T0328313

Design for Additive Manufacturing

Additive Manufacturing Materials and Technologies

Design for Additive Manufacturing

Martin Leary

Elsevier
Radarweg 29, PO Box 211, 1000 AE Amsterdam, Netherlands
The Boulevard, Langford Lane, Kidlington, Oxford OX5 1GB, United Kingdom
50 Hampshire Street, 5th Floor, Cambridge, MA 02139, United States

Library of Congress Cataloging-in-Publication Data
A catalog record for this book is available from the Library of Congress

British Library Cataloguing-in-Publication Data
A catalogue record for this book is available from the British Library

ISBN: 978-0-12-816721-2

For information on all Elsevier publications visit our website at
https://www.elsevier.com/books-and-journals

Publisher: Matthew Deans
Acquisition Editor: Brian Guerin
Editorial Project Manager: Emma Hayes
Production Project Manager: Anitha Sivaraj
Cover Designer: Christian J. Bilbow
Cover illustration: Additive Manufacturing at RMIT University Advanced Manufacturing Precinct: Gyroid structure designed and fabricated by Phil Pille; custom implant designed and fabricated by Milan Brant, Matthew McMillan, Aaron Pateras and Martin Leary

Typeset by TNQ Technologies

Contents

Preface

Additive Manufacturing (AM) is a manufacturing technology that is both maturing and expanding. Commercially mature AM technologies are robust and enable profitable engineering enterprise and innovation in the manufacture of consumer and high-technology products. Concurrently, research AM technologies continue to expand the technical frontiers of manufacturing, including novel material technologies, innovations in manufacturing methods, and technical opportunities to optimise manufacturing rates and minimise production costs. A commonly reported challenge to both the commercial and research facets of AM, is the significant lack of design tools that accommodate the specific manufacturing requirements of AM and enable optimisation of process and product parameters, such that manufacturing costs and effort are minimised. This text provides a timely synopsis of the current and pending state-of-the-art of Design for Additive Manufacturing (DFAM) tools from research and commercial domains. DFAM tools for both metallic and polymeric AM technologies are presented and critically reviewed in the context of existing and emerging AM technologies. Mapping of manufacturing attributes of AM technologies to the associated production economics provides a systematic method for optimising AM geometry, material and process selection. Technical best practice in AM design is espoused through a series of extended Case Studies based on the author's experience. By articulating specific limitations of existing DFAM tools, a clear strategy is provided for DFAM research that responds to a commercial need, thereby increasing the efficiency of commercialisation of technical research.

Intended audience

Both commercial and research AM domains are fundamentally driven by the economics of manufacture. By mapping the production economics of AM with associated DFAM tools, this text provides a methodical and objective summary of the tools and techniques available for commercial success in AM product design. This outcome is of significant value to the design engineer responsible for decision-making that utilises AM technologies. As the text quantifies commercially relevant research strategies, it is of interest to researchers who are intent on developing productive research strategies within the field of Design For Additive Manufacturing. The application focus of the text provides a valuable resource for senior undergraduate students, and the extended Case Studies provide a useful primer for postgraduate students embarking on a DFAM research career.

Author summary

Martin Leary is a Professor of Design for Additive Manufacturing (DFAM) at the Royal Melbourne Institute of Technology (RMIT) University in Melbourne, Australia, and has actively led DFAM research within the RMIT Advanced Manufacturing Precinct since its inception. Within this role Martin has contributed to highly innovative AM research and development with diverse industry partners such as Lockheed Martin, Boeing Australia, Weir Minerals, DST Group, CSIRO Australia, Stryker Corporation, Ford Motor Company, St Vincent's Hospital and the Peter McCallum Cancer Research Institute.

Acknowledgements

This text exists due to the mentorship of Professors Milan Brandt, Ma Qian, Ian Gibson, Colin Burvill, Aleksandar Subic, Peter Choong, Tomas Kron, Dietmar Hutmacher, Adrian Mouritz, Gordon Wallace and Ingomar Kelbassa; the professional support of the Elsevier Staff, Brian Guerin, Emma Hayes, Sheela Josy and Anitha Sivaraj; and especially the encouragement and contributions of my family, Bronwen, Ryan, Sarah and Cheryl.

The quality of the projects summarised in this work is a credit to the calibre of the technical specialists and research engineers with whom I have been privileged to collaborate with since the inception of the Advanced Manufacturing Precinct within RMIT University. In particular, the contributions of the Centre for Additive Manufacturing (DFAM) research theme are acknowledged: Matthew McMillan, Rance Tino, Bill Lozanovski, Tobias Maconachie, Avik Sarker, Sunan Huang, Marcus Watson, Ahmad Alghamdi, Abdu Almalki, Alistair Jones, Tamer Ataalla, Darpan Shidid, Maciej Mazur, Stuart Bateman, Mark Easton, Mahyar Khorasani, Xuezhe Zhang, Elizabeth Kyriakou, Rob Ramsay, Richard Piola, Jeff Shimeta, Kate Fox, Richard Williams, Travis Klein, Omar Faruque, Joy Forsmark, Tony Baxter, Cameron Keller, Gordon New, Stephen Sun, Ian Fordyce, Carl Diver, Jonathan Miller, Jonathan Harris, David Downing, Jonathan Tran, Paul Porter, Phil Pille, Albert Maberley, Alan Jones and Aaron Pateras.

Introduction to AM

1

Additive Manufacturing (AM) refers to a family of manufacturing technologies that sequentially add units of standard input materials to enable the fabrication of discrete physical products. This process is analogous to a printing process but is applied in three dimensions, and results in the commonly applied terminology, *3D printing*.

By this simple definition, AM technologies have existed for centuries, for example, in the form of common brick materials used for the fabrication of complex buildings structures. A commercially relevant definition of AM must also include a restriction that the process be digitally driven; whereby AM is enabled by digital definitions of the intended geometry and associated process parameters. This caveat enables the diverse range of sophisticated design outcomes associated with modern AM technologies, including: inexpensive functional components, high-complexity customised 3D structures, high-value structural systems, and inexpensive patient-specific surgical guides (Fig. 1.1).

1.1 What benefits are enabled by additive manufacture?

AM fundamentally differs from competing manufacturing technologies. In this text, these technologies are collectively referred to as Traditional Manufacturing (TM) methods. In brief, these technologies include (Fig. 1.2): subtractive manufacturing by the sequential removal of bulk input material; and, formative manufacture by the use of some master-reference that physically imparts its geometry to an input material.

The fundamental advantages of additive manufacture can be defined in contrast to the attributes of traditional manufacturing technologies. The nuanced detail of these advantages is discussed in the following Chapters, but can be briefly summarised as:

- ***Reduced production costs:*** The common unit materials and digital design processes inherent to AM enable economic advantages for manufacture, especially for low-volume production, including the manufacture of mass-customised bespoke products (Chapter 7).
- ***Increased complexity:*** The sequential addition of materials that is inherent to AM enables an *outside-in* approach to manufacture; this approach allows the fabrication of high-complexity components that are challenging or even impossible to produce with traditional manufacture, including topologically optimised structures (Chapter 6) and high efficiency cellular structures (Chapter 5).
- ***Custom materials:*** AM processes unit materials in a unique manner. This attribute enables innovation in materials science, including novel polymer chemistries, the manufacture of complex biological constructs and metallurgical properties not feasible with traditional manufacture.
- ***Generative design:*** The digital nature of AM enables an automated approach to the design, documentation and manufacture of complex engineered systems; this opportunity can

Design for Additive Manufacturing. https://doi.org/10.1016/B978-0-12-816721-2.00001-4

Fig. 1.1 A range of commercial AM outcomes: (A). inexpensive functional components (Chapter 8), (B). high-complexity customised 3D structures (Chapter 9), (C). high-value structural systems (Chapter 12), and (D). inexpensive patient-specific surgical guides [1] (Chapter 10).

potentially reduce design cost to the point where AM product can be mass-customised to the specific requirements of a particular design scenario (Chapter 7).

- ***Reduced waste:*** By the efficient utilisation of unit input materials, additive manufacture allows reduced waste and material costs for complex geometries; and paradoxically enables the manufacture of high-complexity product with reduced unit-cost (Chapter 2).
- ***Distributed manufacture:*** Digital files can be shared globally to AM manufacturing centers and fabricated as required. This opportunity enables reduced lead-time and design costs associated with transport and the maintenance of legacy components (Chapter 3).

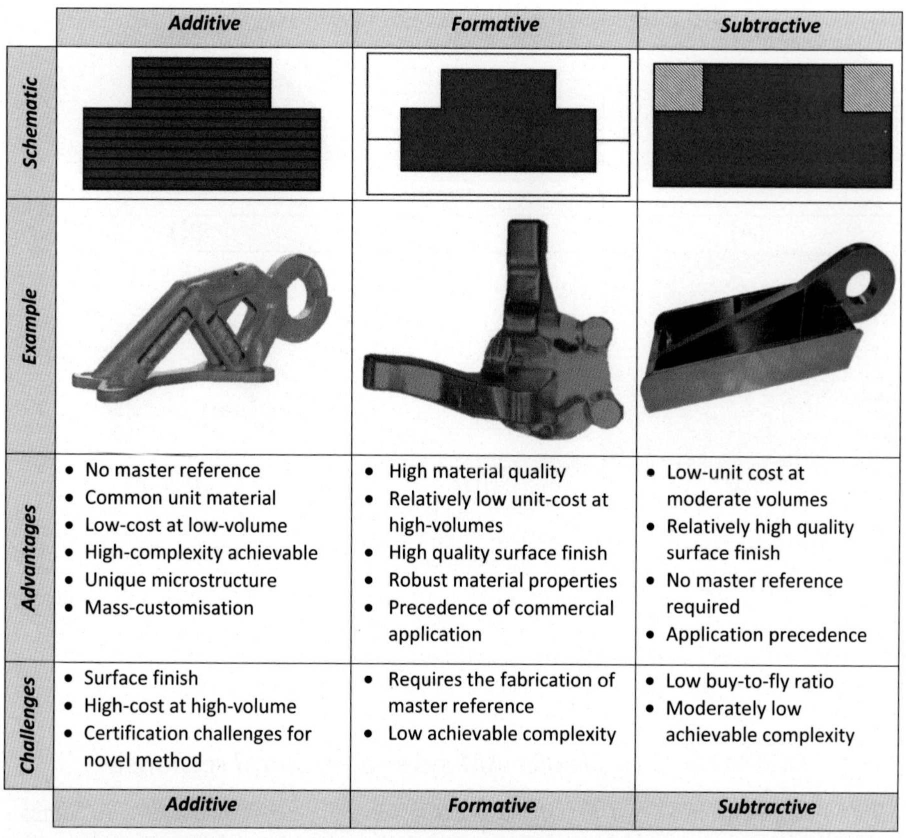

Fig. 1.2 Manufacturing classifications including schematic representation of core advantages and technical challenges.

1.2 Design for Additive Manufacture

The inherent advantages of AM enable significant commercial opportunities, especially for applications that satisfy multiple techno-economic advantages concurrently; for example, the low-volume production of high-complexity components. In fact, AM presents such a profound opportunity for design innovation that it has been labelled as a disruptive technology enabling a fourth industrial revolution [2]. However, the opportunities associated with AM are predicated on the availability of robust Design for Additive Manufacture (DFAM) tools that enable engineers and product designers to (Fig. 1.3):

- Reliably predict the technical response of AM systems
- Define the commercial opportunities of AM systems
- Automate the process of AM product design

These opportunities are elaborated briefly below with specific links to the relevant Chapters and associated case studies.

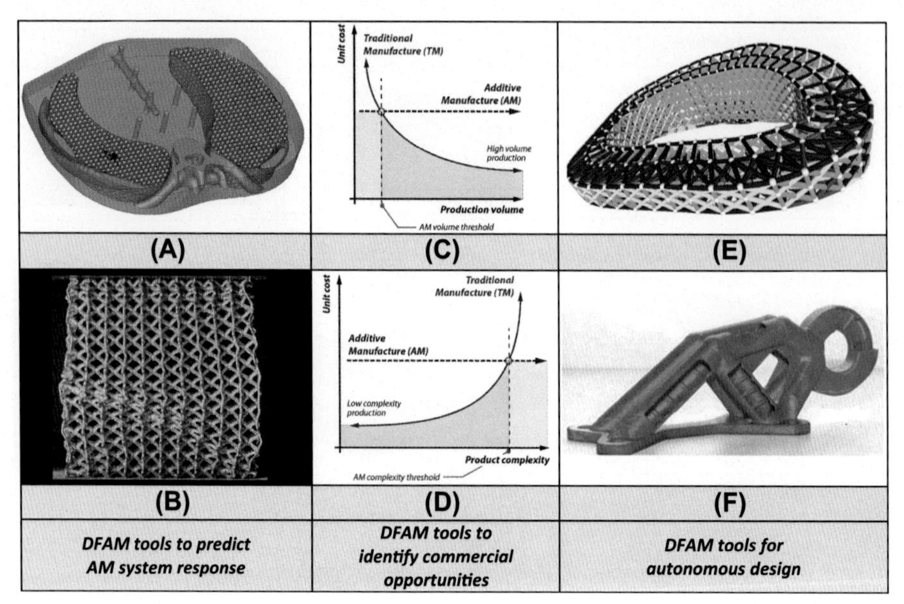

Fig. 1.3 DFAM tool classifications and associated examples developed in this work: methods to characterize the response of (A). polymeric (Chapter 9) and (B). metallic (Chapter 5) AM materials, (C,D). economic DFAM tools for cost prediction (Chapter 2), generative DFAM tools to design high-value (E). medical and (F). aerospace products (Chapter 7).

1.2.1 DFAM tools to predict AM system technical response

DFAM tools that predict *a priori* the effect of design decisions on AM outcomes enable engineers and product designers to minimise the risks of failing to satisfy the technical requirements inherent to a proposed AM technology. This de-risking process enables greater certainty in commercial production, allowing engineers to confidently specify AM product, as well as providing confidence for the manufacturing sector to invest in AM hardware. DFAM tools of this type range from rules-of-thumb on AM manufacturability limits to sophisticated experimental and numerical analysis of AM process parameters (Chapter 4).

1.2.2 DFAM tools to identify commercial opportunities

The evolution of AM technology is driven by a pull from potential customers as well as a push from the design and manufacturing sectors. DFAM tools are required that clearly articulate the commercial opportunities for AM as a function of the geometric, functional and economic requirements of a specific design scenario. These DFAM tools provide a customer-facing tool to identify commercially optimal AM technologies. Alternately, economic DFAM tools provide a manufacturer-facing tool for AM manufacturers to identify potential customers for their specific AM manufacturing specialisation (Chapter 2).

1.2.3 DFAM tools for autonomous design

The complexity enabled by AM processes is not currently matched by the available design tools. A generation of DFAM tools is emerging that enable the complexity of AM processes to be utilised in the design of complex geometries and processes in a highly automated manner. These tools combine manufacturability limits with automated structural optimisation algorithms to algorithmically design AM systems that optimise functional requirements while satisfying the associated manufacturability constraints.

These DFAM tools enable a classification of design known as *generative* design, where the engineer specifies the constraints and objectives that govern the specific product design but is not directly responsible for the specific design outcomes. Generative design algorithms allow the complexity enabled by AM to be systematically applied to optimise the commercial and technical aspects of the product under consideration in a cost-effective manner. Numerous implementations of generative design are under development, including topology optimisation methods that accommodate AM specific attributes. Furthermore, generative design provides an opportunity for the automated self-documentation of design decisions, as is required for the certification and quality control of high-value product (Chapter 7).

1.3 Evolution of AM and technology development trajectories

Novel technologies demonstrate an *S-curve* relationship between implementation (or efficiency) and development time; this curve asymptotes between two limits from initial research to deep commercialisation (Fig. 1.4). A particular challenge associated with this model of technology adoption occurs due to a potential mismatch between research efforts and commercial needs. This mismatch results in a failure for the technology to adequately respond to the end-users technical and economic requirements, leading to a reduced confidence in the underlying technology. This reduced confidence results in a reduced research effort, and reduced levels of implementation. This scenario introduces a real risk that the technology implementation will not exceed the implementation threshold required for technology commercialization.

This text responds to this challenge by promoting DFAM tools that align research with commercial activities, enabling research outcomes to be rapidly industrialised; as well as identifying research opportunities and associated DFAM tools that must be addressed to enable AM commercialisation. It is hoped that these efforts will promote a balance between research and development that maximises commercial uptake.

1.4 The techno-economic motivation for Design for AM (DFAM) tools

Research innovation motivates commercial opportunities, and commercial opportunities motivate research innovation. If this research *push* and commercial *pull* is in

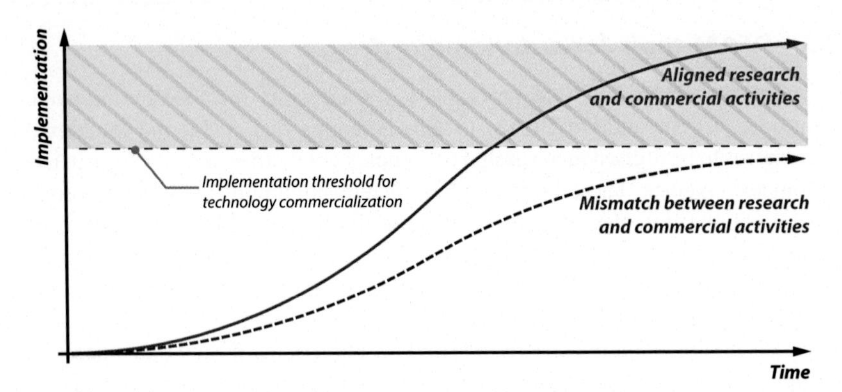

Fig. 1.4 Technology *S-curve*, indicating scenarios where research is aligned to commercial activities, and scenarios where a mismatch exists between research and commercial activities. The latter potentially leading to a failure to achieve the minimum implementation threshold required for technology commercialization.

alignment, the underlying technology matures quickly and evolves to a high level of implementation (Fig. 1.4). Commercialisation of AM technologies requires unambiguous guidance on the associated technical and economic constraints. In this text, economic constraints are given equal priority with technical DFAM requirements. Chapter 2 formalises a series of DFAM tools that can be used to identify scenarios where AM processes are economically optimal. This initial analysis is then refined to address nuanced concepts that affect commercial outcomes in subtle but important ways. Chapter 3 presents technical DFAM tools such that the fundamental technical requirements of AM application are met, as well as tools that enable the complexity of AM to be harnessed without compromising overall design agility and associated costs.

References

[1] Whitley III D, Eidson RS, Rudek I, Bencharit S. In-office fabrication of dental implant surgical guides using desktop stereolithographic printing and implant treatment planning software: a clinical report. The Journal of Prosthetic Dentistry 2017;118(3):256–63.

[2] Berman B. 3-D printing: the new industrial revolution. Business Horizons 2012;55(2): 155–62.

AM production economics

2

The strategic commercialization of AM methods requires that the selection of materials, geometry and AM processes be systematically driven by the associated technical *and* economic constraints, stated here as the *techno-economic* design constraints. It is typical for Design for AM (DFAM) texts and research papers to focus on the technical constraints associated with AM; these are presented in the following chapters and include methods for technical design optimization in the absence of economic considerations. These technical considerations are crucial but must be implemented concurrently with the optimization of economic attributes in order that the manufacture be commercially successful. If the economic aspects of a proposed AM design are overlooked or unknown, the resulting design decisions will be uninformed and may restrict commercial success and stymie future AM applications.

The specific expectations of this chapter are a set of clear definitions for which AM projects are economically optimal. These definitions can be used by engineering designers, business analysts and entrepreneurs to quantify the feasibility of specific projects for AM; and conversely to identify the set of conditions that are optimal for AM, thereby allowing AM service providers to target applications with commercial merit.

This analysis is based initially on elementary philosophies of engineering economics as developed for traditional manufacturing methods. These analyses are then extended based on insights associated with AM production economics. Opportunities for DFAM tools to enhance commercial outcomes are then identified with reference to relevant Case Studies.

2.1 Classical engineering economics

Engineering economics provides a robust tool to quantify the influence of design decisions on costs and to allow the fiscal optimization of an engineered product under development. The charts of Fig. 2.1 are central to classical engineering production economics. Specifically, that for a specified product manufactured with a specified manufacturing process, the associated unit-cost scales exponentially with geometric complexity, and inversely with production volume. The underlying logic being that as the production volume increases, unit-costs are defrayed over an increasing number of manufactured products and that the manufacture of low complexity products requires a lower production effort than for high complexity products.

Costs are categorized as either fixed or variable, depending on whether they are independent, or dependent on the number of products manufactured. Fixed costs include

Design for Additive Manufacturing. https://doi.org/10.1016/B978-0-12-816721-2.00002-6

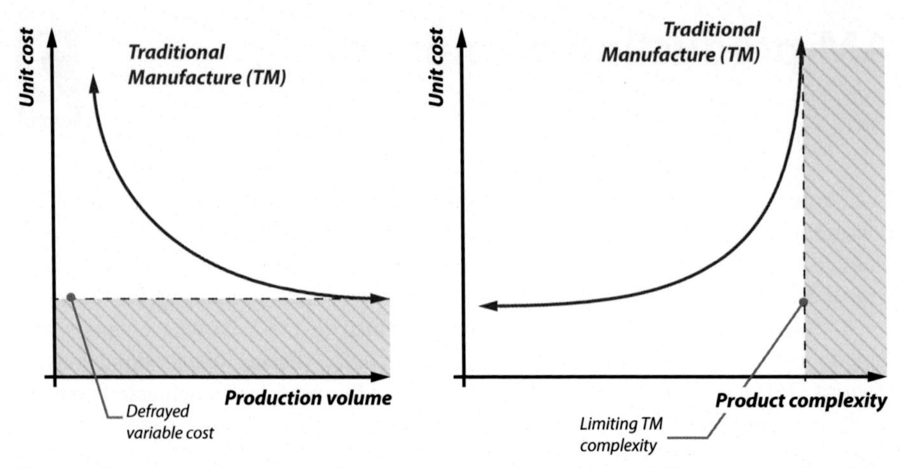

Fig. 2.1 Schematic representation of cost versus production volume (left) and complexity (right) based on classical production economics applied to Traditional Manufacturing (TM) methods.

costs associated with engineering design, including design labor, software and training; sales and marketing; and production hardware and maintenance. Variable costs include materials and utilities consumed during processing, and labor that is directly associated with manufacture. Consequently, for small production runs, unit-cost is approximately equal to the fixed costs incurred with the development of a specific design. As production volume increases, these fixed costs are defrayed over an increasing number of components, and unit-cost tend to the defrayed variable cost (Fig. 2.1, left).

Concurrently, as geometric complexity is introduced to the intended product, production time, sophistication of quality control, and part rejection rate increases. These factors increase production cost in an exponential fashion (Fig. 2.1, right). As product complexity increases, the associated cost asymptotes to infinity at the limiting complexity achievable for the specific manufacturing method selected.

The charts of Fig. 2.1 conclude that for Traditional Manufacturing (TM) of a specified product geometry and manufacturing process, unit-costs are minimized for low complexity components manufactured in large production volumes. This conclusion is valid for traditional manufacturing processes and forms the economic basis for mass production methods. The charts of Fig. 2.2, below, provide a more sophisticated model of TM production economics by indicating a family of curves that represent the impact of *improved production efficiency* and *improved manufacturing capability* on the associated unit cost. Improved production efficiency is achieved by methods including inventory management, efficient supply chains, design for manufacturability and high throughput manufacturing systems—resulting in reduced unit cost for a given production volume. Improved technical capability is achieved by advanced manufacturing methods that reduce costs associated with achieving a specified level of complexity.

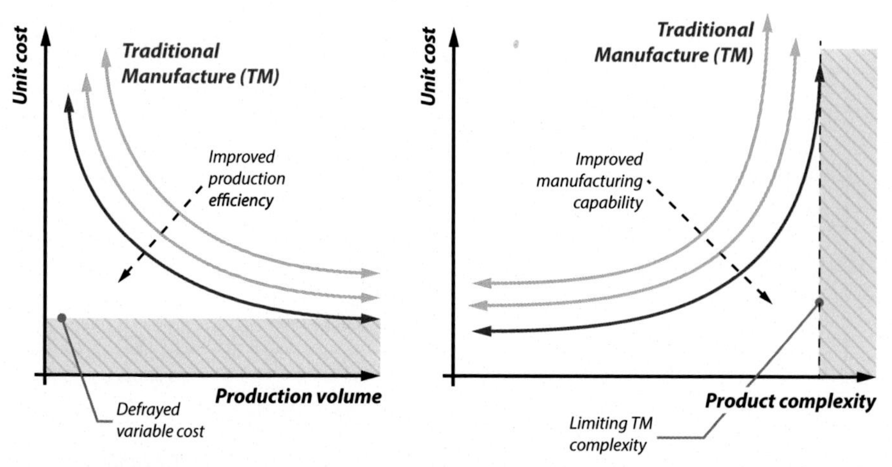

Fig. 2.2 Schematic representation of cost-volume and cost-complexity for Traditional Manufacturing (TM) methods, indicating the impact of improved production efficiency and improved manufacturing capability.

2.2 Additive manufacturing economics

Additive Manufacturing (AM) differs from Traditional Manufacturing (TM) methods in economically critical ways. For example, the addition of a common source material to sequentially generate a final product without the need for a forming master, combined with an inherently digitally driven design and manufacturing process. Because of these differences, the economic conclusions drawn for the engineering economics of traditional methods are not necessarily valid for AM. For example, the cost of high-complexity and low-volume production can be significantly reduced for AM methods (Section 2.3).

To accommodate these inherent economic differences, AM manufacturing unit cost is often modelled as being independent of production volume and part complexity (Fig. 2.3). This simplification of constant unit cost may be approximately valid when robust DFAM approaches are applied, but can introduce non-conservative errors in important design scenarios. These exceptions and associated challenges are discussed in the following Sections. However, the simplification of constant unit cost does allow useful insight into the fundamental economic opportunities associated with AM. For example, it is apparent that AM processes can provide a distinct commercial advantage within the cost-complexity and cost-volume design spaces, and that distinct thresholds exist that delineate the economic optimality of AM and TM methods. Specifically, commercial opportunities for AM over traditional methods exist for scenarios with low-volume production and high-complexity geometries, as defined by the *AM volume threshold* and *AM complexity threshold* (Fig. 2.3).

The specific value at which the AM volume threshold and AM complexity threshold occur depend on the specific characteristics of the AM and TM processes.

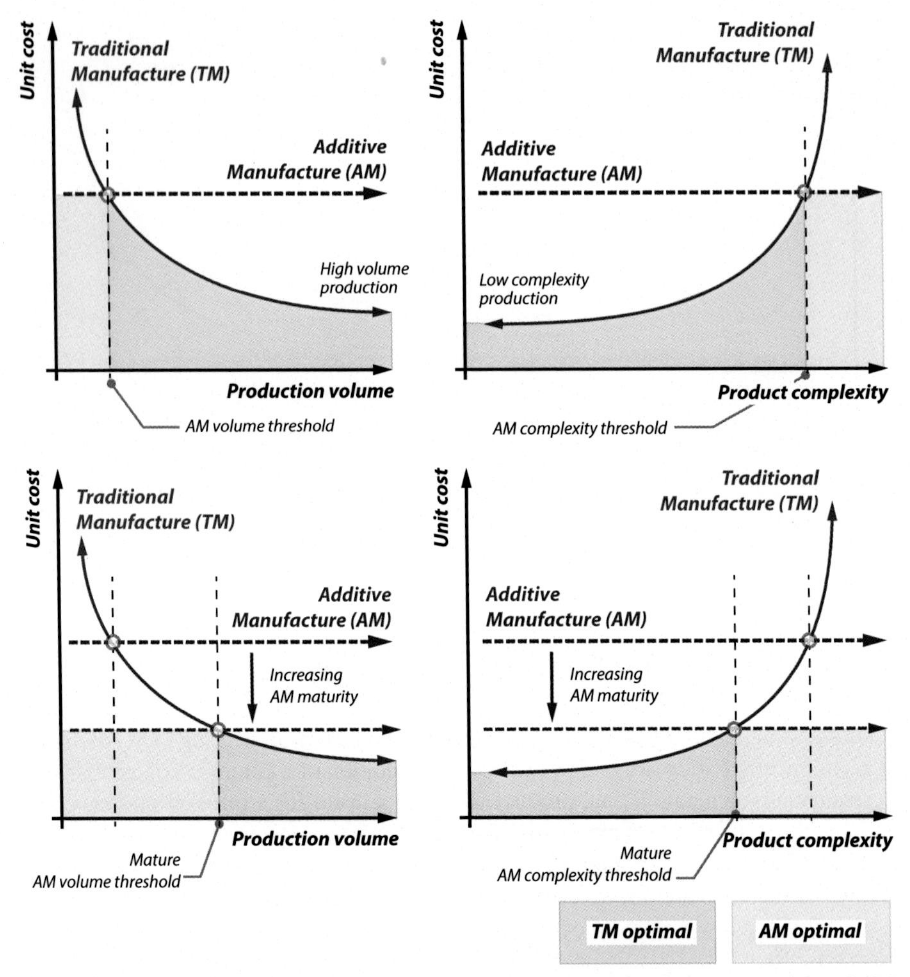

Fig. 2.3 Cost-volume and cost-complexity curves including schematic representations of AM component unit cost for an AM process based on the assumption of AM cost-independence. Upper: Threshold between AM and TM technologies identified based on current AM technologies. Lower: Threshold identified for increasing AM maturity.

When assessing these thresholds for commercial projects, an important variable to consider is the effect of technology maturation. Traditional methods have a high-level of technical maturity when compared to AM. By this logic, maturing AM technologies will increasingly displace traditional methods in terms of economic and technical performance. The eventual techno-economic differentiation between traditional and mature AM technologies is not known; at this stage it is only predictable that AM technologies will become increasingly competitive as the underlying technology matures (Fig. 2.3, lower).

2.3 Economically optimal AM scenarios

The assumption that AM part cost is independent of production volume and product complexity allows useful insights into the relative economic merit of AM and traditional methods. This observation requires that the fixed costs incurred by component design and manufacturing pre-processing be negligibly small contributors to the overall unit-cost. Based on this assumption, key opportunities for AM commercialization can be identified in specific economic regions (Fig. 2.4).

Batch-enabled scenarios

For production volumes below the *AM volume threshold*, unit costs associated with AM are less than for traditional manufacture (Fig. 2.4 – Zone 1). These batch-enabled scenarios include pre-production development of conceptual and technical design models; bespoke design of jigs and fixtures; single unit-production; and short volume serial production. These scenarios are further developed in Section 2.3.1. Generative design methods provide an enabling capability for mass-customized manufacture; these commercial opportunities are presented in Chapter 7.

Complexity-enabled scenarios

AM methods can accommodate increased complexity at lower cost than traditional manufacture. Consequently, the *AM complexity threshold* defines a product complexity, above which AM methods enable lower cost than competing traditional manufacturing methods (Fig. 2.4 – Zone 2). Complexity-enabled scenarios include topologically optimized structural systems, functionally integrated systems and serial production of highly complex geometries. These scenarios are further developed in Section 2.3.2.

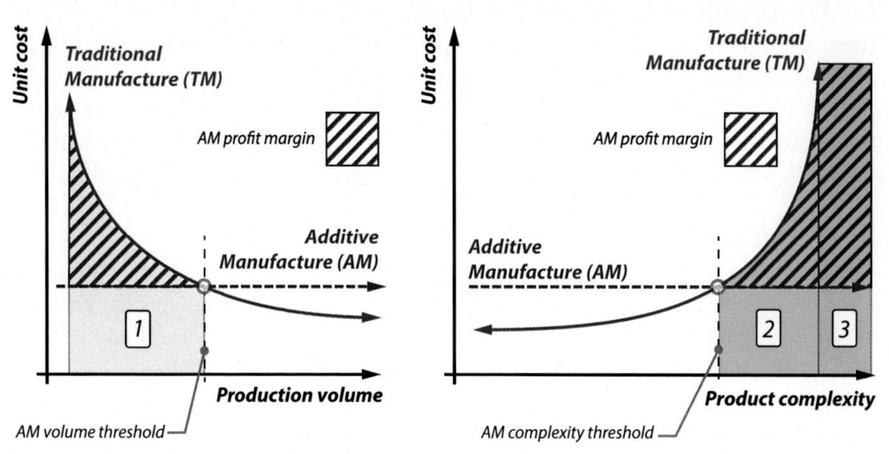

Fig. 2.4 Schematic representation of product unit cost versus production volume and product complexity. Potential AM profit margin over TM, and zones of commercially optimal AM application indicated. Zone 1: Batch-enabled scenarios. Zone 2: Complexity-enabled scenarios. Zone 3: Ultra-high complexity scenarios.

Ultra-high complexity scenarios

The limiting complexity of AM methods often exceeds that of traditional manufacture. These scenarios provide a unique commercial design opportunity for AM, whereby traditional manufacturing methods are technically unable to provide the required complexity at any cost, resulting in a lucrative opportunity for AM design of high-value applications, especially based on highly efficient topologies and generative methods of design (Fig. 2.4 – Zone 3). These scenarios are further developed in Section 2.3.3.

Mass-customisation scenarios

The inherent flexibility of AM enables a new paradigm of *mass-customisation*, whereby the overall production volume is sufficiently high to qualify as mass-production, however, the manufactured product is customized to a product-specific set of design requirements. Mass-customisation is enabled by generative design methods that allow autonomous design and manufacture according to algorithmic expert systems that emulate manual design methods but with such efficiency that their contribution to variable costs is effectively nil. These scenarios are developed from an economic perspective in Section 2.3.4, and from a technical perspective in Chapter 7.

2.3.1 Batch-enabled scenarios

It is apparent that AM processes can provide a distinct commercial advantage within the cost-volume design space. For production volumes below the *AM volume threshold*, the costs incurred by traditional manufacturing methods exceed that of AM processes. These *batch-enabled* scenarios offer a distinct commercial advantage for AM processes (Fig. 2.5).

1. Pre-production conceptual design models

These include geometric and tactile *looks-like* models that are used to validate the intended aesthetics and required customer interaction. These models do not functionally represent the final designed system but can be rapidly manufactured in numerous permutations to accelerate the design process for components that interact with the customer. Case Study: Development of industrial gearbox systems (Chapter 8).

2. Surgical planning models

The use of AM for generating custom surgical models is a special case of a conceptual design model, whereby the AM product is used to represent patient-specific anatomy and clinical features to enhance patient outcomes and reduce risk. Case Study: AM radiation dosimetry phantoms (Chapter 9).

3. Pre-production technical validation models

These include functionally equivalent models that enable pre-production validation of the technical aspects of an intended design. These models do not necessarily represent the aesthetic aspects of an intended product but enable reduced development time by replicating the technical function of products manufactured by traditional methods

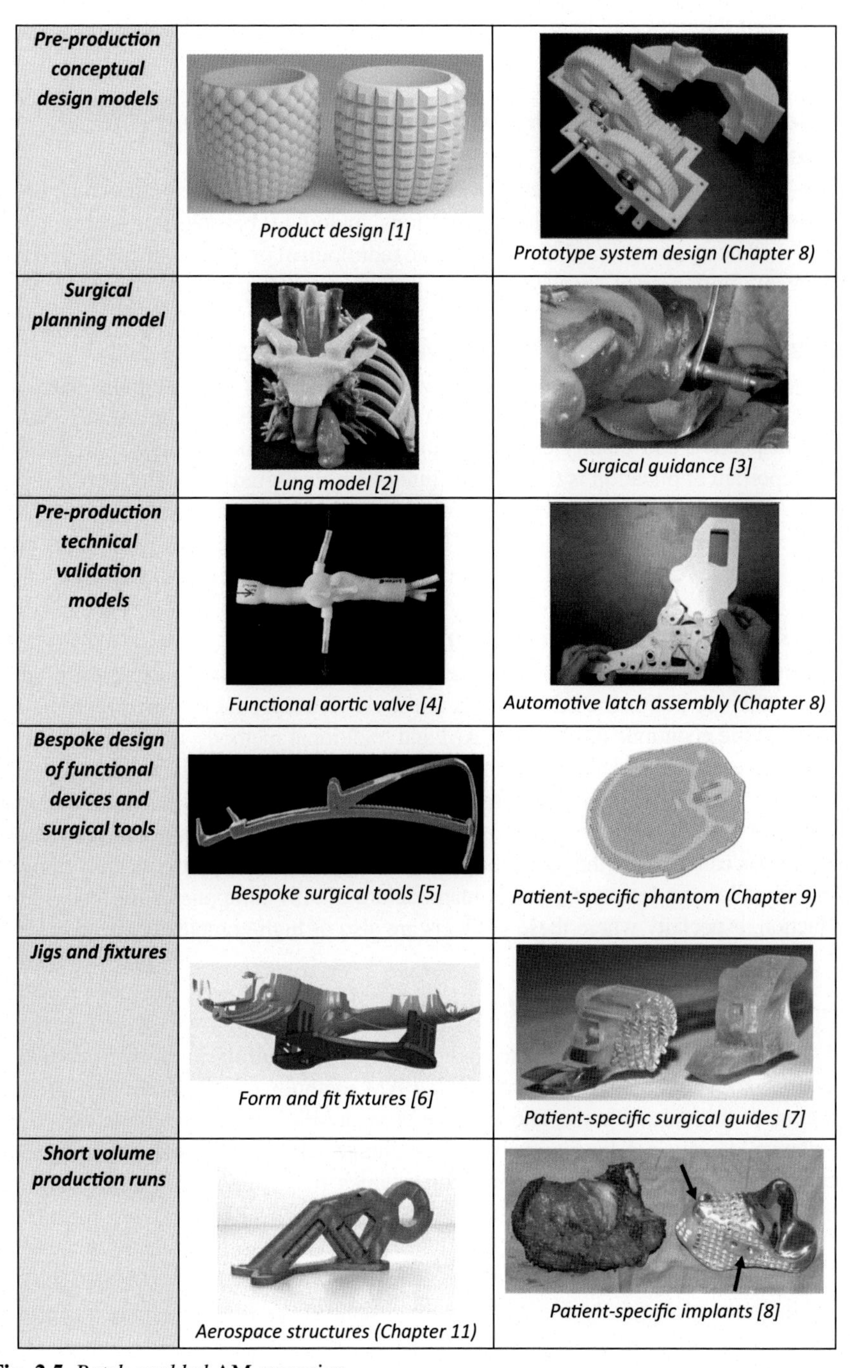

Fig. 2.5 *Batch-enabled* AM scenarios.

without their inherent lead time. Case Study: Validation of kinematic automotive latching assembly (Chapter 8).

4. Bespoke design of functional medical devices and clinical tools

Patient-specific functional medical devices and clinical tools are embodied by a broad range of high-value, batch-enabled applications, including: bespoke surgical tools; radiation dosimetry phantoms that represent the radiological properties of specific tissues (Chapter 9); and, patient-specific orthopaedic implants (Chapter 7).

5. Production hardware – jigs, fixtures and gauges

Jigs and fixtures enable mass production with TM methods by restraining parts and guiding assembling and post-processing. Form-and-fit models and engineering gauges enable inspection and quality control during manufacture and assembly. An increasingly important application is the manufacture of surgical-guides to guide surgical procedures. This critical production hardware can be economically and rapidly manufactured using AM technologies. The end user does not directly interact with such models; they are used to enable manufacture and to confirm manufacturing quality.

6. Short volume production runs

AM provides an economically robust alternative to traditional manufacture for production runs that are below the AM volume threshold. The specific crossover-volume is a function of the economic attributes of AM and traditional methods, as discussed above. A paradox exists associated with this threshold, whereby the use of AM for pre-production prototyping and initial series production, especially when sales volumes are difficult to predict, can result in an increase in the effective AM volume threshold. This paradox can increase the production volume for which AM is economically optimal, potentially allowing for the economic manufacture of AM components for low volume production, especially where these products are also of high-complexity (Section 2.6).

2.3.2 Complexity-enabled scenarios

AM methods readily accommodate high complexity, resulting in an *AM complexity threshold*, above which *complexity-enabled* AM methods enable lower cost production than traditional methods. The inherent compatibility of AM methods with high-complexity design is a key competitive advantage when compared with traditional manufacturing methods. This observation enables numerous economically optimal design scenarios (Fig. 2.6).

1. High efficiency topologies

These include topologically optimized structures and space frames that are technically challenging to fabricate with traditional methods. Such applications are important for aerospace and medical design applications where structural optimization is a critical design objective. Case Study: topologically optimized structures for aerospace applications (Chapter 6).

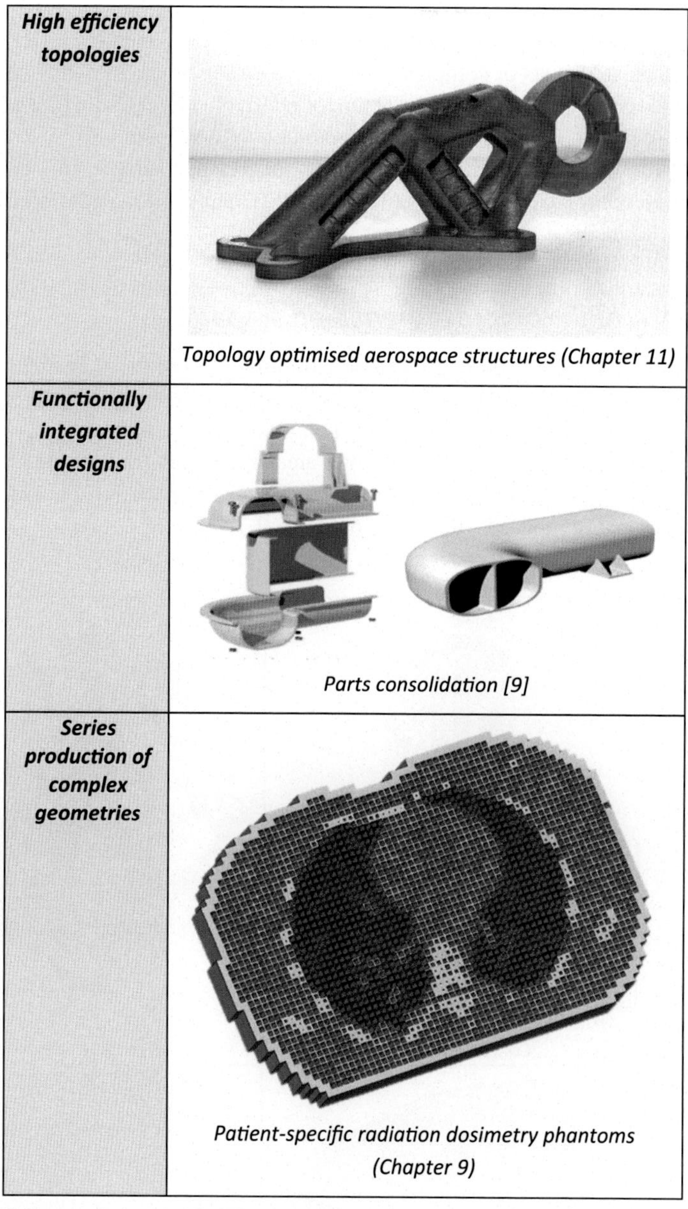

Fig. 2.6 *High-complexity enabled* AM scenarios.

2. Functionally integrated designs

Functional integration provides an opportunity to increase commercial value by utilizing the capability of AM for increased complexity. Functional integrated design allows parts consolidation, thereby reducing part count and associated inventory as well as reducing the stress concentrations and potential failure modes associated with component interaction. Examples of functional integration include integrated cooling and structural parts consolidation (Chapter 11).

3. Series production of complex geometries

Complex geometries are often incompatible with traditional manufacturing methods due to the associated costs of complex designs. For these scenarios, it may be economically feasible to use AM for the manufacture of relatively large production volumes of a common design, known as *series production*. AM provides a commercial opportunity for series production of such components; for example, the manufacture of structurally optimized aerospace componentry (Chapter 6) and the fabrication of patient-specific medical devices (Chapter 7).

2.3.3 Ultra-high complexity-enabled scenarios

Furthermore, AM methods typically enable a limiting complexity that exceeds that of traditional manufacturing methods. Consequently, AM provides *ultra-high complexity-enabled* scenarios where the manufactured complexity is sufficiently high that traditional manufacturing methods are not technically feasible. These scenarios provide a monopoly opportunity for commercial application of AM, as traditional manufacturing methods are infeasible (Fig. 2.7).

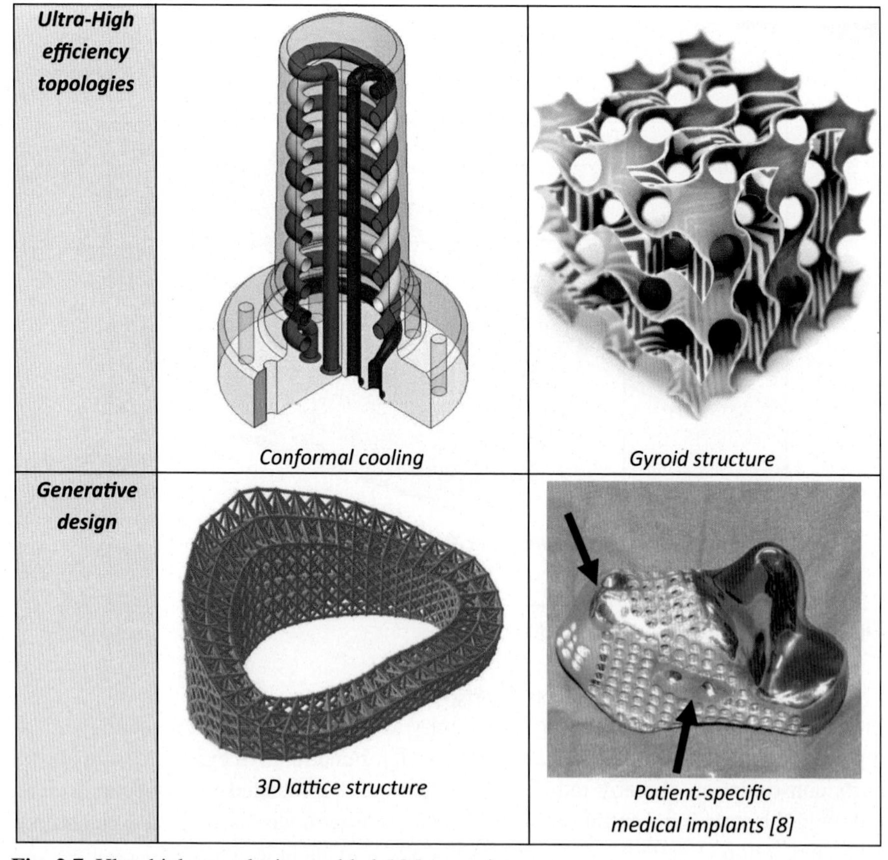

Fig. 2.7 Ultra-high-complexity enabled AM scenarios.

1. *Ultra-high-efficiency topologies*

The high limiting-complexity associated with AM processes enables the fabrication of ultra-high complexity structures with exceptional mechanical efficiency. This category of structures includes conformal cooling systems and zero-mean curvature structures that provide unique technical advantage but are exceptionally challenging for traditional manufacture (Chapter 5).

2. *Generative design*

The complexity associated with sophisticated AM structures introduces a significant challenge for timely and economical design. Increasing complexity results in an exponential increase in the required design effort, this potentially debilitating challenge is referred to as the *curse of dimensionality*.[1] For high complexity design scenarios, generative design techniques are required, whereby the designer specifies the design intent, and the algorithm efficiently deploys the specific structural details. Successful generative design requires a robust understanding of AM manufacturability limits and associated DFAM tools.

2.3.4 Mass-customized scenarios

Production costs are categorized as being either fixed or variable according to whether they are independent, or dependent on the production volume (Section 2.1). This representation provides insight into the relative magnitude of these costs as a function of production volume. For example, at small production volumes, fixed costs dominate the overall production costs. As production volume increases, variable costs become increasingly important, and for large production volumes the variable costs typically dominate (Fig. 2.8). When defrayed by the production volume, this observation leads to the unit cost versus production volume relationship of Fig. 2.3.

This representation is useful to provide insight into the influence of fixed and variable costs on unit cost as a function of production volume. However, the underlying assumptions are not fully compatible with AM production economics and can lead to erroneous predictions of unit cost as well as missed commercial opportunities or unanticipated commercial failures. In particular, the model does not accommodate the nuances associated with the cost of product design and whether these costs are correctly described as being fixed or variable. The following text allows these design costs to be more correctly understood such that commercially relevant insights for AM design can be made.

The costs associated with product design are typically assigned as being fixed. This notion is compatible with TM methods where an *initial design effort* is made prior to manufacture and this design remains unchanged for the entire production volume. This initial design cost is correctly represented as a fixed cost, where the magnitude of this cost is typically smaller for evolutionary design activities than for revolutionary design activities[2] (Fig. 2.8). An alternative scenario is highly relevant for AM design;

[1] The concept of dimensionality and its effect on problem solvability is introduced in Chapter 7.

[2] In concise terms, evolutionary and revolutionary design refers to the design of product that is respectively well understood or novel, and is actively discussed in Chapter 3.

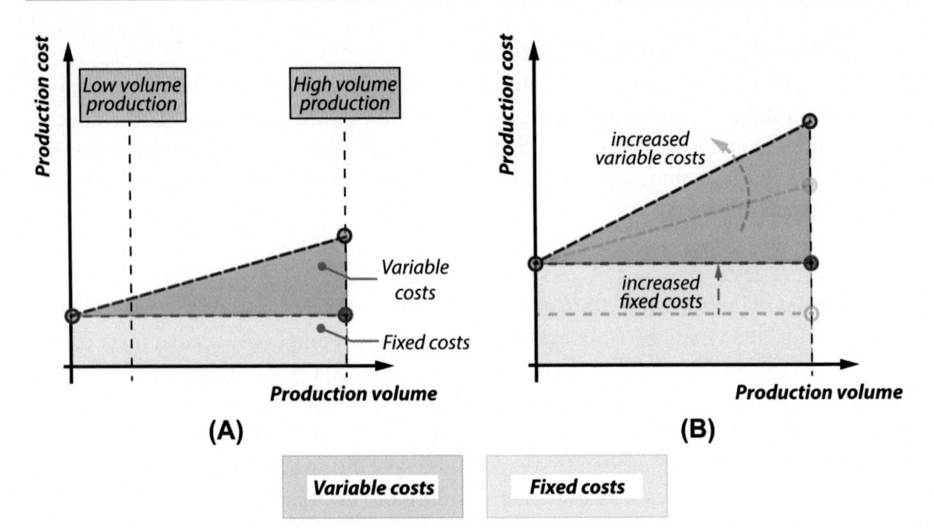

Fig. 2.8 Simplified representation of the effect of fixed and variable costs on production cost as a function of production volume. (A) Low fixed and variable costs, (B) Increased fixed and variable costs. Dashed lines indicate relative magnitude of these costs for low and high production volumes.

specifically, where *ongoing design effort* is required during production in order to customize the product according to the requirements of particular design scenarios. For these scenarios, the *ongoing design effort* is a variable cost and, as production volume increases, so does the associated design cost.

When the ongoing design effort is of high complexity (for example, as is required for sophisticated high efficiency structures) or when this effort is frequently required (for example, in the design of patient-specific products), the high ongoing design costs associated with manual design are likely to be economically infeasible (Fig. 2.9). For these scenarios it may be economically advantageous to invest in *generative design methods* that algorithmically assist the designer to customize the design with relatively low design effort (Chapter 7). Generative methods increase the initial design costs due to the effort required to develop and validate the algorithmic design tools but reduce the ongoing design costs by reducing the manual effort of customization. The overall effect is to introduce a *production volume threshold for generative methods* above which generative methods are economically optimal in comparison with equivalent manual design (Fig. 2.9).

If utilized effectively, this economic insight can enable the *mass-customisation* of AM product, whereby highly efficient generative design methods are applied to autonomously customize every manufactured product according to unique design requirements. This capability is challenging for TM methods but can be enabled by the manufacturing flexibility inherent in AM methods. For example, for the generative design of customized structural bracketry and patient-specific implants (Fig. 2.10).

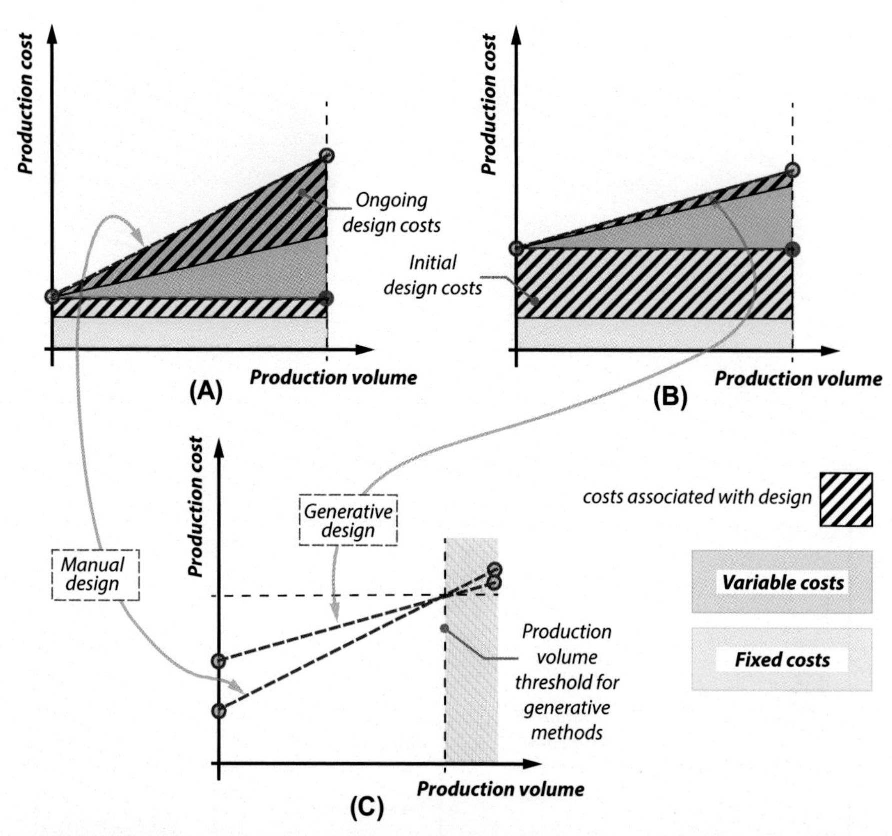

Fig. 2.9 Left: Simplified representation of the effect of increased initial design costs and ongoing design costs on overall production costs. Right: Effect of generative design methods on overall production costs (Chapter 7).

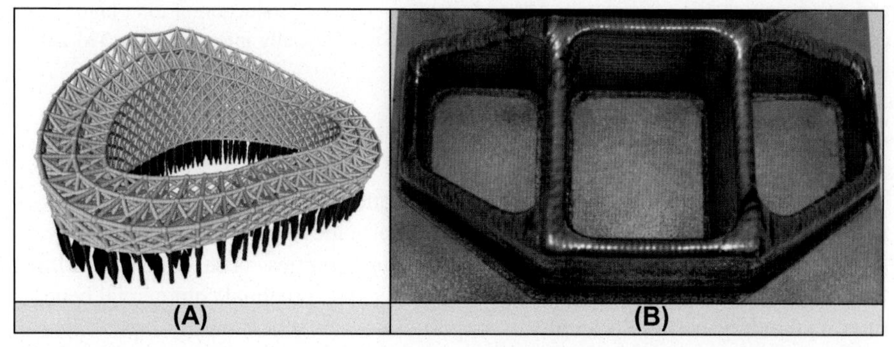

Fig. 2.10 Sample componentry enabled by mass-customization. (A) Patient-specific medical implants, and (B) Structural aerospace bracketry. These components are generatively designed in various permutations to accommodate manufacturability and structural requirements (Chapter 7).

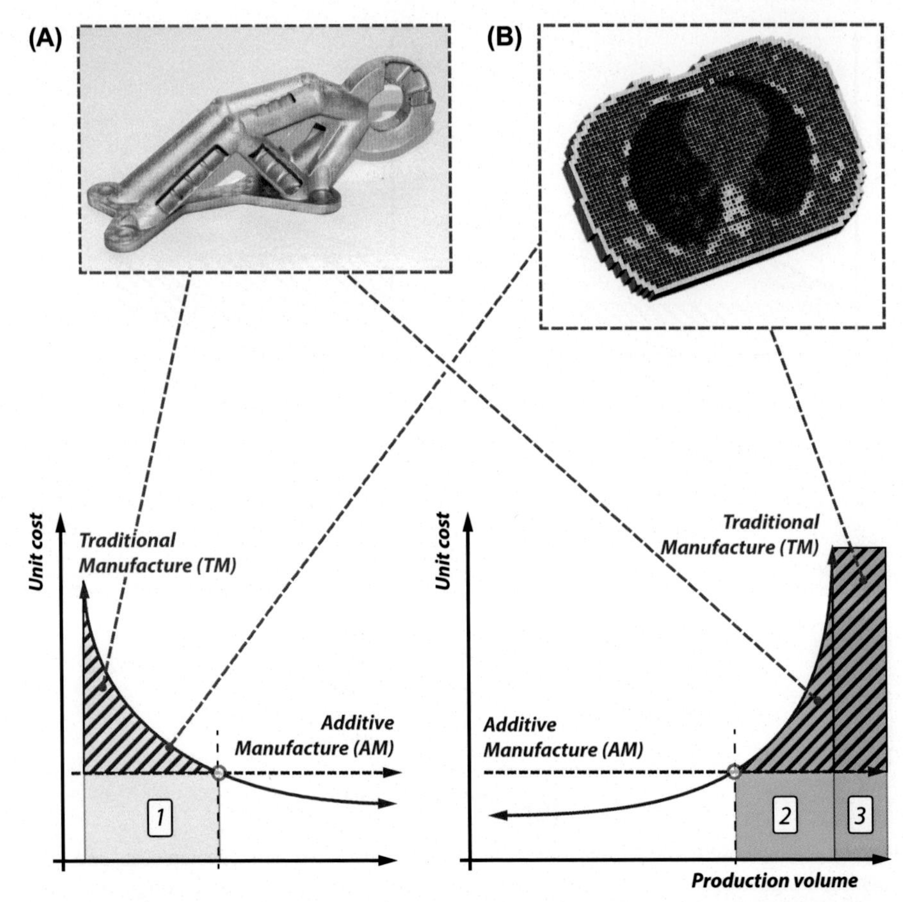

Fig. 2.11 Schematic representation of enhanced commercial opportunities for AM. (A) Low-volume production of high complexity structural aerospace component. This geometry is potentially feasible with TM (for example by investment casting) but has a production volume and complexity that is optimal with AM and, (B) Low-volume production of ultra-high complexity patient-specific radiotherapy phantom that is technically infeasible with TM and provides a significant AM commercial opportunity.

2.4 Enhanced commercial scenarios for AM

The economic analysis of AM and traditional manufacturing techniques enables systematic insight into the scenarios for which AM provides a robust commercial advantage. An amplification of the commercial opportunities occurs when multiple AM-optimal regions are addressed simultaneously. The potential commercial benefits of compounding cost and complexity opportunities are schematically represented in Fig. 2.11. Specifically, these opportunities occur for the following scenarios:

Low-batch and high-complexity scenarios

Low-batch production of high-complex geometry is technically feasible with traditional manufacturing methods; however, AM potentially provides a reduced cost alternate. This opportunity is suitable for serial production of high-complexity components with production volumes that are expensive for traditional manufacture at small volumes; and is especially suited to high-value applications such as medical devices and aerospace components (Case Study, Chapter 6).

Low-batch and ultra-high complexity scenarios

Low-batch production of ultra-high complexity components provides an exceptional opportunity for the economic manufacture of components with a complexity that is not technically achievable with traditional manufacturing methods. Examples of this opportunity include: the design and manufacture of patient-specific custom implants, where the specific implant is tailored to the patient-specific geometry and material properties (Case Study, Chapter 7); and the design of advanced rocket motors that allow structural and thermodynamic optimization that is not feasible with traditional manufacturing methods.

The complexity of these applications will typically necessitate some level of generative design procedures that allow complex geometry to be deployed algorithmically in response to a specified engineering requirement. The application of generative design methods to implement complexity without compromising the available design time and budget are described in detail in Chapter 7.

The simplified analysis of engineering economics developed for Fig. 2.4 and ensuing sections provides useful economic design guidance for many AM applications; however, certain caveats exists that can modify this simplified analysis, and must be considered when analysing the economic profile of a proposed AM scenario. These caveats include the:

- paradox of reduced cost with increased complexity
- economies of high-volume AM production
- paradox of volume intersection for optimal cost
- flawed cost-independence assumption of AM

These caveats and associated case studies are introduced in the following sections.

2.5 The paradox of reduced cost with increased complexity

AM provides a beneficial exception to the traditional economic relationships of cost and complexity. For traditional manufacture, an increase in complexity incurs an exponential increase in associated costs. For AM, this relationship does not hold, and unit cost of AM products is often simplified as being independent of complexity. This simplification does not capture a core technical advantage of AM in being able to manufacture highly complex geometries, and in particular, that these complex geometries enable functional efficiencies that reduce the physical volume of the manufactured component, and therefore reduce variable costs by reducing material use and associated manufacturing time.

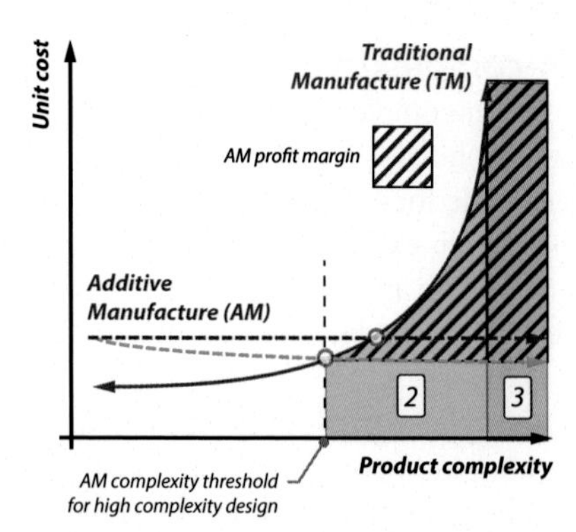

Fig. 2.12 Schematic representation of unit cost versus product complexity indicating the effect of reduced cost with increased complexity.

Fig. 2.12 schematically indicates the phenomenon of increased complexity resulting in reduced AM unit cost. When designed according to robust DFAM strategies, AM componentry can achieve high technical efficiency that reduces material consumption and associated manufacturing time. Because of this phenomenon, higher complexity AM components can potentially be manufactured with reduced cost; a response that is entirely opposite to that of traditional manufacture. This outcome highlights a strategic commercial opportunity for AM. Specifically that, not only does AM enable geometric design opportunities that provide highly optimized structural response, but also that such designs can potentially be implemented with a cost-benefit over lower complexity designs. This opportunity is so profound and counter-intuitive that it will be referred to as *the paradox of AM cost and complexity*. This paradox enables significant commercial opportunities and motivates many of the commercially oriented opportunities identified in this text.

The paradox of AM cost and complexity is predicated on an understanding that increased complexity does not significantly increase the costs associated with the design process. This outcome can be achieved by following the DFAM strategies presented in Chapter 4 such that robust AM design decisions are implemented. As well as by utilizing the generative design strategies of Chapter 7 such that complex design outcomes are efficiently implemented and that the *production volume threshold for generative methods* is satisfied (Section 2.3.4).

2.6 Economies of high-volume AM production

According to simplified economic analysis, the unit cost of AM is independent of production volume. This simplification is useful in that it characterises the fundamental advantage of AM in enabling efficient production at low volume, whereby a digital design is fabricated in a single process using a common source material. This simplification, however, does not accommodate the practical influences of production volume on the costs of manufacture. When production volume increases, various commercial advantages come into play, even for AM systems. These include

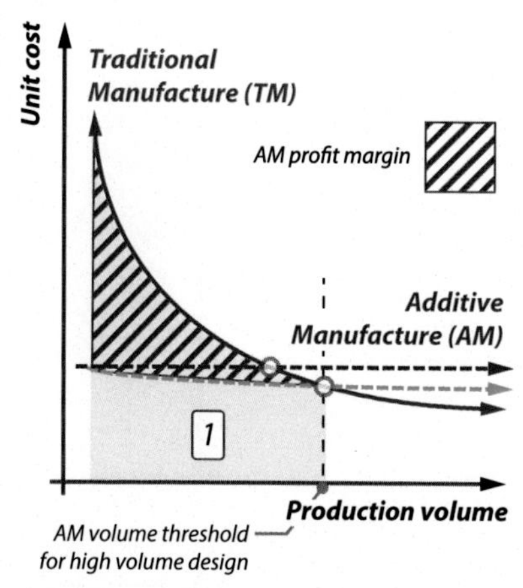

Fig. 2.13 Schematic representation of unit cost versus production volume indicating the effect of reduced cost with increased production volume.

opportunities to purchase input material at reduced rates; opportunities to maximize machine yield and production rates; reduced maintenance costs associated with material change-over and cleaning; and the potential to optimize workflows and post-processing that comes with consistent manufacture of a specific product at high volumes. These factors together tend to enable AM unit costs to be reduced with increasing production volume. This reduction may be less pronounced that for high-volume TM production methods but it does allow an enhanced economic advantage that must be considered for higher production volumes (Fig. 2.13).

2.7 The paradox of volume intersection for optimal cost

According to the presented economic analysis, AM is cost-optimal for production volumes below the *AM volume threshold* (Fig. 2.4, left). However, this outcome may be oversimplified in some scenarios, resulting in non-conservative estimates of the cost-optimality of AM. These scenarios are especially associated with the use of AM for pre-production validation, and for scenarios where product sales volumes are largely unknown or difficult to predict.

AM is fundamentally compatible with pre-production validation for both aesthetic and functional attributes, known respectively as *looks-like* and *works-like* products. Consequently, if AM has been used for pre-production validation, the *margin cost* (the cost of adding an additional unit of production) associated with manufacturing the final product with AM methods is essentially nil. Conversely, commissioning production with TM methods incurs additional costs and risk. *Consequently, if AM is used for pre-production validation, there may be economic benefit in applying the same AM method for initial series production.* This insight is especially relevant for low production volume or when production volume is difficult to predict.

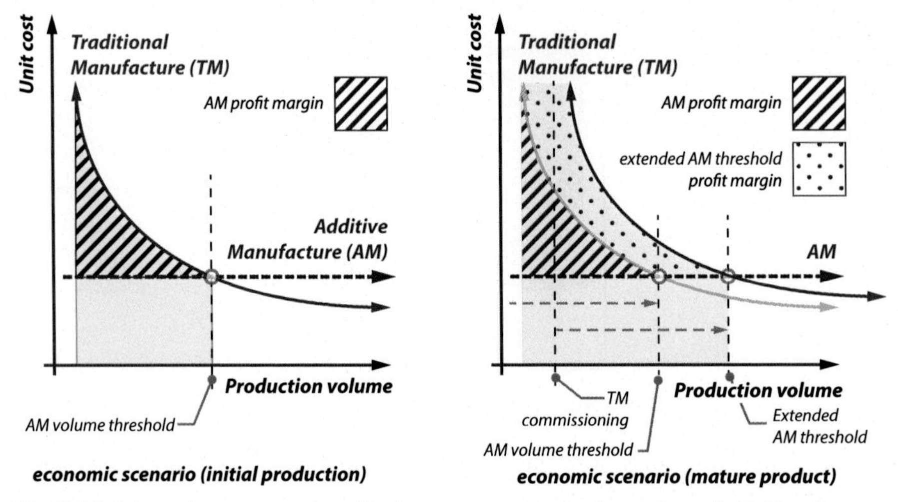

Fig. 2.14 Schematic representation of unit cost versus production volume, indicating the paradox of volume intersection for optimal costs; whereby an extended AM threshold occurs in the favour of AM for a mature product when AM is used for initial production.

For scenarios where sales volumes are unknown or difficult to predict, it is challenging to know whether the *actual* production volume will exceed the AM volume threshold. For these scenarios, a conservative approach is to proceed with AM for short-to medium-production volumes. If the production volume is below the AM volume threshold, the AM process is cost-optimal. If future sales volumes are predicted to exceed this threshold, production can either migrate to traditional manufacture or remain with AM. If production remains with AM, the margin cost is low, alternatively, the margin cost for migrating to TM may be relatively high, such that these additional funds may be more effectively utilized within other aspects of the business, for example in new product development.

When predicting economic impact, it is critical to acknowledge that if AM is used for initial production, the production economics of TM manufacture must be recalculated from the production volume associated with *TM commissioning* (Fig. 2.14, right). This results in a resetting of the TM cost curve such that an *extended AM threshold* occurs in the favour of AM production. This scenario demonstrates the *paradox of volume intersection for optimal costs*, whereby if AM methods are used for pre-production validation and initial production, an extended AM threshold exists that dramatically extends the production volumes for which AM is economically optimal[3] (Fig. 2.14).

By this logic, AM provides a mechanism to manage financial risk for enterprises embarking on product development where final production volumes are difficult to predict; as well as providing a low economic barrier to entry for enterprises with limited capital to invest in TM methods. This strategy also provides an opportunity

[3] Note that these considerations can be combined with the economies of high-volume AM production (Section 2.7) to potentially further increase the AM volume threshold.

to mitigate technical risks by allowing the product design attributes to be modified during production with little influence on profit margin. For example, structural safety factors can be initially high to allow robust technical outcomes and can potentially be reduced as in-field product behaviour is better understood. This approach is especially compatible with scenarios that apply AM for pre-production validation, which can then seamlessly transition to serial production.

2.8 The flawed cost-independence assumption of AM

According to the previous analysis, product cost for traditional manufacture scales inversely with production volume and exponentially with complexity (Fig. 2.2). Conversely, AM is typically modelled as presenting a cost curve that is independent of production volume and complexity. This simplified analysis is useful; however, it includes idealisations that may result in erroneous predictions of AM unit cost, especially for low volume and high complexity production. For low volume production the initial design costs are defrayed over a relatively small number of production units, thereby inflating unit costs. Similarly, for high complexity production, the design effort required to overcome the technical challenges of manufacture are potentially high. These increased design costs are especially high for designs that are revolutionary in nature (Chapter 3), whereas for evolutionary design the AM manufacturability challenges are well understood, and the associated design costs are relatively low.

Fig. 2.15 indicates a modification of the previously developed relationships between volume, complexity and cost, indicating schematically the contribution of increasing design costs to unit cost. It is apparent that increasing design costs directly translate to a decreasing economic feasibility of AM methods. In the limit, increasing design costs contribute so significantly to unit cost that AM methods lose their competitive advantage over traditional manufacturing methods. This under prediction of cost

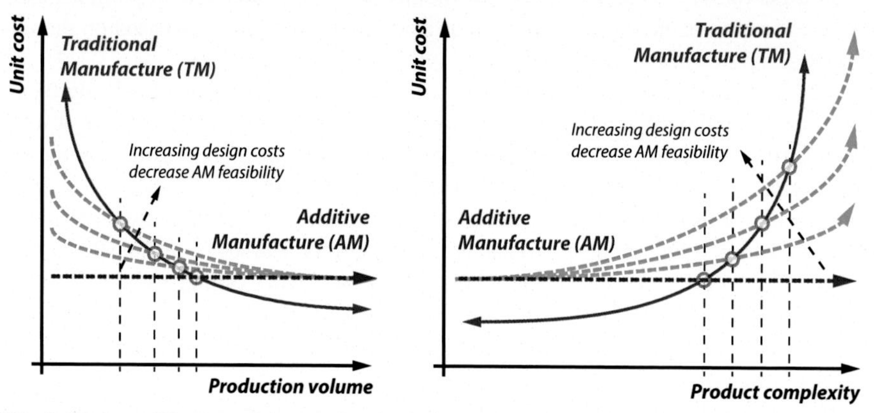

Fig. 2.15 A modification of the previously developed correlation between production volume, complexity and cost, indicating schematically the contribution of increasing design costs. Reduced design costs, as can be achieved by DFAM methods, result in a larger domain for which AM is cost-optimal.

is a potential pitfall that must be understood by those involved in AM design, especially for low volume and high complexity production. The following section describes the necessity for DFAM tools to minimize design costs such that failure to achieve intended commercial outcomes is avoided.

2.9 The economic necessity of design for AM

The economic Design for AM (DFAM) insights developed in this chapter provide a systematic approach for identifying scenarios for which AM methods are optimal, as well as caveats to these scenarios that may result in either opportunities to increase commercial benefit, or potential pitfalls that may compromise commercial success (Sections 2.6–2.9).

Based on the analysis of Fig. 2.15, it is apparent that increasing design cost constitutes a potential pitfall to successful AM commercialization. Specifically, that increased design costs diminish the cost-benefit of AM in comparison with traditional methods. In order to reduce design costs, systematic approaches are required to more comprehensively understand the manufacturing attributes of a potential AM technology, input material, and associated geometry as early in the product-design phase as possible. This understanding is aided by (DFAM) tools that accommodate the specific technical attributes of AM. These *technical DFAM tools* are categorized in this text as either: DFAM tools to predict AM system response, or DFAM tools for autonomous (generative) design.

DFAM tools that predict system response provide *a priori* guidance on the effect of relevant design variables, including material selection and geometry, on technical performance for a specific AM technology. These technical DFAM tools are invaluable for de-risking commercial AM projects and increasing profitability. These tools exist in multiple forms, varying from rules-of-thumb on AM manufacturability limits based on direct experimental observation to non-linear numerical analyses based on the underlying physics of the AM process.

AM enables the design of complex structures; however, to be economically beneficial this complexity must be technically useful. The addition of technically useful complexity is increasingly challenging. DFAM tools for autonomous design enable the engineer and product designer to engage with the complexity enabled by AM without undue increase in associated costs. For example, Fig. 2.11 represents potential responses to a structural design objective with various degrees of technically useful complexity. As complexity increases, the cost of achieving further technical benefit is subject to a diminishing return. The application of DFAM tools for autonomous design enables a higher degree of complexity to be accommodated within the available time resource (Fig. 2.16).

A similar logic can be applied to the reduction of design costs for batch-optimized design, especially for low-volume production. As production volume increases, design

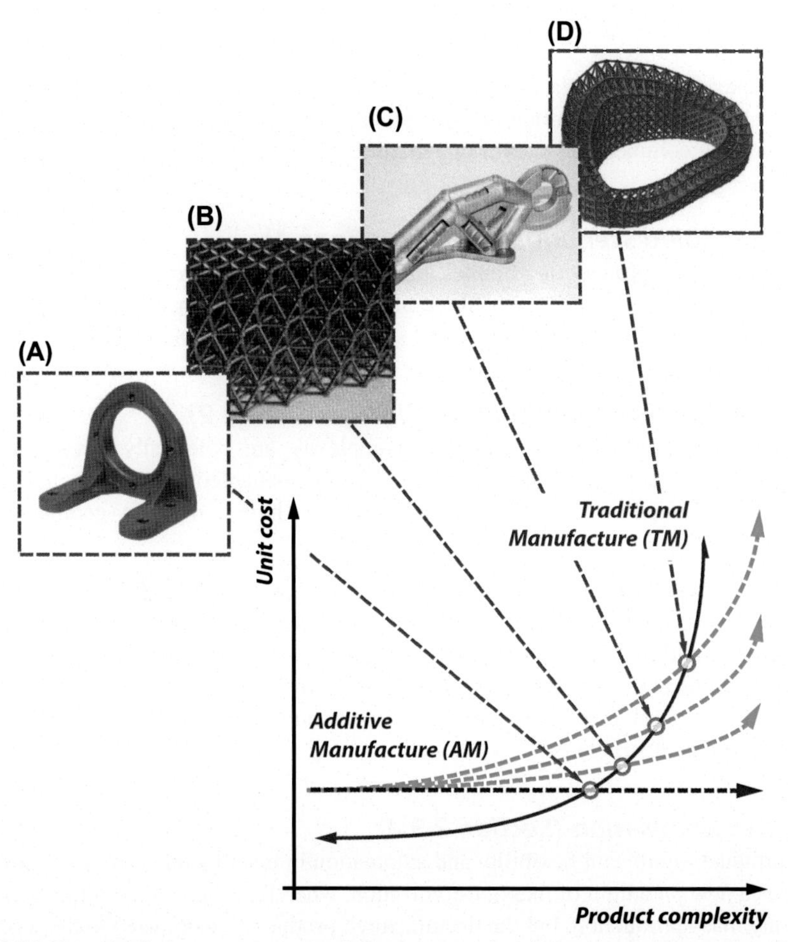

Fig. 2.16 Potential design responses to a defined structural optimization requirement. (A) First-principles structural analysis. (B) Periodic lattice structure. (C) Topological optimization. (D) Conformal lattice structure. Optimization objective (mass minimization) is subject to diminishing return with effort.

costs are defrayed over a larger production run, and are therefore less critical to the associated unit cost. However, series volume production is predicated by successful concept development prior to approval. Therefore, excessive cost of concept development presents a barrier to commercialization, even if the eventually serialized part could be cost-competitive. It is critically important that DFAM be applied to minimize design costs, especially for entrepreneurial activities and new product development that requires production to be economically sustainable at all stages of the product life-cycle from pre-production prototyping to serial production.

2.10 Summary of chapter outcomes

This chapter distils elementary philosophies of engineering economics with fundamental technical and economic attributes of AM production. Based on this simplified analysis, the following commercially robust opportunities for AM application are identified.

Batch-enabled scenarios (Fig. 2.4 – Zone 1)

At relatively low production volumes, the unit costs of AM are lower than for traditional manufacture. Batch-enabled commercial scenarios include pre-production development, single unit production and short volume serial production.

Complexity-enabled scenarios (Fig. 2.4 – Zone 2)

The costs incurred for high geometric complexity are potentially lower for AM methods than for traditional manufacture. Complexity-enabled scenarios include topologically optimized systems and serial production of high complexity geometry.

Ultra-high complexity applications (Fig. 2.4 – Zone 3)

The limiting complexity of traditional manufacture is less than for AM methods. These scenarios provide a unique opportunity for the manufacture of high-value structures where traditional manufacturing methods are not feasible at any cost. Applications include high complexity lattice structures, functionally integrated designs and generatively optimized structures.

Mass-customisation (Section 2.3.4)

AM provides significant flexibility and automation in manufacture. These capabilities enables a new paradigm of mass-customization, whereby production volume qualifies as being mass-production, but the manufactured product is customized to the product-specific design requirements. The commercial implementation of mass-customization requires that the economics of generative design methods be understood such that the investment in autonomous DFAM tools is economically rewarded.

These identified opportunities are not mutually exclusive and can be combined for enhanced commercial benefit. Specifically, these enhanced commercial opportunities include low-batch manufacture for scenarios that are either of high- or ultra-high complexity. These scenarios benefit from the economic advantage of both low-volume and high-complexity and can potentially generate significant commercial benefit, for example, patient-specific medical implants (Chapter 7) and generatively designed radiation dosimetry phantoms (Chapter 9).

This chapter begins with a simplified analysis of engineering economics (Section 2.1). This simplification provides a useful economic DFAM tool for guiding the economic aspects of many AM applications. A more nuanced understanding of AM identifies certain caveats to these outcomes that must be considered when analysing the economic performance of a proposed AM scenario. These caveats include the following.

The paradox of reduced cost with increased complexity

In response to increased complexity, AM can actually provide a reduction in product cost. This cost reduction occurs due to a reduction in production costs, especially associated with reduced material usage and machine time. The behavior is paradoxical to that of traditional manufacture and can provide economic advantage for components with significant design complexity. This opportunity should be considered in parallel with considerations of potentially flawed cost-independence of AM (Section 2.5) such that the overprediction of the cost of high complexity AM products be avoided.

The paradox of volume intersection for optimal cost

AM provides significant opportunities for pre-production validation and conceptual design. For such scenarios where the actual production volume is not clearly known, there are economic benefits for remaining with AM manufacture as production grows, including: reduced economic barrier to entry for manufacturers with limited start-up capital; economic risk-management for uncertain production volumes; and the potential for product design attributes to seamlessly evolve during production. Furthermore, if a transition to TM methods is considered, the associated economics must be recalculated at the time of *TM commissioning*, resulting in an extended threshold for which AM methods are optimal (Section 2.7).

The flawed cost-independence assumption of AM

The simplified representation of engineering economics for AM and traditional methods is predicated on design costs providing a negligible contribution to unit cost (Section 2.8). Cases for which this assumption is valid do exist, however the absence of these idealized conditions can result in an underprediction of manufacturing costs, and can be avoided by employing technical DFAM tools and generative design methods to manage the contribution of design costs, as presented in Chapters 3 and 7, respectively.

This chapter provides a strategic overview of the *techno-economic* design constraints associated with AM methods. The underlying intent is that designers and commercial managers be systematically forewarned on the technical *and* economic constraints inherent to the robust commercial implementation of AM technology. It is important that simplified economic analysis be understood such that the inherent opportunities of AM be clearly articulated to potential customers, thereby enhancing the commercial *pull* for the application of AM. However, it is imperative that AM practitioners be aware of the limitations of this simplified economic analysis such that they can prevent flawed commercial decisions and promote advanced opportunities for AM technologies.

The criticality of insightful *techno-economic* design understanding is high, as without this insight commercial ventures fail or are only sporadically successful, leading to a collapse in industry confidence in the underlying AM technology. Despite this criticality, relatively few research contributions in the field of techno-economic design exist, possibly due to the multidisciplinary nature of the technical and economic expertize required. For those research teams that are capable of contributing to this requirement, there exist significant opportunities for the generation of techno-economic DFAM tools, including accountancy expert systems that embody nuanced understanding of the opportunities and potential pitfalls for potential AM applications; systems that illustrate (and quantify) the relative economic risks and opportunities associated with evolutionary and revolutionary design activities; published case studies on experiences in successful (and unsuccessful) commercial AM ventures; and research investigation on the dynamic effects of relevance to commercial AM application, including for example the paradox of volume intersection that occurs when contemplating the migration of AM production to TM.

References

Rapid prototyping in new product development

[1] Van Rompay TJ, Finger F, Saakes D, Fenko A. "See me, feel me": effects of 3D-printed surface patterns on beverage evaluation. Food Quality and Preference 2017;62:332–9.

Three-dimensional printing of anatomical models for complex surgical cases

[2] Gillaspie EA, Matsumoto JS, Morris NE, Downey RJ, Shen KR, Allen MS, Blackmon SH. From 3-dimensional printing to 5-dimensional printing: enhancing thoracic surgical planning and resection of complex tumors. The Annals of Thoracic Surgery 2016;101(5): 1958–62.

In-house manufacture of patient-specific guides for planning and surgery

[3] Whitley III D, Eidson RS, Rudek I, Bencharit S. In-office fabrication of dental implant surgical guides using desktop stereolithographic printing and implant treatment planning software: a clinical report. The Journal of Prosthetic Dentistry 2017;118(3):256–63.

3D printing for enhanced treatment of cardiovascular disease

[4] El Sabbagh A, Eleid MF, Al-Hijji M, Anavekar NS, Holmes DR, Nkomo VT, Oderich GS, Cassivi SD, Said SM, Rihal CS, Matsumoto JM. The various applications of 3D printing in cardiovascular diseases. Current Cardiology Reports 2018;20(6):47.

Additive Manufacture of bespoke surgical tools

[5] Kontio R, Björkstrand R, Salmi M, Paloheimo M, Paloheimo KS, Tuomi J, Mäkitie A. Designing and additive manufacturing a prototype for a novel instrument for mandible fracture reduction. Surgery 2012;1. pp. 2161–1076.

3D printing of custom jigs and fixtures

[6] Krznar N, Pilipović A, Šercer M. Additive manufacturing of fixture for automated 3D scanning–case study. Procedia Engineering 2016;149:197–202.

A review of the application of AM in medicine including patient-specific surgical guides

[7] Malik HH, Darwood AR, Shaunak S, Kulatilake P, Abdulrahman A, Mulki O, Baskaradas A. Three-dimensional printing in surgery: a review of current surgical applications. Journal of Surgical Research 2015;199(2):512–22.

Additive manufacture of patient specific medical implant

[8] Imanishi J, Choong PF. Three-dimensional printed calcaneal prosthesis following total calcanectomy. International Journal of Surgery Case Reports 2015;10:83–7.

Parts consolidation using AM

[9] Yang S, Tang Y, Zhao YF. A new part consolidation method to embrace the design freedom of additive manufacturing. Journal of Manufacturing Processes 2015;20:444–9.

Digital design for AM

3

The Additive Manufacturing process is inherently digital. Production geometries are specified as 3D digital representations of the intended solid geometry, often using generative methods to algorithmically specify complex designs based only on *digital data representations*. This design data is then *digitally processed* to generate layerwise 2D data including AM toolpaths and associated process parameters. This data is then *digitally queued* for production according to the available AM hardware. During production, logs of processing data are *digitally stored and analyzed* to provide in-situ quality control and process parameter feedback. Post-production metrology consists of *digital records* of actual manufactured geometry, which can then be assessed according to the intended solid geometry for quality control verification.

AM enables unique commercial opportunities, including the economical production of ultra-high complexity geometry, as well as high-value products inexpensively manufactured at extremely low production-volume. These scenarios are entirely incompatible with the constraints of traditional manufacture and enable profound opportunities for techno-economic production advantage. However, these opportunities are *entirely predicated on the robust and efficient flow of digital data* within the AM design process.

Despite the criticality of interconnected and efficient digital design capabilities for AM design, this particular DFAM aspect has arguably received little research attention; at least in comparison with the geometric and process parameter DFAM tools introduced in Chapter 4. This deficiency is likely due to a mismatch between research *push* and market *pull*; whereby the research focus is primarily on lower Technical Readiness Level (TRL)[1] technical activities (such as fundamental material and process optimization) and is less focused on the higher TRL commercial activities (such as economic analysis and commercial implementation tools) (Chapter 1).

The limited availability of digital DFAM tools presents a significant opportunity for informed designers to generate commercial advantage, while managing the risks and costs associated with digital AM design. Digital DFAM limitations also provide a strategic research opportunity for those interested in developing commercially useful innovation tools. This chapter provides a comprehensive summary of the digital aspects of AM, as well as the associated opportunities and challenges for commercial outcomes and associated research opportunities.

As the design process applied for AM is so inherently digital, this chapter on digital design begins with a review of the *engineering design process*; including concepts ranging from the paradox of engineering design and the present value of knowledge,

[1] The Technical Readiness Level (TRL) and associated Manufacturing Readiness Level (MRL) are systematic measures of the *technical* and *manufacturing* maturity of a specified technology.

Design for Additive Manufacturing. https://doi.org/10.1016/B978-0-12-816721-2.00003-8

to the distinct design strategies appropriate for evolutionary and revolutionary product design. From this review, opportunities for digital AM design are systematically identified in the context of commercial best practice AM design processes.

3.1 Digital data and engineering design

The inherently digital nature of the AM design process enables significant opportunities for innovation in design. To maximize this opportunity requires an understanding of philosophies of *engineering design*; whereby the enabling attributes of design excellence are formally defined to provide a roadmap to commercial best practice design methodologies. Of the extensive available literature, a selection of engineering design concepts that are pertinent to optimized AM design are introduced in the following sections.

As a design progresses toward an implemented solution, product attributes such as geometry, materials and manufacturing processes are formalized. Consequently, the flexibility to implement change within a specific design rapidly diminishes, thereby restricting flexibility in design decision-making. Conversely, design understanding is initially low but increases rapidly as the design progresses. The strategic utilization of AM provides a robust opportunity to positively influence design understanding and flexibility in many unique ways (Section 3.1.1). These opportunities are predicated on the ability for AM to be implemented with certainty, repeatability and high-efficiency; and, especially for high-value and safety-critical applications, robust validating documentation.[2] These requirements *can* be enabled by digital AM, in particular with the benefit of generative design methods (Chapter 7); however, this requires that designers and production managers be aware of the inherrent technical requirements of Digital AM design and available DFAM tools. The selection of appropriate DFAM tools is strongly influenced by the classification of the associated design, as either as *evolutionary* or *revolutionary* (Section 3.1.3).

From this review of relevant concepts of engineering design, opportunities for digital DFAM design, both available and emerging, are systematically identified and reviewed in the context of commercial best practice AM design processes.

3.1.1 The present value of knowledge

A fundamental concept of engineering design is associated with *the paradox of design flexibility and understanding* [1–4]; whereby for a specified design activity of fixed duration, the understanding of the underlying physics, customer response and associated economics progressively increases as the design evolves. Paradoxically, the flexibility of the designer to integrate the insights gleaned by this understanding

[2] The design of highly complex product is enabled by digital generative methods that are inherently compatible with the automated generation of documentation to assist in achieving regulatory compliance (Chapter 7).

continuously decrease due to the irrevocable nature of manufacturing decisions. This paradox is analogous to the more commonly understood fiscal concept of the *present value of money*; whereby the financial value of a given resource is greater the earlier it is available, due to its ability to generate a revenue return over time. Analogously, a given resource of design understanding is more valuable the earlier it is available, due to its ability to positively influence design decisions while there remains sufficient flexibility for these decisions to be acted on. Consequently, in this text, this concept will be referred to as the *present value of knowledge* (Fig. 3.1).

AM can utilize the present value of knowledge to enable profound and disruptive change in the engineering design process (Fig. 3.2). These opportunities are associated with the ability of AM technologies to positively influence the present value of knowledge, either by increasing the level of design understanding, or by increasing the flexibility to respond to this increased understanding (Fig. 3.3); in particular, by enabling the following opportunities.

1. The opportunity for AM to enable rapid insight into a specific technical challenge, i.e. by allowing the rapid manufacture of functionally and aesthetically equivalent products more quickly than would be possible with traditional manufacturing (TM) processes. Such products are informally known as *works-like* and *looks-like* products, respectively. This application of AM can be classified as *rapid-prototyping* and provides an economical opportunity to rapidly increase design understanding.
2. The opportunity to rapidly deploy commercial product using AM technologies, with an understanding that TM technologies will ultimately be used to fabricate the mature product. This strategy of *in-situ product development* allows a market share to be nurtured by being first-to-market, while allowing modifications to the design to be rapidly implemented during production; an infeasible outcome with TM. Furthermore, this strategy allows AM product to be brought rapidly to market, while TM facilities (with longer lead time) are brought online.

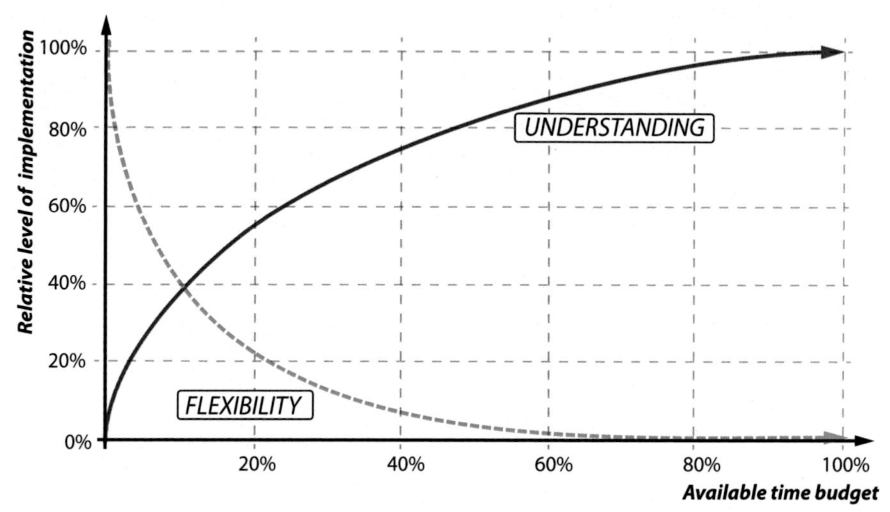

Fig. 3.1 The present value of knowledge. Paradoxically, as the understanding of the design problem increases, the flexibility to exploit this understanding diminishes.

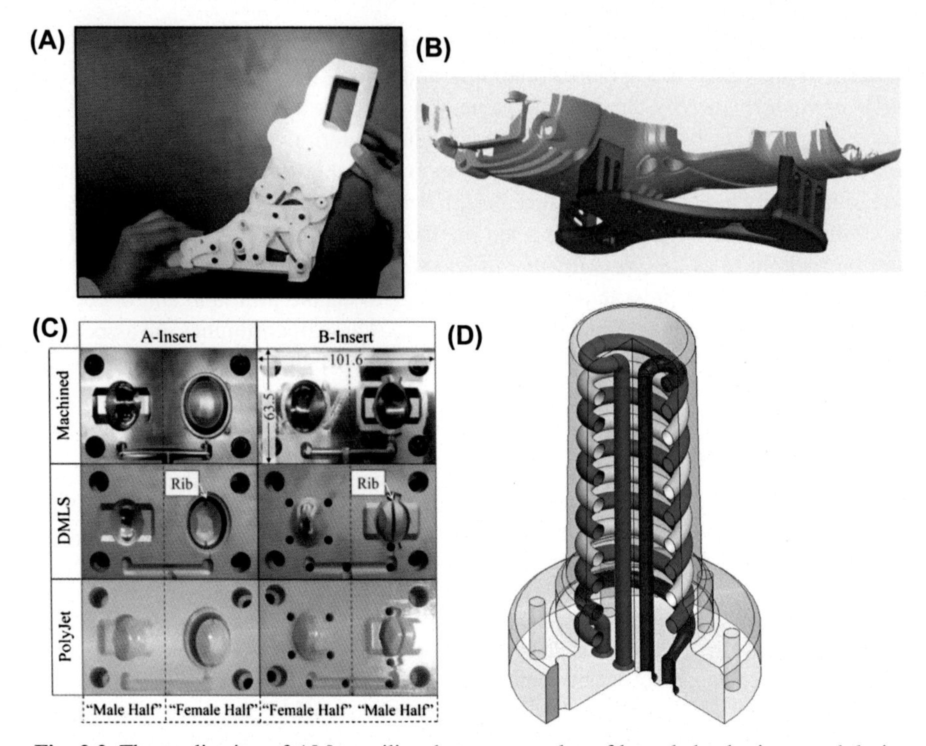

Fig. 3.2 The application of AM to utilize the present value of knowledge by increased design understanding and increased flexibility, for example by: (A) the rapid manufacture of technically functional products (Chapter 8), (B) the fabrication of fixtures and jigs to assist TM methods [5], (C) the rapid manufacture of tooling with polymer jetting and powder bed technologies (Chapter 9) [6] and (D) the manufacture of high-value AM tooling with conformal cooling channels (Chapter 11).

Once the associated TM technologies are available, the product can then be manufactured in the low-cost, high-volume regime, although with the low flexibility inherent to TM processes. This application of AM enables increased design flexibility, as illustrated by the product development examples of Chapter 2, where product development and commercial production are directly integrated.

3. Reduction of manufacturing lead-time by utilizing AM to complement TM technologies. For example, AM can be used to rapidly fabricate fixtures and jigs to assist TM methods. This strategy allows for cost-reduction and reduced time-to-market by allowing technically robust manufacturing hardware to be rapidly fabricated at low cost directly from the existing digital design data.
4. The opportunity to retain flexibility in decision making for a greater proportion of the available time budget. For example, by utilizing AM to reduce the lead-time for tooling manufacture, the available time for design changes can be extended. For example, rapid AM tooling using polymeric or powder fusion methods, or by the rapid manufacture of high-value AM tooling with conformal cooling channels.

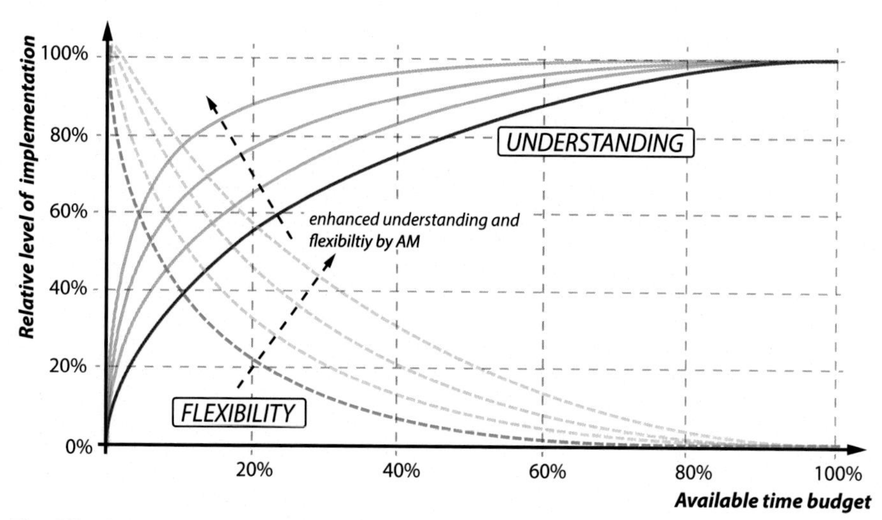

Fig. 3.3 The application of AM to utilize the present value of knowledge by increasing both understanding and flexibility.

3.1.2 Parallel set narrowing

Furthermore, these strategies are compatible with design philosophies of *parallel set narrowing*, a strategic approach to time-constrained product design in the presence of uncertainty (Fig. 3.4). For these scenarios, multiple independent solutions are developed in parallel. At various times, these independent solutions are reviewed, and infeasible or sub-optimal designs are cancelled, allowing the associated design resources to be then applied to the remaining feasible solutions. Parallel set narrowing can be advantageously integrated with AM opportunities to utilize the present value of knowledge in order to de-risk time constrained engineering design projects.

3.1.3 Evolutionary versus revolutionary design

Revolutionary design refers to the class of designed products that are implemented for the first time, or with little relevant prior experience.[3] In these cases, there is no available precedent for the optimal design architecture, and significant effort is required to clarify the associated design requirements and develop a feasible solution. The effort and risk required to commercialize revolutionary projects is potentially high, as are the opportunities for financial return if such a design can be successfully commercialized. Conversely, if a design has been previously commercialized, the product understanding is relatively high, and the risk of technical failure is commensurately low, as is the potential for commercial return. Such design scenarios are referred to as *evolutionary*. Relevant design strategies and associated DFAM tools depend strongly on whether a

[3] For a design bureau engaging in the design of a product that is novel according to their experience, the design should be categorized as revolutionary, even if other bureaus are experienced in similar designs.

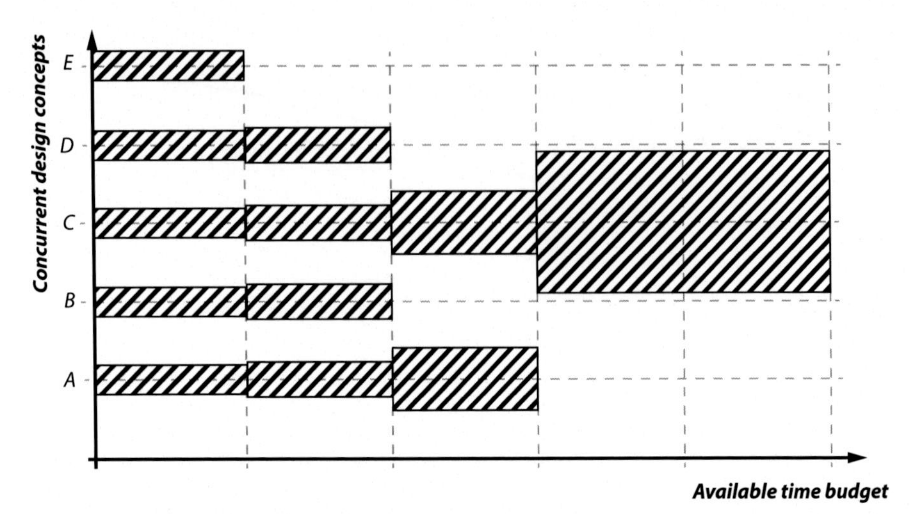

Fig. 3.4 Schematic representation of the parallel set narrowing process for risk-managed time-critical engineering design. Multiple design concepts (A–E) are initially developed in parallel. These concepts are periodically reviewed, and poor performing concepts are cancelled with the design resources redeployed to the remaining concepts. Band width indicates the design resource applied to a specific design concept.

design scenario is fundamentally evolutionary or revolutionary. Appropriate DFAM tools and strategies to assist in managing the data associated with these scenarios are presented in the following sections.

By matching the identified concepts of robust engineering design (i.e. the present value of knowledge; challenges for evolutionary and revolutionary design; and parallel set narrowing) with the specific challenges associated with digital AM design, the criticality of these challenges can be formally evaluated. These challenges are then presented with reference to existing DFAM tools and commercial best practice AM design to identify nascent commercial research opportunities associated with the field of digital DFAM.

3.2 Digital data flow within AM design and production

Engineering design refers to a set of problem-solving strategies that are applied with the objective of specifying a formal solution to a specific technical problem. Various schematic representations have been proposed to quantify the flow of information and associated decision-making within the engineering design process. A schematic representation of the AM design process is presented in Fig. 3.5 as an amalgam of these representations, including the sequential design phases[4] of

[4] This sequential representation is a useful simplification of the actual design process. In practice, design processes include feedback loops that enable review of design decisions as the design progresses. The simplifications applied in this text allow less-distracted focus on the digital AM aspects of the associated design phases.

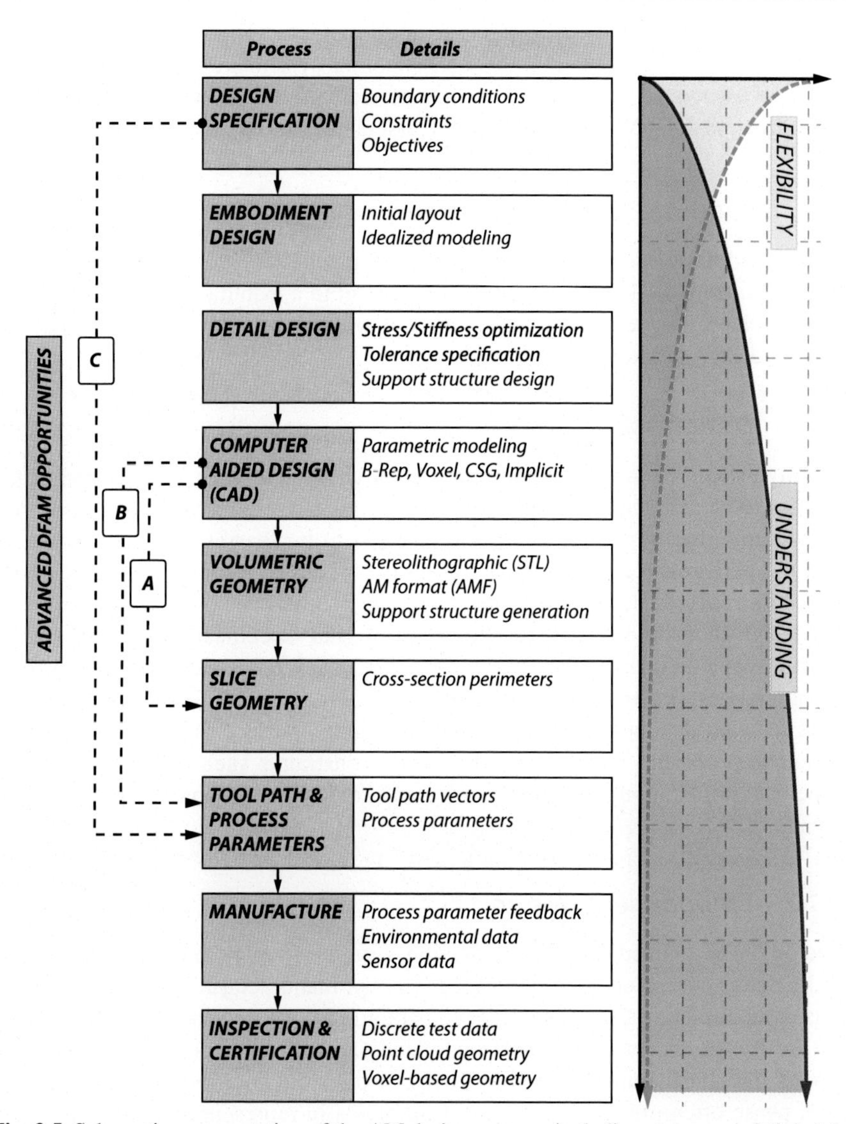

Fig. 3.5 Schematic representation of the AM design process, including summary of digital data and transformations associated with each design phase. Solid arrows indicate the transfer of data. Dashed lines indicate opportunities for enhanced digital DFAM by the feed-forward of digital data (Section 3.4).

relevance to AM design, and the design management attributes associated with these phases. This schematic representation of the design process has been annotated to include the associated digital inputs, transformations and outputs; as well as the potential for advanced DFAM opportunities by the feed-forward flow of digital data to subsequent design phases (Section 3.4). This schematic representation serves as a focus for this chapter, enabling systematic identification of the challenges associated with digital design as well as the associated opportunities for commercial DFAM applications and associated research.

3.2.1 Digital AM design sequence

The proposed design sequence of Fig. 3.5 presents a systematic strategy for the conversion of a formally-defined customer requirement to a manufactured, inspected and commercialized product. In summary, the proposed design sequence embodies the following design phases:

3.2.1.1 Design specification

The design specification should provide a clear and unambiguous[5] definition of the outcomes to be achieved. Design requirements are formally defined; often in terms of constraints that must be met for the solution to be feasible, and objectives by which the performance of a feasible design is ranked. Digital data associated with the design specification phase may include numerical quantities associated with design constraints and objectives; a digital specification of the spatial constraints and associated boundary conditions; and point cloud reference data. Commercial best practice for AM design specification should also include reference to AM relevant design details; for example, required surface finish and directional mechanical properties and associated material constraints such as biocompatibility for medical applications or flame, smoke and toxicity requirements for aerospace design.

3.2.1.2 Embodiment design

Based on the associated design specification, a preferred embodiment is defined that broadly defines the intended topology,[6] geometry, manufacturing processes and materials of manufacture such that design constraints are satisfied, and the design objectives are (to an acceptable degree) optimized. To achieve timely outcomes while allowing the consideration of a broad array of potential solutions, the embodiment design phase often utilizes significant idealisations in representing the design problem. The embodiment design phase is completed in the absence of a complete understanding of the implications of these decisions on the subsequent design phases, as

[5] In this text we will assume that customer requirements are unambiguously defined, although in practice this is often not the case. Vast tomes of engineering design philosophy are dedicated to the resolution of this challenge (see references 1–4); no further effort will be made to resolve this here.

[6] The terms *topology* and *geometry* may convey confusingly similar meanings. To clarify, topology refers to the connections within a structure, where geometry refers to parametrically measurable dimensions of a structure.

indicated by the concept of the present value of knowledge (Fig. 3.1). Digital design methods can be utilized in concert with AM to rapidly increase both design understanding and flexibility (Fig. 3.3), thereby increasing the likelihood of successful design outcomes. These digital design methods include topology optimization, numerical simulation of structural response, simulation of AM manufacturability, generative design, and formal processes for material and process selection (Section 3.3.1). Commercial best practice for AM embodiment design will consider preferred build orientation, associated supporting structures, and AM process parameters at this stage.

3.2.1.3 Detail design

Once the design embodiment phase has been formalized, specific design details are then defined such that engineering production documentation may be specified, including: nominal and toleranced geometries; primary and finishing processes; and specific inspection details. Digital design methods of interest during the detail design phase include CAE analysis, parametric optimization, and generative design. During the detail design phase, fewer design permutations are considered than for embodiment design, and modelling analyses are more sophisticated and less idealized. At this stage AM can be utilized to experimentally validate CAE predictions (for example as discussed in the safety-critical design of AM medical devices, Chapter 7), as well as to experimentally confirm predicted performance and increase the understanding of engineering performance. Commercial best practice for AM detail design will specify preferred build orientation; preferred support structure type and location; and relevant AM process parameters at this phase. However, these parameters may be formalized in later design phases as appropriate.

3.2.1.4 Computer aided design

The Computer Aided Design (CAD) phase refers to the generation of digital geometry data that formally defines the component specified in the detail design phase. CAD often occurs in tandem with detail design; however, it is enumerated here as a discrete phase as the generation of CAD data is often a distinct responsibility to detail design, and there exist distinct opportunities for enhanced CAD methods that are optimized for the specific requirements of AM or are generatively implemented (Section 3.3.3).

3.2.1.5 Volumetric geometry

Based on the formally defined CAD of the intended geometry, a series of data files are generated to represent the external envelope of the nominal AM geometry (Section 3.3.4). It is desirable that this volumetric geometry data be compatible with the subsequent design phases (i.e. slice geometry data) and provide acceptable computational and storage efficiency. Various volumetric geometry formats are available, each with its own set of compatibility, resolution and file size attributes. These include re-purposed data formats initially developed for alternative data storage purposes, as well as custom data formats developed specifically for the geometric, processing and storage requirements of AM. The efficient management of the digital data is crucial to enabling efficient and robust AM outcomes. The formal specification of AM support structures is typically implemented concurrently with the volumetric geometry phase, although this is not strictly necessary, especially for enhanced digital DFAM opportunities that omit the formal definition of volumetric geometry (Section 3.4). The generation of support structures is an active area of DFAM research that enables significant opportunities for commercial benefit by the use of optimized digital design strategies (Chapter 4).

3.2.1.6 Slice geometry

In this phase, the volumetric representation of the intended AM geometry is converted to the discrete layerwise representation of individual *slices* as is required for the specific AM process.[7] Depending on the intended application, this data can be processed automatically, either by proprietary software over which the user has little direct control, or according to explicit design specifications, or by custom or open-source algorithms. The data sets generated at this stage can take various forms depending on the specific processing methods, but they essentially define the geometry of each discretely manufactured layer (Section 3.3.5).

3.2.1.7 Toolpath and process parameters

This phase involves the specification of the end effector trajectory, known as the *toolpath*, and the associated process parameters such that material addition is achieved for the layerwise slice data (Section 3.3.6). Toolpath optimization is a challenging AM design problem, and numerous DFAM strategies have been proposed to generate toolpaths that minimize fabrication time and associated fabrication defects; as discussed in Chapter 4. Selection of toolpath generation strategy also requires consideration of the available computational time and required digital storage space. The process parameters defined at this stage vary in sophistication, from open-loop prescriptive settings based on experimental test coupon data, to complex closed-loop protocols that accommodate predictions of optimal process control and potentially include

[7] AM is typically based on layerwise addition of source material, although material addition may be made according to any convenient coordinate system. The term *slice* is strictly true for Cartesian systems only, but is used here for convenience with no loss of generality.

the capacity to provide feedback process control. The output of this design phase is a data set that unambiguously defines the toolpath and associated process parameters for each stage such that AM manufacture can occur.

3.2.1.8 Manufacture

The digital data generated during manufacture varies significantly depending on the specific AM system used. For example, low-cost systems may generate no formal data during manufacture, whereas for more sophisticated systems extensive data may be generated. Digital data generated during manufacture may include environmental variable sensor readings and process parameter feedback that logs the performance of relevant variables during manufacture. These data sets provide invaluable insight into process behaviour; however, unless computational costs and file size are managed, this can lead to data overload and efficiency challenges. These data challenges are especially challenging if in-situ generated data are intended to be available sufficiently quickly to provide process feedback control (Section 3.3.7).

3.2.1.9 Inspection and certification

This phase intends to assess the quality of the manufactured product and includes the metrology data required to confirm correct dimensional outcomes, and proof testing (either destructive or non-destructive) to confirm mechanical response (Section 3.3.8). This inspection data can take the form of discrete-point data from CMM or tensile testing; point-cloud data from surface-scanning technologies; or volumetric data from micro-CT scanning. The complexity of this data can readily expand such that inspection processes dominate production cycles and associated costs. To avoid undue costs, it is necessary that the functional intent of an inspection process be matched to the associated inspection cost (Chapter 4). For example, the initial qualification of an intended AM process for high-value applications requires that the influence of relevant process variables on mechanical, dimensional and aesthetic properties be robustly quantified; and, for these scenarios relatively high inspection costs can be borne. Conversely, for production inspection processes, the available inspection budget is typically lower, and reference may be made to less expensive methods such as Coordinate Measuring Machine (CMM), vision systems and coupon testing.

These phases define a sequential series of requirements for AM design, specifically by identifying the associated digital data sets that are associated with each design phase (Fig. 3.5). In the following sections, opportunities for enhanced digital data management are identified for each design phase; these opportunities are elaborated with examples of the use of DFAM tools that provide faster outcomes and reduce unnecessary data storage. DFAM methods that allow computationally seamless integration of design phases are of particular importance, as these are necessary for the implementation of generative design methods (Chapter 7).

-	INVALID ZONE
M	TYPICALLY MANUAL
A	TYPICALLY SEMI-AUTOMATED
X	ADVANCED DFAM OPPORTUNITY

		Design Specification	Embodiment Design	Detail Design	CAD Data	Volumetric Geometry	Slice Geometry	Tool path & Process Parameters	Manufacture	Inspection & Certification
		1	2	3	4	5	6	7	8	9
Design Specification	1	-	M							
Embodiment Design	2	-	-	M				X		
Detail Design	3	-	-	-	M					
CAD Data	4	-	-	-	-	A	X	X		
Volumetric Geometry	5	-	-	-	-	-	A			
Slice Geometry	6	-	-	-	-	-	-	A		
Tool path & Process Parameters	7	-	-	-	-	-	-	-	A	
Manufacture	8	-	-	-	-	-	-	-	-	M
Inspection & Certification	9	-	-	-	-	-	-	-	-	-

Fig. 3.6 Review of state-of-the-art of AM design practice, highlighting phases that are typically implemented by manual or semi-automated DFAM tools and nascent opportunities for *advanced DFAM tools* (Section 3.4).

In addition to these opportunities for the application of commercial best practice DFAM design tools, the subsequent section provides opportunities for new DFAM outcomes by taking advantage of *advanced* DFAM opportunities (Section 4). These advanced methods allow the commercial best practice AM design methods to be either bypassed or automated, thereby providing opportunities for highly efficient AM outcomes, but requiring additional awareness and strategic planning on the part of the design team. These opportunities are symbolically identified in Fig. 3.6.

3.3 Digital DFAM opportunities for digital design and data management

The design phases identified in Fig. 3.5 are presented in greater detail in the following sections to allow discussion of the available digital DFAM tools, and to present commercial best practice design and opportunities for commercially relevant DFAM research.

3.3.1 Embodiment design

The embodiment design phase involves the high-level generation and evaluation of potential solutions to a specific statement of requirements. Embodiment design is therefore less relevant to evolutionary design programs (Section 3.3.1) as the preferred design embodiment is already defined (although specific aspects of the design need to be formalized in the detail design phase). However, for revolutionary design, the selection of a preferred embodiment is highly challenging for two reasons. Firstly, it involves the selection of an optimal solution among a parametrically vast number of unique permutations.[8] Secondly, the present value of knowledge is such that, at this critical stage, when much flexibility exists to influence the design outcomes, the problem understanding is relatively low (Fig. 3.1). These challenges can together result in a very real risk of failure to identify an acceptably high performing solution within the (limited) available design time. As previously identified, the task of design embodiment is not a core focus of this text. However, several opportunities of particular relevance to AM exist and are discussed below in the context of digital design, including: the present value of knowledge and parallel set narrowing; topological optimization to identify efficient structures; and systematic methods of material selection.

Embodiment design is a critically important phase of engineering design as it is endowed with significant design flexibility. This flexibility can be utilized to investigate a variety of solutions that address the associated design requirements. AM provides a valuable opportunity to utilize the present value of knowledge in order to improve the understanding of these potential solutions early in the embodiment phase when there remains flexibility to implement these solutions (Section 3.1.1). Specific DFAM techniques include the rapid fabrication of technical and aesthetic prototypes; the efficient fabrication of tooling and manufacturing systems; and initial series-production using AM methods. These techniques are directly compatible with the philosophy of parallel set narrowing, which allows multiple potential solutions to be considered while managing the risks associated with timely product delivery.

Design embodiment typically involves the accommodation of structural requirements within the available design volume in such a manner that the intended method of manufacture is accommodated and the available budget is not exceeded. Structural optimization is a subject to multiple constraints and optimization variables. Consequently, obtaining a high-performing optima within the available design budget can be extremely challenging. Topology optimization provides a systematic method for identifying structurally robust design embodiments for a specific set of design requirements and can therefore provide a profoundly useful DFAM tool. Furthermore, the topologies generated by topological optimization routines are often not compatible with traditional manufacturing methods, but are potentially more suited to AM. The synergies between AM and topology optimization are explored in Chapter 6 with a focus on commercial best practice and associated research opportunities.

[8] The computational and intellectual challenge of optimizing such dimensionally large problems is referred to as the *curse of dimensionality* and is presented in detail in Chapter 7.

In addition to the being highly compatibility with AM, topology optimization is directly compatible with digital design methods. Topology optimization methods, summarized in Chapter 6, are based on a digital definition of boundary conditions, loads and allowable spatial envelope. This data is digitally processed by a topology optimization algorithm to generate an optimized material distribution (Fig. 3.7).

Although topology optimization is more compatible with AM than traditional methods, it is not inherently compatible, and topologically optimized structures can fail to be directly manufacturable by AM methods. These failures can be due to any of the AM manufacturability challenges identified in Chapter 4, including excessively acute inclination angle and failure to accommodate thermal load paths or entrapped source material. DFAM tools that can integrate AM manufacturability requirements with the fundamental topology optimization routines provide a significant opportunity for commercial advantage. Such DFAM tools are currently under development and the field is highly research active (Fig. 3.8). Commercial best practice, DFAM tools and research foci for the topological design of AM systems are presented in detail in Chapter 6.

Material selection is a similarly challenging aspect of embodiment design, particularly because the available library of candidate materials is vast, multiple disparate material properties are of relevance to performance, and the designer's expertize in the available materials is potentially limited or biased by their prior experience [12]. The field of systematic material selection is a relatively recent discipline of engineering design; and, in particular, consists of formalized procedures for predicting the response of the associated structural element when subject to a library of candidate material properties [13]. These allow systematic and efficient identification of feasible and optimal materials for a specified engineering objective. Fig. 3.9 indicates a potential application of these methods for a specific combination of material related objectives. The extension of these methods to the specific manufacturing requirements of AM remains an open research opportunity, including the development of alloys that are optimized to the thermal cycles of metal additive manufacture as well as the design of polymers that enable novel functionality.

Furthermore, although systematic material selection methods are fundamentally compatible with digital DFAM methods, there remain significant research opportunities to implement DFAM tools that accommodate material selection in a practically useful manner. Specifically, the documentation of AM-specific manufacturability requirements remains *ad-hoc* and without consistency in reporting, and the current generation of DFAM tools, particularly topology optimization methods, do not typically accommodate material selection as part of their standard optimization strategy. The integration of topology optimization and material selection is exemplary commercial practice and can yield significant performance enhancement (Fig. 3.10); however, systematic DFAM tools to support this practice are not well implemented.

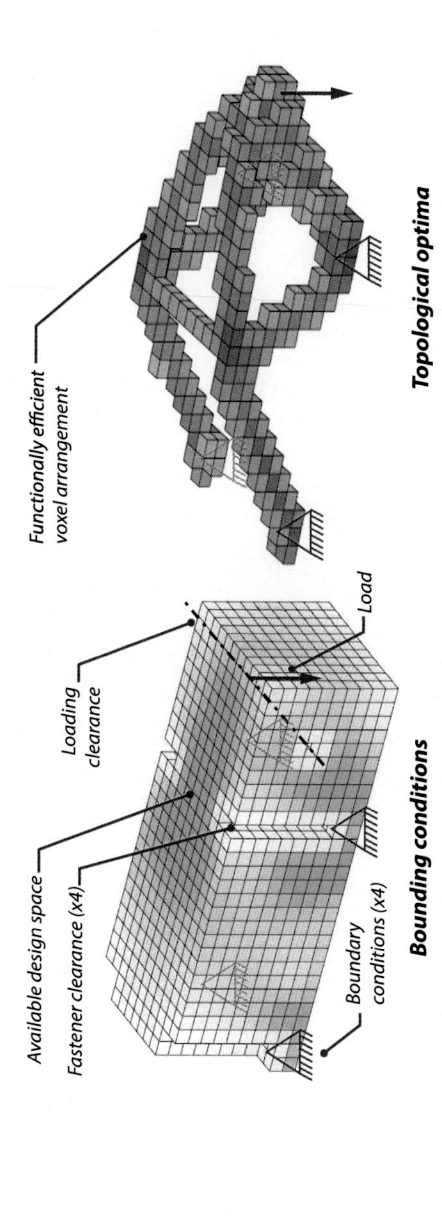

Fig. 3.7 Topology optimization methods are compatible with digital design processes. A digital specification of boundary conditions, loads and allowable spatial envelope is algorithmically processed to generate a digital representation of the optimized material distribution.

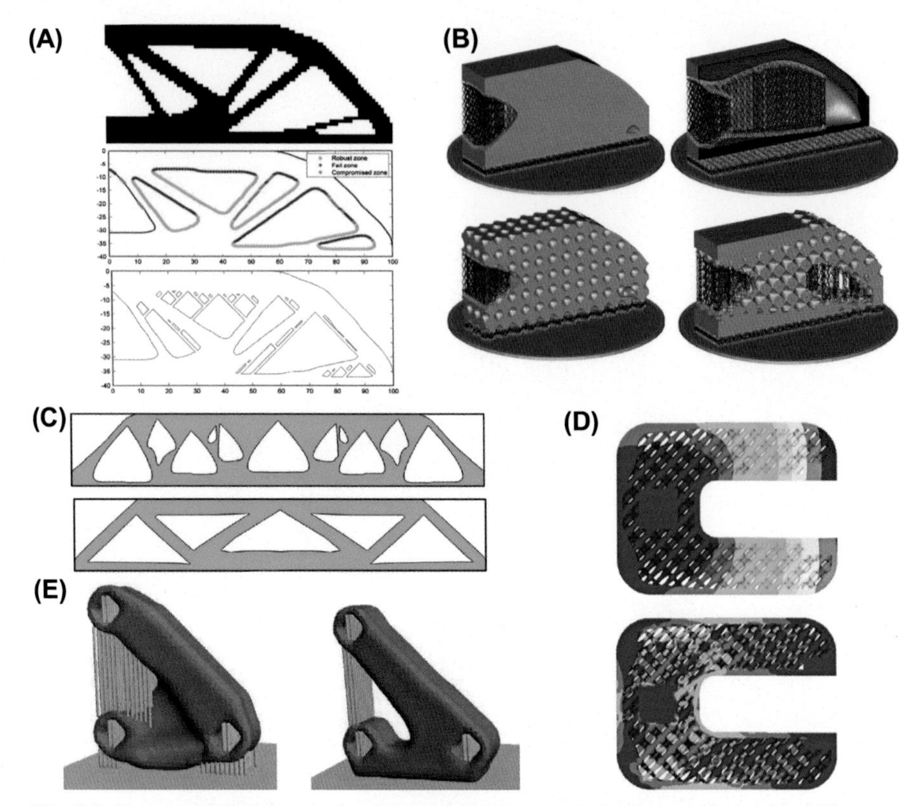

Fig. 3.8 Proposed DFAM tools that integrate topology optimization routines with AM manufacturability constraints. (A) [7]. (B) [8]. (C) [9]. (D) [10]. (E) [11].

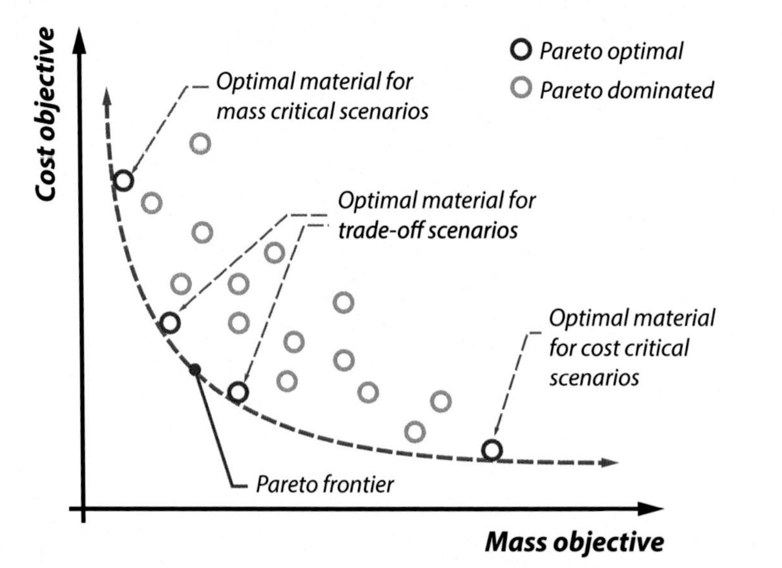

Fig. 3.9 Material selection chart example indicating a Pareto frontier for mass and cost objectives (performance increases as objective tends to zero).

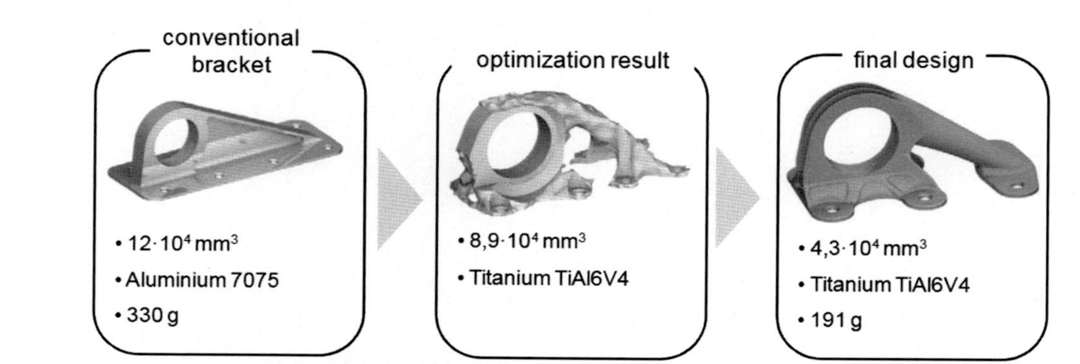

Fig. 3.10 Commercial best practice example of the integration of topology optimization with material selection for the design of aerospace bracketry to achieve a mass reduction of 42% [14]. Note that in this image the optimization result (middle) is the output of the embodiment phase, whereas the final design (right) is the output of the detail design phase.

3.3.2 Detail design

At the detail design phase, the system embodiment is finalized, and specific parametric details are then defined with the intent of optimizing system performance according to the design specification. This phase typically requires parameterization of the associated embodiment: a technically challenging and potentially labour-intensive process. The embodiment is often defined as some level-set function (e.g. for gyroid scenarios, Chapter 5), voxel field (for topology optimization scenarios, Chapter 6), or continuum methods (for example, medical implant applications, Chapter 7). These data sets are typically not directly compatible with the requirements of parametric optimization.

Despite significant research effort, the robust conversion of embodiment data to a parametric representation is typically achieved manually, based on the experience and interpretation of the human designer (Fig. 3.11). This manual intervention provides opportunities for the engineer's creative input, experience and high-level project understanding to add value to the manufactured outcome. However, it does impose a potentially crippling bottleneck to the digital AM workflow, i.e. that manual intervention introduces a possibility for uncertainty based on the specific designer's preference and has the potential to stymie process automation. The former issue is potentially challenging for safety-critical scenarios where process documentation and justification of design decisions is paramount; the latter issue is of critical importance to the opportunities of generative design (Chapter 7). In response to this challenge, a significant opportunity for enhanced digital design exists, whereby the transition from embodiment to detail design is digitally automated, as detailed in the

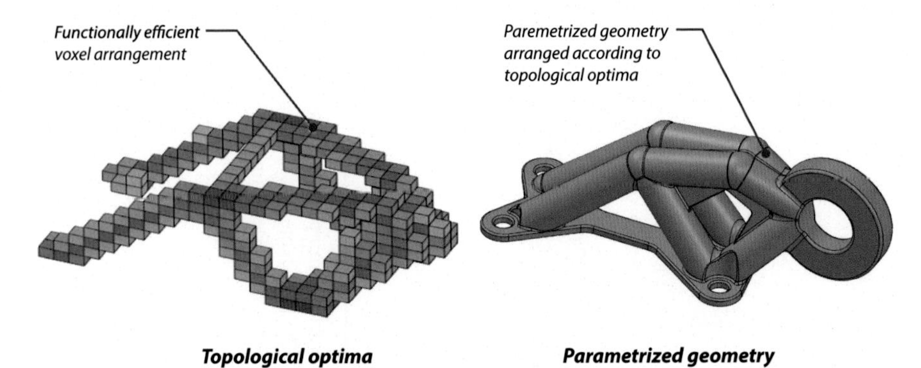

Fig. 3.11 Geometric representations appropriate for embodiment (left) and detail design phases (right). Embodiment design is often compatible with spatial partitioning methods that efficiently describe the connectivity of points within a specific volume, such as the voxel output of topological optimization methods. Whereas, detail design phases typically require functionally modifiable parameterized geometry. Advanced digital DFAM opportunities to automate the transition between embodiment and detail phases are discussed in Section 3.4.

enhanced digital design methods of Section 3.4. The merit of automated methods depends on the complexity and frequency of the conversion from embodiment to detail design. For example, when detail design occurs infrequently (as is typical in evolutionary design) the optimization penalty associated with the manual generation of parametric design data is less critical than for revolutionary design scenarios where the conversion from embodiment to detail occurs for each unique product.

Commercial best practice for AM detail design will include the optimization and specification of build orientation and associated support structures. These details are presented here as they are of critical influence on the quality of the manufactured component and therefore must be understood early in the design process such that their influence can be robustly controlled and accommodated[9]; in practice however, these attributes may be formally refined in later design phases as appropriate.

3.3.2.1 Build orientation

The feasibility and performance (both economic and technical) of an AM product is highly dependent on its orientation within the build volume. Optimization of AM part orientation was one of the first fields of formal DFAM research, and a significant range of associated DFAM tools have been developed. DFAM guidelines and algorithmic tools for optimal part orientation are readily available and are typically implemented as commercial best practice. However, the optimization of AM part orientation is highly multivariate and must consider many relevant design variables, including manufacturing time; local roughness; necessity and location of support structures; local temperature profile; and microstructure. Many of these variables are poorly understood or computationally challenging to acquire, such as temperature field data for thermal AM systems (Chapter 11). A selection of commercial and research implementations of DFAM orientation optimization are introduced in Fig. 3.12 and are further discussed in Chapter 4.

For various scenarios, especially for thermal AM systems with complex manufacturability characteristics, the designer's skill, experience and understanding of specific product function is invaluable in manually optimizing part orientation. Manual optimization of orientation can provide the most agile commercial solution to the orientation optimization problem. In particular, for scenarios that are so computationally demanding that no feasible DFAM tools exist; for example, the nesting of multiple components that are subject to thermal constraints. However, manual orientation presents a commercial challenge, in that it presents a potential workflow bottleneck and potentially introduces uncertainty in orientation according to user-preference: a significant challenge to the commercial application of AM to medical and safety-critical applications.

[9] Failure to accommodate the performance-critical attributes of orientation and supporting structures at the design phase can result in failure to meet the associated design requirements, thereby necessitating costly redesign at a late design phase.

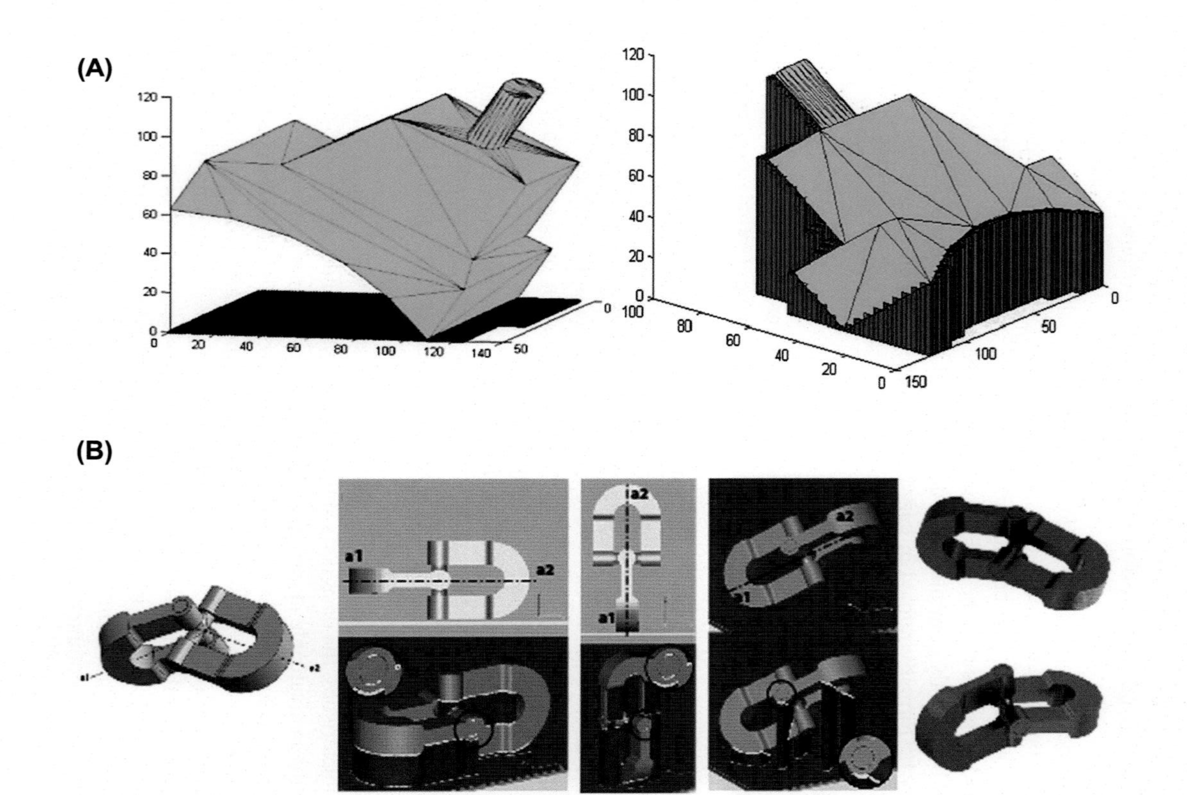

Fig. 3.12 Selected DFAM approaches for optimization of part orientation. (A) Orientation to minimize part error and support structure volume [15], (B) optimization to maintain technical function [16].

3.3.2.2 Support structure generation

Support structures are a core AM design capability that can be used to offset some of the manufacturability constraints inherent to AM processes, thereby dramatically increasing the scope for AM solutions to technical challenges. The opportunities for support structures to address specific AM manufacturing limitations are typically classed as follows (Figs 3.13–3.16).

- Providing physical support for AM processes that cannot naturally accommodate overhanging geometry, as occurs due either to an accute inclination angle, α, or an unsupported region within a build layer. For example, in Material Jetting processes (MJT, Chapter 9), overhanging materials cannot be added to the build without direct physical support from the supporting materials on the previous layer.

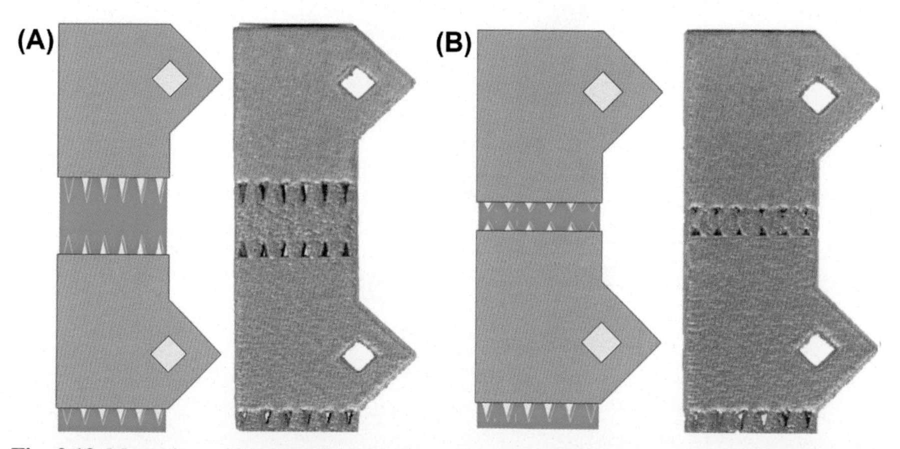

Fig. 3.13 Manual specification of supporting structure and as-manufactured specimen using Powder Bed Fusion (PBF, Chapter 11) AM methods. (A) Relatively sparse support with large vertical offset, (B) relatively dense support with small vertical offset.

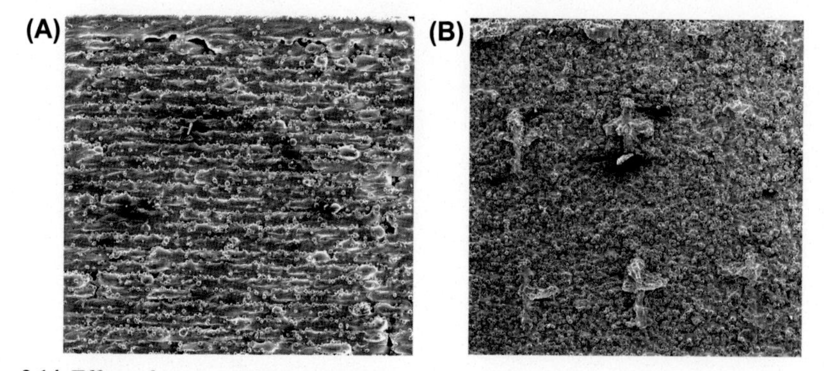

Fig. 3.14 Effect of support structure removal on Powder Bed Fusion (PBF, Chapter 11) specimen surface (image side length is approximately 5 mm): (A) upward facing surface with no support structure, (B) downward facing surface showing physical damage associated with support structure removal.

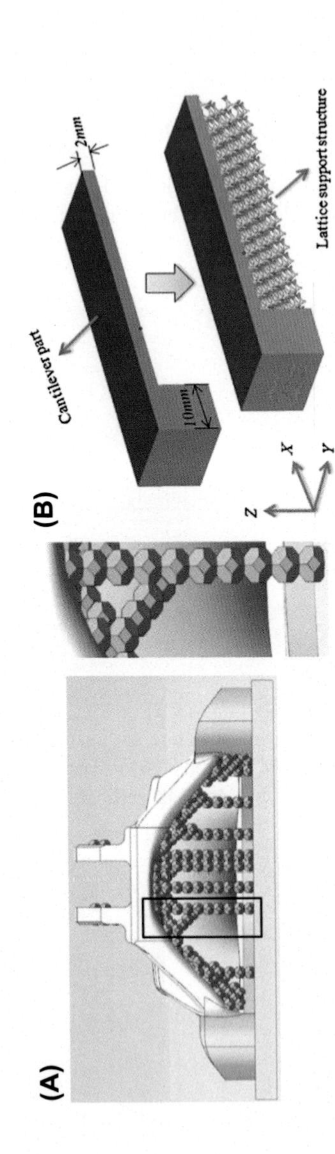

Fig. 3.15 Manufacturing limitations addressed by the use of support structures generated by: (A) algorithmically deployed unit cells [17], (B) lattice support structure [18].

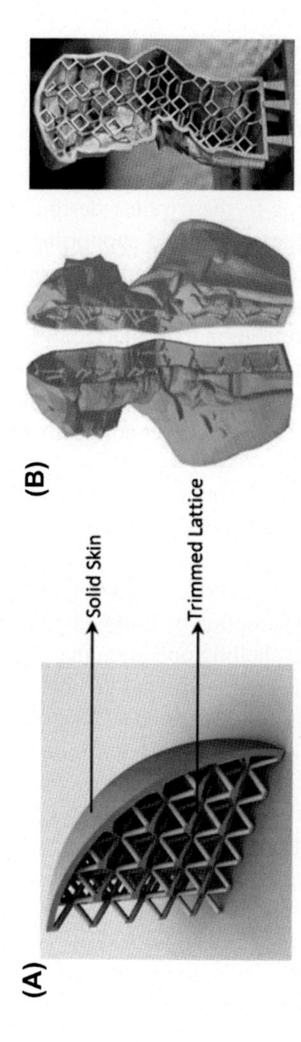

Fig. 3.16 DFAM approaches to hollowing and then internally supporting AM geometry. (A) Functionally graded internal lattice structure [19], (B) hierarchical supporting structure [20].

- Ensuring manufacturability of overhanging materials. For example, as occurs in Fused Filament Fabrication (FFF, Chapter 8) in the presence of acute overhangs. This classification is distinguished from the previous, as the material requires support only during solidification. In addition to the use of support materials, alternative methods, such as active cooling, optimized process parameters, and polymers with a narrower transition temperature can also be used to address this issue.
- Addressing fusion challenges due to heat transfer paths. For example, Powder Bed Fusion (PBF, Chapter 11) processes require heat flow to solidify the meltpool. The use of support material can introduce additional heat transfer paths that can increase manufacturability by influencing the heat transfer resistance.
- Anchoring fused materials within the bath of curable resin. For example, as is required by Vat Polymerization (VPP, Chapter 10) methods.

Despite the geometric opportunities enabled by the use of support structures, there remain open-research questions as to their robust design and optimization. Support structures are specified using either automated supporting structure generation algorithms or by the manual deployment of standard supporting geometries (Fig. 3.13). Support structures can become problematic if entrapped within the associated geometry and can result in physical modification to the surface structure, which may be functionally or aesthetically undesirable (Fig. 3.14). Furthermore, support structures consume input material and increase manufacturing time (both for direct manufacture and subsequent removal) and are therefore economically significant for commercial AM scenarios.

As the size of a three-dimensional structure is increased, the fraction of internal volume scales to the side length cubed. Therefore, especially for larger structures, the potential for increased manufacturing time and material consumption increases significantly with structure size. For these scenarios, algorithmic DFAM tools are available that reduce material consumption by hollowing the intended geometry. Often these tools involve some Boolean shelling operation that removes internal material according to some defined wall thickness. Internal geometry that does not satisfy AM manufacturability requirements is then identified and accommodated by some supporting structure that is typically a self-tessellating infill, or some hierarchical or fractal structure that is recursively applied (Fig. 3.16).

The generation of support structures for commercial AM design is challenging as it is a highly multivariate optimization problem; there exist somewhat limited data on the practical effect of AM design variables on manufacturability; and the simulation of relevant effects can be challenging, especially for physically complex AM technologies. Consequently, most of the available DFAM tools for support generation are based on simple proxies for manufacturability, notably feasible build angle, which is commonly implemented for FFF systems (Chapter 8). Alternatively, DFAM tools exist that generate support structures in a semi-automated fashion according to the designer's experience and requirements – as is commonly implemented for physically complex AM systems such as PBF (Fig. 3.13). The optimization of support structures themselves, and the associated DFAM tools for their deployment remains a valuable and active research area (Chapter 4).

3.3.3 Computer aided design

Computer Aided Design (CAD) data provides unambiguous representations of the geometric envelope associated with a specific intended geometry. Numerous formal protocols for CAD representation exist and are defined for various purposes according to their specific capabilities and attributes. These attributes include the fundamental method of geometric representation, either as a volume or external surface; the representation of geometry as either explicit or implicit data; the associated data storage protocol; and the representation of either toleranced or nominal geometries. As noted earlier, although CAD is typically generated at the detail design phase, this is not strictly necessary, and it is considered as a unique design phase here due to its importance to the digital aspects of AM.

CAD data can be generated according to numerous protocols. These protocols consist of both proprietary and open-source formats and can be classified according to the method of geometry generation. Prominent *explicit methods* for data generation include *Constructive Solid Geometry* (CSG) and *boundary representation* (B-rep). *Implicit methods* for geometry representation include voxel methods, level sets, and scalar fields that indirectly represent the geometry of interest. Each of these methods has distinct opportunities and challenges for AM application (Fig. 3.17).

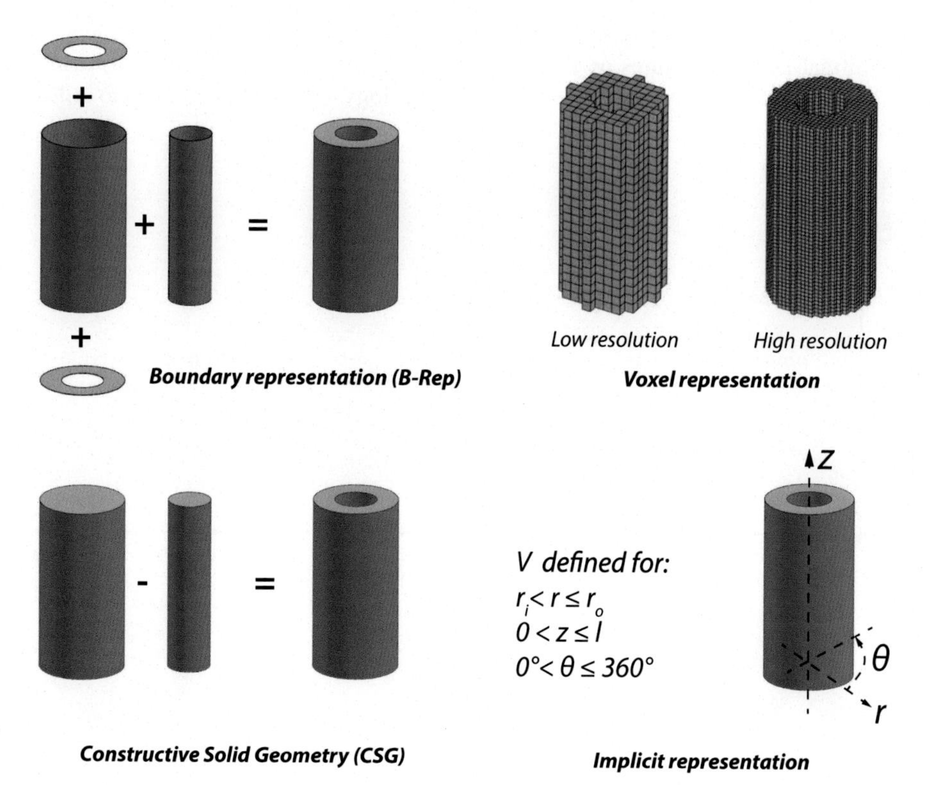

Fig. 3.17 Candidate methods of CAD representation.

Constructive Solid Geometry (CSG) represents the intended component geometry with Boolean operations applied to primitive geometry structures, such as spheres, cylinders and cubes. These CSG methods are therefore eminently compatible with algorithmic methods, as are required for generative AM design (Chapter 7). However, these representations are not necessarily compatible with curvilinear geometries, which are not readily constructed from the available library of primitive structures.

Boundary representations (B-rep) consist of surface elements that interconnect to define the volume of interest. B-rep is robust and flexible, especially for curvilinear geometries including complex variable radius fillets and sweeping blends. Consequently, B-rep is often the preferred data representation used in manual CAD software. The explicit data representation of B-rep and CSG methods can be data-inefficient for repetitive geometries, including for self-tessellated lattice structures and the complex geometry that is often preferred for high-value AM applications. For these scenarios, the necessary file size can rapidly expand beyond the feasible data processing limit (Fig. 3.18).

This data storage challenge can be potentially offset by efficient CAD protocols that allow geometry to be defined in terms of the repetitions of a unit cell, rather than requiring that this geometry be re-stated for each repetition. This approach is feasible for self-tessellating geometry and can achieve a significant file size reduction. However, this data compression approach may be incompatible with functionally graded or conformal lattice structures as for these structures each lattice cell is geometrically unique, although it may be based on a similar topology. Another approach is to utilize

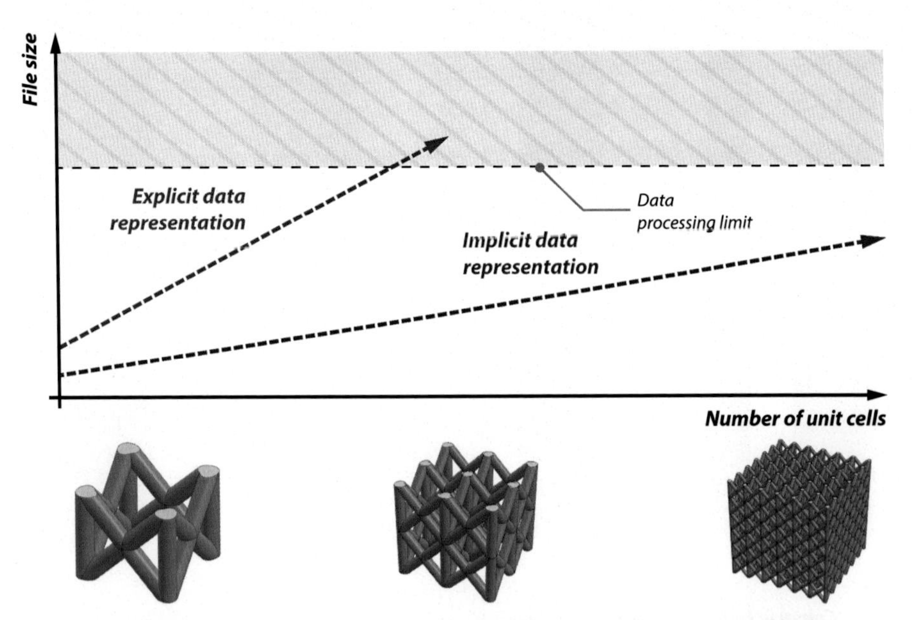

Fig. 3.18 Schematic representation of explicit and implicit data representation of lattice structure. Data generated by commercially available tools with no data compression for repetition elements.

implicit CAD representations to define complex geometry. These implicit representations may provide data size efficiencies for a specific geometry complexity, and consequently such implicit representations are emerging within the commercial DFAM tools (Fig. 3.18).

Topological optimization is an enabling design method for the commercialization of AM technologies. The geometric output of topological optimization methods is typically in the form of a *voxel field* that implicitly defines the volume of interest by discrete cubic sub-volumes (Fig. 3.17). The application of implicit voxel field data to represent AM geometry presents a research opportunity for the development of data structures that represent topologically optimized geometry with high data efficiency.

As well as accommodating digital processing and storage requirements, CAD data must be generated with an understanding of the technical requirements and associated computational costs of the subsequent design phases, particularly the conversion of CAD data to volumetric and layerwise data as required for AM processing.

3.3.4 Volumetric geometry

The available CAD data is converted to a data set that represents the volume of the intended component in a manner that is compatible with the subsequent design phases; in particular, the slice geometry phase, whereby volumetric data is converted to discrete planar slice data (Section 3.3.5). As discussed in Section 3.4, enhanced digital DFAM methods can make this design phase redundant: it is included here because it is representative of current commercial best practice, and to allow insight into opportunities for DFAM research contributions. Various methodologies are available to represent the volumetric envelope of the intended geometry; these are either based on existing data formats that are re-purposed for AM or are custom developed AM formats. Of these available formats, the stereolithographic (STL) format is the most commonly adopted in commercial practice. In response to identified limitations of the STL for AM application, alternative data formats, such as the Additive Manufacturing Format (AMF) have been proposed. These volumetric data representations are presented below.

3.3.4.1 Stereolithographic file format

At the time of writing, the standard volumetric AM format is the stereolithographic file (STL). This format was originally generated by 3D Systems for use with their steriolithographic methods of 3D printing. The STL format represent the surface of the intended geometry with a finite number of adjoining triangular facets that are assigned by a standard format consisting of three vertices per facet and a normal vector specified to define the outward direction of the surface being defined (Fig. 3.19). These facets are presented in unstructured manner with no specific scale data. Facets are defined uniquely and therefore data repetition is inherent in the STL format as facets share vertices and this data is redundantly repeated for each facet. STL data can be stored as either as ASCII or (more efficiently) binary data (Table 3.1).

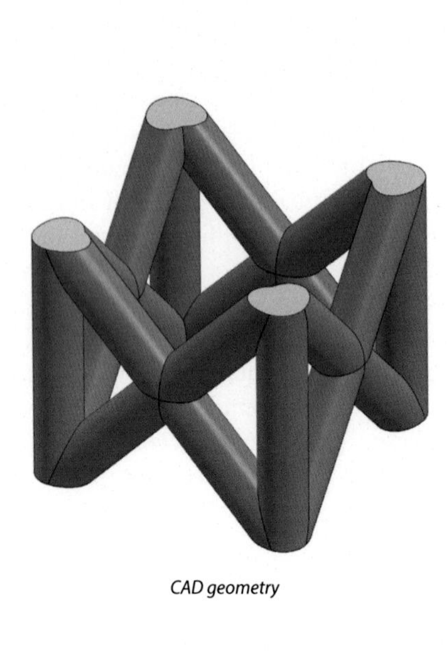

CAD geometry

```
solid FCZ_ASCII
   facet normal 7.879406e-001 -4.354019e-001 -4.354019e-001
      outer loop
         vertex 5.871572e+000 4.192288e+000 8.345653e-001
         vertex 5.933013e+000 4.396447e+000 7.500000e-001
         vertex 5.871572e+000 2.500000e+000 2.526853e+000
      endloop
   Endfacet
   facet normal 9.975989e-001 4.897104e-002 4.897104e-002
         ...

      ...
      endloop
   endfacet
endsolid
```

STL file structure

STL geometry visualisation

Fig. 3.19 CAD geometry and associated STL format representation in steriolithographic (STL) file format, including a truncated sample of the ASCII STL file used to generate this data.

Table 3.1 Standard STL formatting: ASCII (left) and binary (right). While either file type contains the same information and is acceptable for most AM pre-processors, the binary option offers a significant reduction in file size over ASCII files.

ASCII	Binary	
solid (name)	UINT8	*%header*
facet normal $(n_1)(n_2)(n_3)$	UINT32	*%total number of facets*
outer loop	FLOAT32	*%facet normal*
vertex $(v_{1-1})(v_{1-2})(v_{1-3})$	FLOAT32	*%vertex 1*
vertex $(v_{2-1})(v_{2-2})(v_{2-3})$	FLOAT32	*%vertex 2*
vertex $(v_{3-1})(v_{3-2})(v_{3-3})$	FLOAT32	*%vertex 3*
endloop	UINT16	*%free attribute byte count*
endfacet	...	
...	end	
Endsolid (name)		

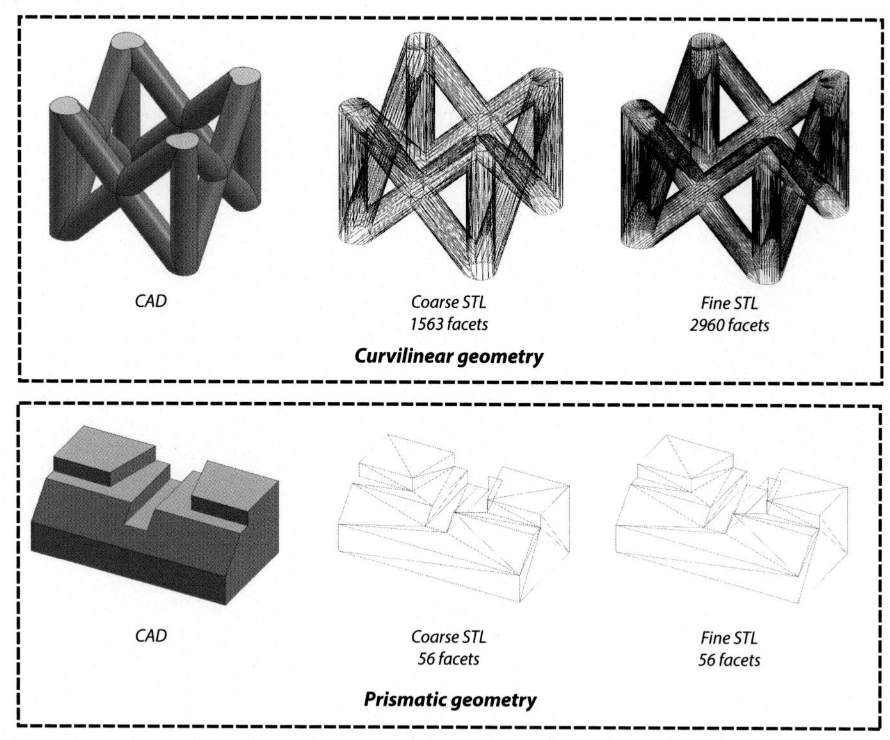

Fig. 3.20 Representation of curvilinear and prismatic geometries with STL format at coarse and fine resolution. The STL represents prismatic geometry without error, but can suffer from poor data efficiency for the representation of curvilinear geometry.

The STL data structure provides a straightforward and technically robust mechanism for defining volumetric data. For prismatic geometries, this structure is computationally efficient, as few facets are required to unambiguously define the component volume. However, the representation of curvilinear geometry with discrete planar facets is a particular challenge associated with the STL format (Fig. 3.20). There exist multiple strategies for measuring and accommodating the associated *faceting error*, including the angular and chordal deviation associated with a specific facet in representing the intended geometry. The typical response to the challenge of faceting error is to manually prescribe an allowable deviation from the original CAD representation: this is often qualitatively described in CAD export options as, for example, either *fine* or *coarse* resolution (Fig. 3.20). It is apparent that an infinite number of planar facets are required to represent a curvilinear geometry without geometric error (Fig. 3.21). It is in response to this challenge that custom AM volumetric data storage formats have been developed.

The data storage challenge for the STL format exists only for curvilinear geometries. Unfortunately, many of the geometries of interest to AM generate their technical function and structural integrity by their local curvature. Two examples of this locally curved

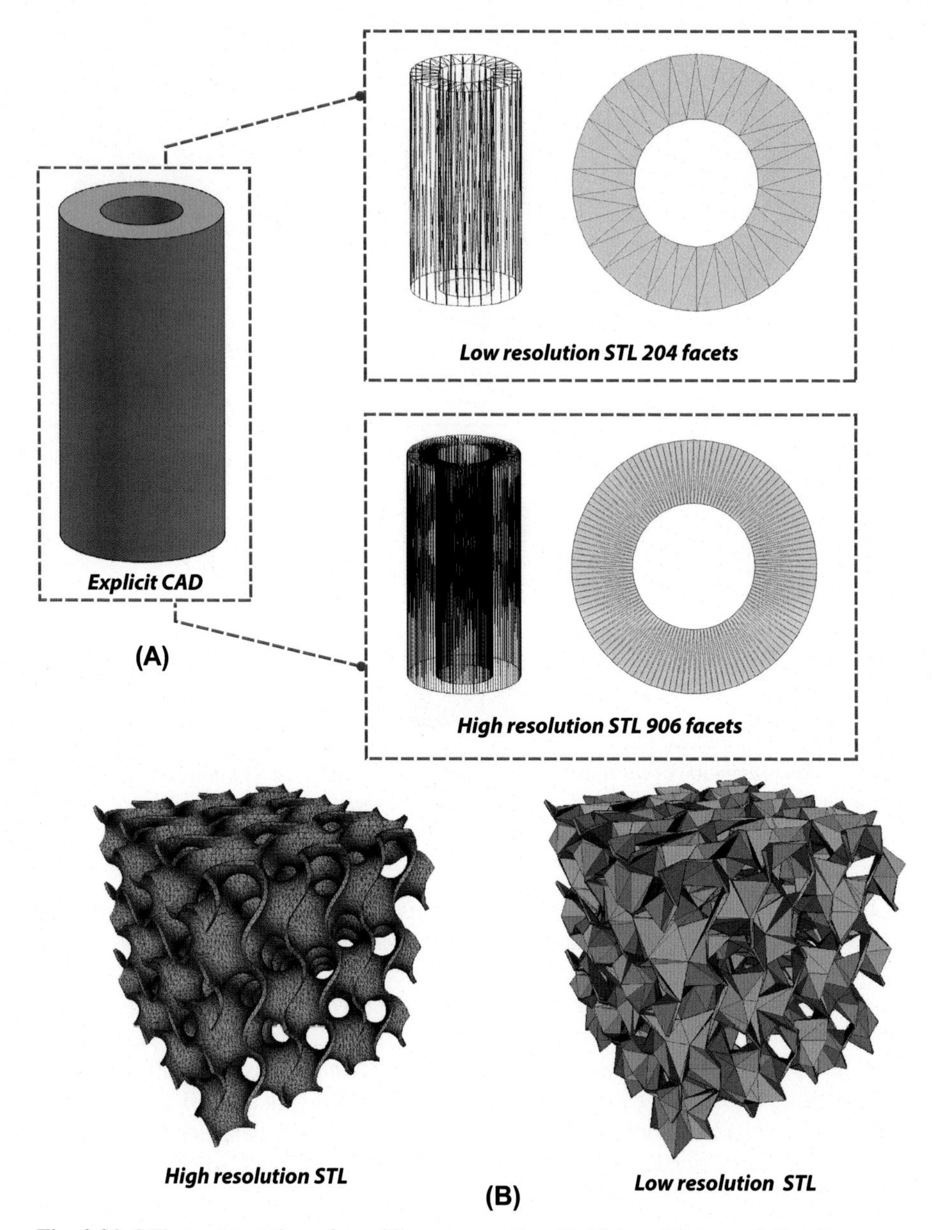

Fig. 3.21 STL representation of curvilinear geometry with high and low resolution for (A) explicitly defined cylinder and (B) implicitly defined gyroid.

geometry are the Triply Periodic Minimal Surface (TPMS) and the lattice structure, both of which are of significance to optimal AM design (Chapter 5) and can be mechanically and aesthetically compromized by *under-sampling* of the STL (Fig. 3.22). Under-sampling occurs when the STL facet representation provides insufficient resolution

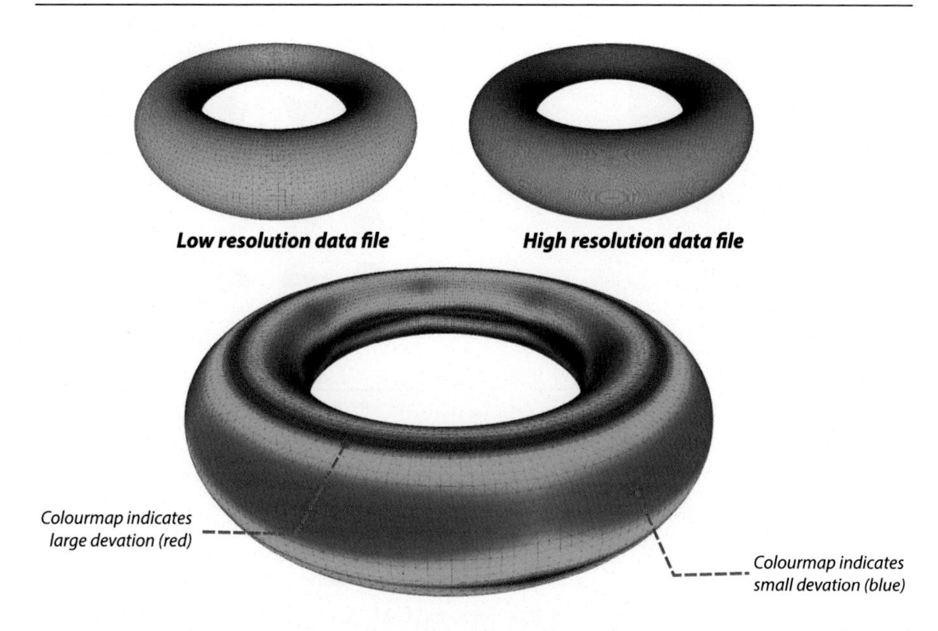

Fig. 3.22 STL representations of toroidal structures with high and low facet resolution, colourmap indicates the discrepancy between these data representations.

for the specific AM process, resulting in measurable facets in the manufactured part. This insight provides a little-utilized DFAM design opportunity for the optimization of STL data sets – i.e. that the facet frequency and associated geometric error is not technically compromising if its effects on the manufactured geometry are understood and actively accommodated. There are two associated DFAM opportunities: experimentally quantifying the effect of faceting error on the as-manufactured geometry, and utilizing the STL file data format to more efficiently represent the required geometric data.

Faceting error incurred by STL data representation can compromise data quality, resulting in visible geometric artefacts[10] and compromised structural response (Fig. 3.23). The specific features of these artefacts are not only dependant on the geometric faceting effect but are also influenced by all phases of the AM design process, including the local geometry, orientation within the build-envelope, material feedstock specifications, prescribed toolpath and process parameters. The practical effect of these errors can be actively mitigated by experimentally quantifying the effect of these variables on the as-manufactured geometry and associated mechanical properties. Various experimental methods exist for quantifying this experimental error, depending on the

[10] In this text, the term *artefact* is used to refer to geometric features that are a predictable and inherent attribute of AM manufacture, whereas the term *defect* is used to refer to a random or unpredictable attribute that is fundamentally avoidable.

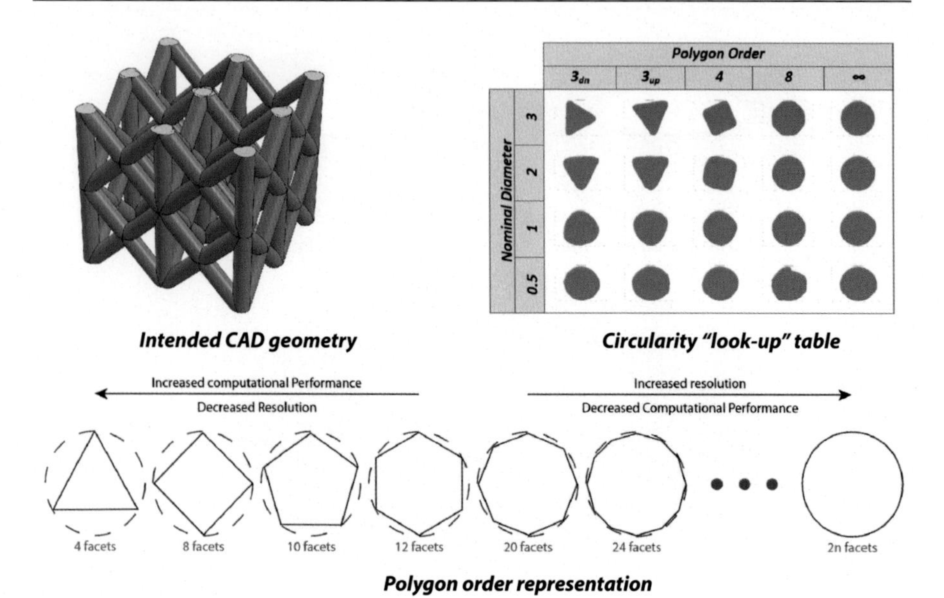

Fig. 3.23 Informed STL generation based on intended CAD geometry and experimental data on strut circularity for a given polygon order representation.

specific geometry of interest. By way of example, an experimental method to quantify the error introduced in AM cylindrical lattice strut elements is presented below.

Lattice strut elements are an example of a technically important AM geometry that is challenging to implement in the STL format. Fig. 3.23 demonstrates that the digital file size exponentially increases as the allowable error decreases. For this scenario, the relevant functional quality is the associated strut circularity, as measured by the *isoperimetric quotient*, Q, which is the ratio of cross-sectional area, A, to the area of a circle with equal perimeter, p. It is apparent that, as the number of facets used to represent a strut element increases, the isoperimetric quotient tends to unity – implying that, for a given geometry and material and process parameters, a threshold exists whereby further refinement of the STL data will render no further benefit to technical performance. For a particular AM scenario of interest, the relationship between the STL polygon order and the measured isoperimetric quotient provides an explicit and systematic DFAM tool to quantify the technical limit to the required volumetric data resolution (Fig. 3.24).

$$Q = 4\pi A/p^2 \tag{3.1}$$

A particular criticism often levelled at the STL file format is a supposed inability to accommodate surface specific data; for example, to identify characteristics such as allowable roughness, intended fit, colour and other surface functions. It is correct

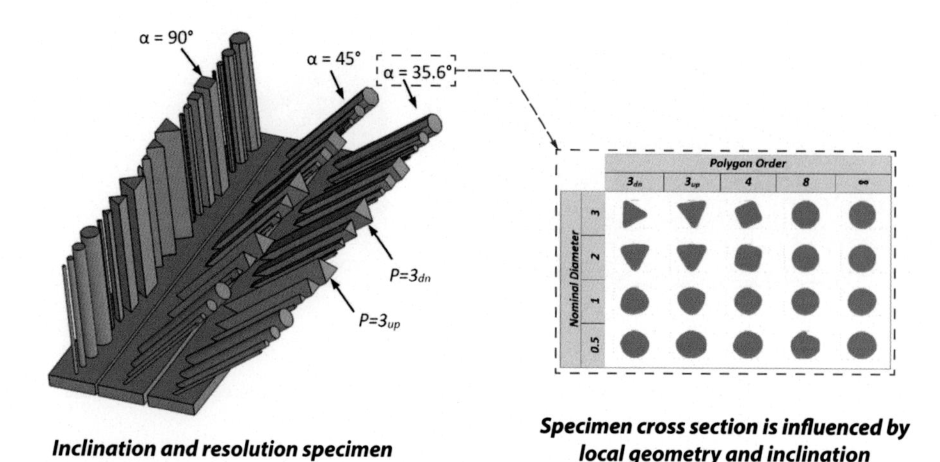

Fig. 3.24 A potential experimental approach to quantify the technical limit to the required STL resolution.

that the STL does not explicitly accommodate such data; however, a relatively simple workaround allows this problem to be largely circumvented. As identified in Table 3.1, the standard binary STL format accommodates a user-definable *header* (UINT8), as well as a free *attribute byte count* (UINT16) of data per facet. These user-definable data structures provide an informal mechanism for the accommodation of relevant surface data. For example, the header can be used to specify the global classification of surface functional requirements (such as nominal colour or preferred layer thickness) and this global classification can then be locally modified at local facets by modifying the attribute byte count data. The attribute byte count is specified by an unsigned 16 bit integer (UINT16) and thereby allows 65,536 unique combinations of surface data inputs. Hybrid data, for example integrated colour and allowable roughness, can be specified by a look-up table.[11] A potential limitation of this method is that additional facets may need to be defined to identify surface feature attributes that are incompatible with the facets required for geometric specification (Fig. 3.25).

Another often-cited challenge associated with the STL file format is its inability to accommodate multiple volumetric qualities. For example, the STL format cannot formally accommodate build and support materials concurrently within a single data file. This is technically correct; however, informal methods exist to accommodate this challenge. One method is to generate a unique STL file for each unique volume

[11] Note that the STL file does not accommodate the integration of a look-up table within the formal STL structure. This additional data must be formalized outside of the standard STL format, in an acceptable manner for the intended users. This informal data is not characterized by the STL format and can potentially result in data misinterpretation, but can enable technical innovation in the use of STL with surface specific data.

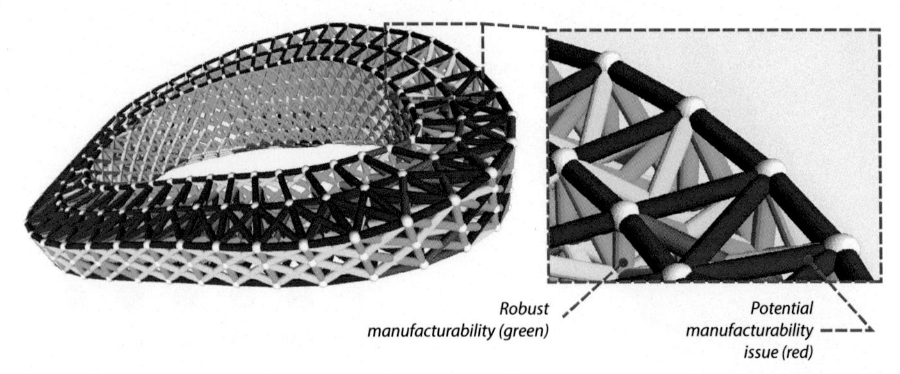

Fig. 3.25 Informal method to integrate surface specific data (in this case, colour) within the STL format.

required. The function of these unique files can then be informally identified within the file header data (UINT8, Table 3.1) as to their intended function. Custom scripts or manual processing can then be used to integrate these data sets as required,[12] for example, in the specification of support structures and manufactured geometry with unique process parameter attributes (Fig. 3.26).

Overall, the STL format has proven remarkably compatible and resilient to the needs of AM. This is not particularly surprising, as, although the technical requirements for AM systems have considerably evolved, the format was initially developed as a fundamental surface data format for the stereolithographic AM processes. To provide an informed review of the compatibility of the STL format for the requirements of current AM technologies, the associated advantages and disadvantages are enumerated below.

Advantages of the STL file format include:

- the simplicity of data structure definition
- an industry standard input for the slicing phase (Section 3.3.5)
- both ASCII and binary versions accommodate readability and storage efficiency, respectively
- compatibility with slicing in any orientation and with any layer thickness
- opportunities for informally accommodating surface and volumetric data.

Disadvantages of STL file format include:

- potentially large file size, especially for curvilinear geometries
- challenges in incorporating additional surface and volumetric functionality, for example, formally accommodating surface and volumetric data
- challenges in accommodating multiple independent volumes, and associated volume specific data
- computational challenges in generating slice data.

[12] An alternative method involves generating multiple unique watertight volumes within a single STL file. Facets associated with each volume are then informally labeled according to the associated attribute byte count. This method has advantages as it involves only one STL file, but requires careful post-processing and may result in compatibility issues if examined with commercial software.

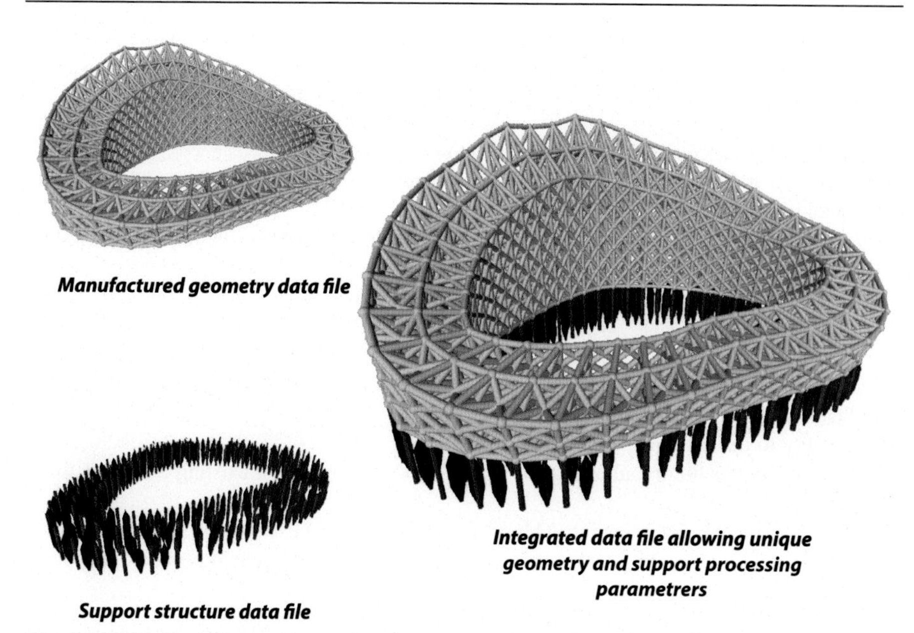

Fig. 3.26 Informal method to accommodate distinct volumetric data within the STL format. Support structure and manufactured geometry are informally distinguished by unique STL files that are then integrated with custom scripts.

In response to the challenges associated with the available data files, significant research effort has been directed towards the development of novel DFAM formats for surface specification. Of these developed formats, of interest are the Polygon File Format (PLY), Additive Manufacturing Format (AMF) and 3D Manufacturing Format (3MF). These formats were developed to overcome specific limitations of STL including the accommodation of multiple unique volumes, formal definitions of surface attributes, and unlimited character header files and comments. The PLY format was primarily developed to enable the storage of sophisticated data sets generated by 3D scanning. The AMF and 3MF format represent similar efforts to enhance data storage for AM. Industry consortia developed the 3MF format; whereas international standards organizations developed the AMF, which is presented in detail below.

3.3.4.2 Additive manufacturing format

It is apparent that the stereolithographic format is useful, but not necessarily optimal for AM applications. In response, significant research and development effort has been applied to develop custom file formats that are tailored to the specific requirements of current and evolving AM technologies. The AMF is an example of such a DFAM tool. To encourage application and research contributions in the use of AMF, the important AM data structure attributes are described in detail below.

The AMF format was formally developed based on recommendations by ASTM[13] Committee F42 on Additive Manufacturing Technologies[14] with the following considerations:

- open source to encourage universal application
- technology independence such that the data format can be compatible with any AM technology
- process independence such that the data does not restrict process settings such as slice thickness
- simple data structures such that understanding is promoted, while data repetition is to be avoided
- efficient data storage, including efficient scaling of file size with geometric complexity, increasing AM resolution, periodic geometric structures, and multiple concurrent structures
- efficiency of data read and write operations
- capability to describe geometric units
- backwards compatibility, allowing for example, AMF versions of existing STL data to be generated
- forwards compatibility, allowing AMF to be converted to STL for use in legacy systems
- future-technology-proofing such that future generations of AM technology can be supported by the standard AMF format.

The AMF format was initially released in 2011, and in 2013 the management of AMF became the joint responsibility of ASTM and ISO.[15] AMF is defined in XML format and is compressed according to the ZIP compression protocol. AMF may be defined in binary or ASCII by a data structure including the following five high-level elements:

- object – definition of volumes of specific materials
- material – specific printing material with unique identification
- texture – images or textures for surface mapping with unique identification
- constellation – hierarchical combination of objects in a relative pattern
- metadata – additional information on objects to accommodate future AM technology needs.

Geometry is then specified for each object as a face-vertex polygon mesh as follows:

- vertices – sequentially listed (from zero) in order of their definition to define the object mesh
- coordinates – a child of the vertices defines the position of each vertex in Cartesian 3D space

[13] ASTM International (formerly the American Society for Testing and Materials) is an international standards organization that publishes a range of technical standards for materials, products, systems and services. Voting rights are conferred to voluntary personal members, who may volunteer to be part of committees of interest.

[14] ASTM Committee F42 on Additive Manufacturing Technologies was formed in 2009 with the scope of promoting knowledge, research and implementation of AM technologies through the development of relevant standards.

[15] International Standards Organization (ISO) is an international standards-setting body. Voting rights are conferred to representative standards bodies of member nations.

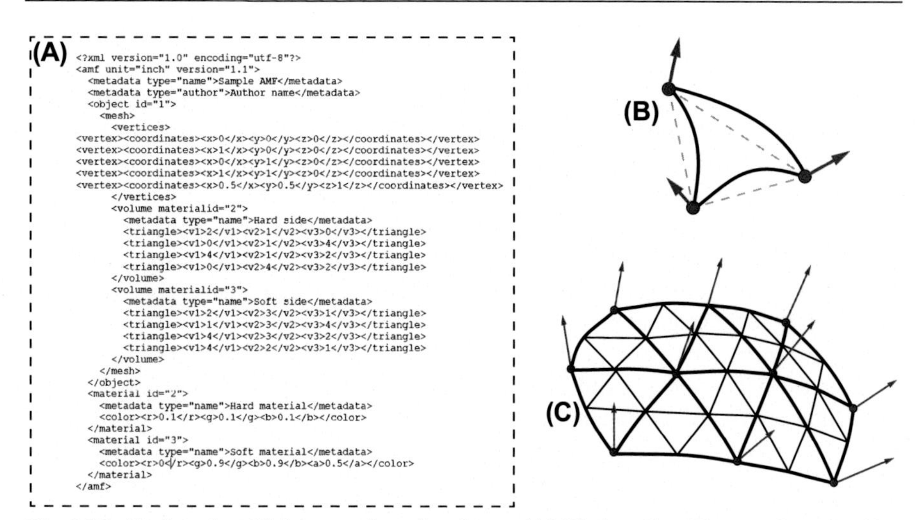

Fig. 3.27 (A) Sample additive manufacturing format (AMF) data file, (B) curved patch with tangent normals, (C) recursively defined self-tessellation.

- volume – encapsulates an enclosed volume of vertices (vertices can be shared, but volumes must not overlap)
- triangle – a child element of a volume that specifies triangles by reference to three vertices.

The triangle geometry interlocks to define a continuous surface but differs from the STL facet. Each triangle is numbered from zero in the order of definition and is defined according to the right-hand rule in clockwise order when viewed from outside of the surface; this method eliminates the need for the specific surface normal of STL. Furthermore, these triangles may be curved to increase the associated geometric resolution (Fig. 3.27).

Additional attributes of interest to the AMF format include the explicit accommodation of: colour data, texture data, graded materials, periodic structures, multiple independent geometries, comments and unlimited metadata for the accommodation of future AM digital requirements. Selected attributes and their specific DFAM opportunities are described below.

Periodic and gradient structures: Periodic and gradient structures enable significant innovation in the AM design space; however, these structures potentially compound the complexity of the surface data file. For example, STL file size rapidly escalates when attempting to generate high-resolution data for even relatively small periodic lattice structures. The AMF format allows periodic data to be embedded as a classification, eliminating the need for the redundant re-definition of geometric data. Furthermore, AMF provides a capacity for the definition of stochastic attributes within periodic structures by allowing the use of pseudo-random variables. These capabilities enable the freedom to design ultra high-complexity structures (Chapter 6) without the data management burden inherent to the STL format; an emerging research and commercial DFAM opportunity, especially when combined with the algorithmic design capabilities of generative design (Chapter 7).

Multi-Materials: When the STL data format was initially defined, monolithic single material structures were the cutting-edge of AM technologies. Current AM technologies enable the routine fabrication of complex multi-material structures, however, the technical bottleneck to such capabilities is often the limitations of the data format. AMF explicitly allows for multiple material types, each associated with a specific *volume* definition. Furthermore, material type may be defined with reference to a *composite* data type that allows materials to be blended according to a constant value, or as a function of local coordinates. As for periodic structures, multi-material definitions allow the use of pseudo-random variables, thereby enabling stochastic attributes to be defined to enable the implementation of novel material properties.

Nesting: Part nesting is a standard commercial consideration, whereby unique parts are arranged within the available build space in two-dimensions, and increasingly, in three-dimensions. AMF facilitates this DFAM imperative by the *constellation* element, which allow multiple unique parts to be defined concurrently within a single AMF data file both in terms of their position and rotation. Arrays of identical elements are also accommodated. This capability of the AMF format is of commercial significant for manufacturers that intend to maximize their production yield by maximizing the production volume per cycle. As well as for the designers of performance-critical AM systems (such as aerospace componentry and medical implants) where it is critical that the nesting attributes, including build orientation, are formally documented and implemented such that process control is assured.

Efficient facet implementation: The fundamentally planar facet representation of the STL format inherently introduces inefficiency in representing curvilinear geometries. This challenge can be computationally debilitating, even for relatively small structures of the type of interest to commercial AM such as lattice and zero-mean curvature cellular structures (Chapter 5). The AMF format enables curvature to exist in the definition of tessellating elements. The function of this curvature is to reduce the number of elements required to accurately describe a curvilinear surface. Curvature is enabled by the specification of an outward-facing *normal* element as a child to specific *vertex* elements of interest. Accordingly, the triangular element is then defined according to a circular arc that maintains this normal condition. A curved triangle is then approximated by decomposing the defined *triangle* element into four triangular sub-elements (Fig. 3.27). This decomposition is recursively made for five generations during the slicing phase, resulting in a total of 1024 flat triangles representing the curved triangle. Where curvature at a point is not defined, for example at an edge or corner, an *edge* element can be specified to clarify local geometry. The AMF methodology for accommodating curvature reduces the file size required to represent a curved surface to a given accuracy; however, it still requires that the explicit linear data be provided at a resolution that is known *a priori*.

Future proofing: In an attempt to accommodate the future needs of the AM format, the *metadata* hierarchical element can be used to define additional information. A metadata element defined at the global level indicates information that is relevant to the entire data structure; when defined locally the metadata is applied only to the locally relevant

Table 3.2 Attributes of AM volumetric data storage formats.

	STL	*PLY*	*3MF*	*AMF*
Formal method for specifying surface attributes	N	Y	Y	Y
Informal method for specifying surface attributes	Y	–	–	–
Formally accommodates surface curvature	N	N	Y	Y
Accommodates periodic structures	N	N	Y	Y

attributes. This metadata data provides useful functionality that can be formalized within the ISO definition as is required by the evolution of AM technology.

File compression: AMF file compression is achieved by the lossless open-source *zip* compression format.

Table 3.2 summarizes the various attributes of AM surface data formats considered in this brief review. It is apparent that the AMF format formalizes many of the limitations of STL format. However, the STL does provide the capacity for many of these attributes in an informal manner. The AMF format provides higher storage efficiency for curvilinear geometry, in particular by allowing the formal definition of curved triangular patches. However, this advantage is tempered by the inherent computational challenges associated with characterizing the intersections of this curved patch with layerwise slices. Intersections can be identified by recursively representing the patch by successively smaller sub-patches up to the required resolution (Fig. 3.27), although this may incur high computational cost at the slice data phase. Associated DFAM tools are emerging in response to the need for advanced understanding and utilization of second-generation AM data structures such as the AMF.

Although AMF has addressed some of the challenges associated with AM data formats, it is apparent that open DFAM research questions remain regarding volumetric data storage, and that these challenges must be intelligently accommodated by the designer in order to enable commercially robust design outcomes.

Despite formally defined specifications for AM data formats, it is commonly observed in commercial practice that volumetric digital data is corrupt, incorrect or incomplete; potentially resulting in flawed outcomes in the following design phases. In response to this potential risk, numerous commercial and open-source DFAM tools have been developed that attempt to identify and correct these flaws; which typically include incorrectly defined surface normals, invalid facet definitions, and non-manifold (watertight) surface definitions. These tools include methods that range from autonomous systems that attempt to algorithmically resolve these flawed surface representations, to manual methods that allow the designer to provide input on the preferred technical response. Although technical DFAM tools exist, the challenges associated with ensuring robust surface data representations and the associated tools to correct these identified flaws remains an open research question.

3.3.5 *Slice geometry*

AM technologies typically fabricate the intended three-dimensional component volume by the addition of materials within consecutively offset layers.[16] The slice geometry phase refers to the digital generation of component perimeters at these consecutive layers.

A brute force slicing approach for STL/AMF can be implemented as follows. A plane is defined at an offset from the build platen that corresponds to the local layer thickness. For this defined plane, each triangular facet is sequentially examined, and recursively generated AMF facets must be explicitly assessed (these can be either unpacked within a reference data repository or can be calculated *on the fly* for each facet of interest). For both the STL and AMF formats, each facet is identified by three spatial coordinates. If these coordinates are ordered according to their height from the build platen (Z-value), the existence of an intersection and the associated segment is then readily characterized (Fig. 3.28). Identifying the intersection of each facet with the plane of interest allows a set of polygons to be generated that defines the associated layer *slice* [21].

This brute-force approach is mathematically robust but can become computationally challenging for relatively large input data files. An alternative approach is to pre-process the associated facets such that they are then categorized according to which layers they are associated. This pre-processing requires an initial processing step but can reduce the overall computation time for large data sets. Such data management tools for slicing optimization remain an open DFAM research opportunity. For those seeking to enhance the computational performance of slicing operations, it is important to note that, computationally, the surface volume data file (either STL or AMF) can be processed in parallel for each layer, as the associated slice data is independent of other layers.

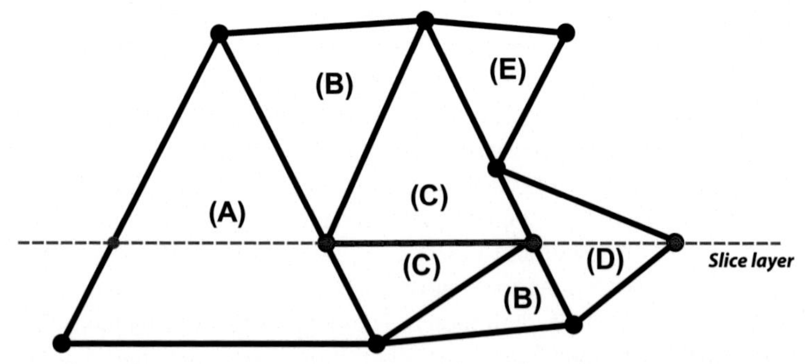

Fig. 3.28 Possible intersection scenarios for slice layer and facet intersection: (A) slice plane intersects two facet edges, (B) plane intersects one vertex only, (C) plane is collinear with facet edge, (D) plane intersects one vertex and one facet edge, (E) no intersection.

[16] AM processes are not necessarily planar, alternative geometries including cylindrical and robotic are feasible (Chapter 4); nonetheless, planar geometries are examined here due to their prevalence in AM.

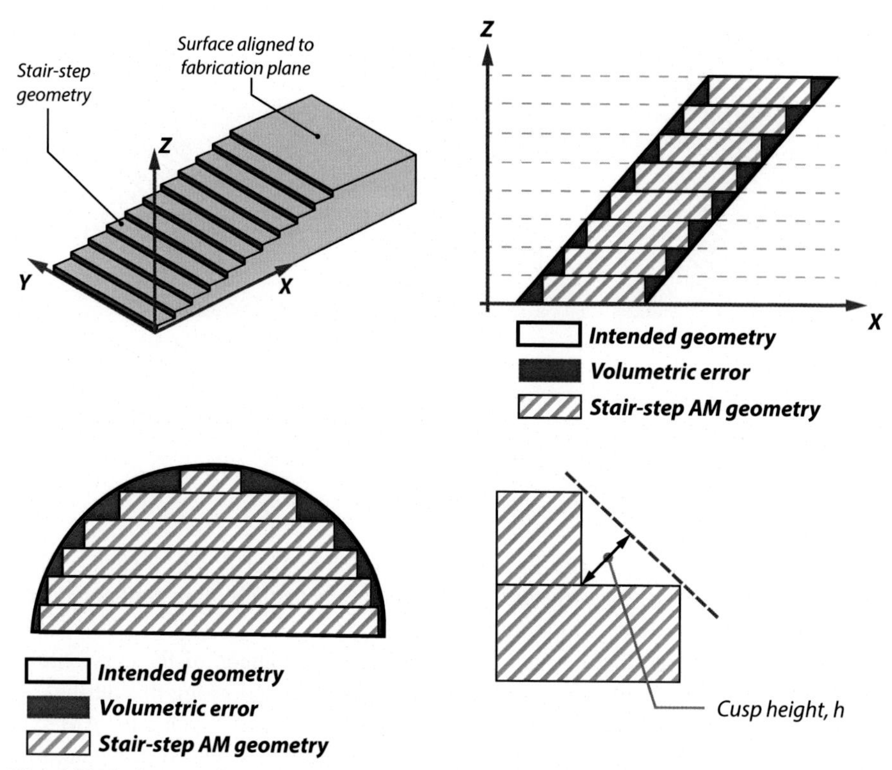

Fig. 3.29 Stair-step effect associated with a specific inclination angle (upper), and associated error measures, including volumetric error and cusp height.

The slicing of a continuous geometry into a finite number of prismatic volumes inherently introduces geometric error, specifically known as the *stair-step effect* (Fig. 3.29). The stair-step effect is especially significant as the *inclination angle*, α, becomes increasingly acute, and is zero for structures that are either horizontal or vertical.

Metrics introduced to assess the error associated with a slicing operation include the cusp height and volumetric error (Fig. 3.29). *Cusp height error* is measured as the maximum distance measured by a normal chord between the sliced geometry surface and the design surface. Alternatively, *volumetric error* may be computed between the sum volume of all associated prismatic sections and the intended component volume. Depending on the specific scenario, either of these slicing error estimates may be more appropriate.

DFAM methods have been proposed to mitigate the measured slice error. For example, *adaptive slicing* refers to the slicing of AM part geometry with slice thickness varying according to allowable values of local slice error (Fig. 3.30). Adaptive slicing requires that the feasible slice thicknesses for the intended AM technology be known *a priori*; however, it can provide significant technical advantages including a reduction in manufacturing time without compromising slice error, as well as a potential for reduced slice file size. A variant of adaptive slicing allows the processing parameters to be locally

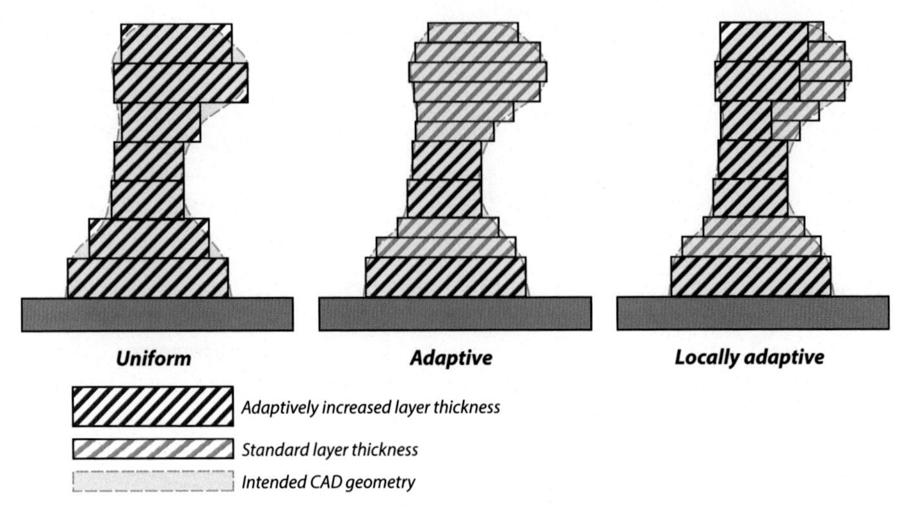

Fig. 3.30 Adaptive slicing methods to mitigate slice error.

modified such that the externally facing surfaces to be fabricated with a smaller slice thickness that the internal core structure; this method can potentially allow further reduced manufacturing time without compromising surface error.

A slice data attribute that is often overlooked is the resolution of the curve-fitting polygon used to represent the slice data. The rationale for overlooking this attribute is likely due to the significantly higher resolution inherent in the XY-plane for AM systems. However, for high-resolution volumetric data with small feature size, the number of facet interactions, and consequently the number of data points within the slice polygon, can become large. In practice the resolution of this polygon may exceed the manufacturable resolution of the intended AM technology; for these scenarios it is useful to reduce the polygon resolution according to the AM manufacturability to avoid excessive file size. Alternatively, the polygon can potentially be replaced with a spline or other curvilinear geometry format, with the objective of reducing the data file size required to achieve an allowable error in the manufactured geometry. These methods are compatible with advanced design methods that omit traditional phases for AM slicing, especially those that generate slice data directly from the original CAD (Section 3.5).

3.3.6 Tool path and process parameters

The layerwise slice data generated in the previous phase define the external boundary of specific manufactured layers. For most AM systems, further data on the specific toolpath are required. The following summary describes the standard data used to define typical toolpaths, with associated optimization strategies discussed in Chapter 4.

For AM systems with a point wise *end-effector*, for example, Fused Filament Fabrication (FFF) systems that lay filament (Chapter 8) and Powder Bed Fusion (PBF) systems that apply thermal energy to selectively fuse regions in a powder bed (Chapter 11), it is necessary to fill the interior polygon space with a toolpath trajectory. The allowable toolpath is restricted by a series of technical constraints; for example, melt-pool geometry and allowable filament extrusion profiles. Consequently, the geometric challenge of filling space subject to these technical constraints is a challenging problem. To alleviate this space-filling complexity, and to reduce the computational cost associated with toolpath definition, a number of space filling patterns have been proposed. These are typically implemented with a *perimeter-and-raster* strategy. This strategy allows for *perimeter* toolpaths that are visibly continuous and provide aesthetic and low-roughness surfaces, while an internal *raster* of closely spaced parallel toolpaths allows for relatively effective space-filling with low computational cost.

Note that the perimeter-and-raster terminology prevalent for FFF (Chapter 8) systems is compatible with the *skin-and-core* terminology commonly applied for PBF (Chapter 11) systems. For systems that incur thermal stresses, the thermal field that occurs during manufacture is challenging and computationally demanding to predict. Therefore, toolpath strategies exist to reduce overheating and associated thermal distortion. For example, the island strategy consists of geometrically isolated islands that are sequentially fused such that thermal distortion due to local overheating is reduced (Fig. 3.31).

Advanced DFAM tools also accommodate the influence of toolpath on the mechanical and functional properties of the manufactured component. It is apparent that the geometric features of a proposed structure are challenging to accommodate without introducing geometric defects including linear fill defects, island voids and intersection voids (Fig. 3.32). These geometric defects introduce structural discontinuities that result in sub-optimal mechanical performance. Typical AM products consist of thousands of unique slice layers, each with unique toolpath data. Consequently, manual generation or inspection of these unique layers is infeasible. DFAM tools that actively minimize defects within these layers, as well as tools that can predict and quantify their effect, are much needed for structurally critical commercial AM applications.

Adaptive layer methods with variable skin-and-core thickness can be utilized at this phase. This variant of the adaptive slicing concept allows the external skin surfaces to be processed at a finer layer thickness than the core, thereby allowing

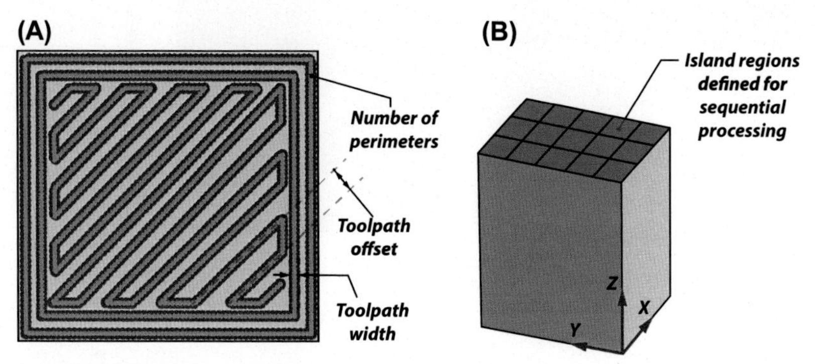

Fig. 3.31 (A) Standard toolpath geometry and associated terminology, (B) island tiling patterns for space filling without introducing excessive local temperature.

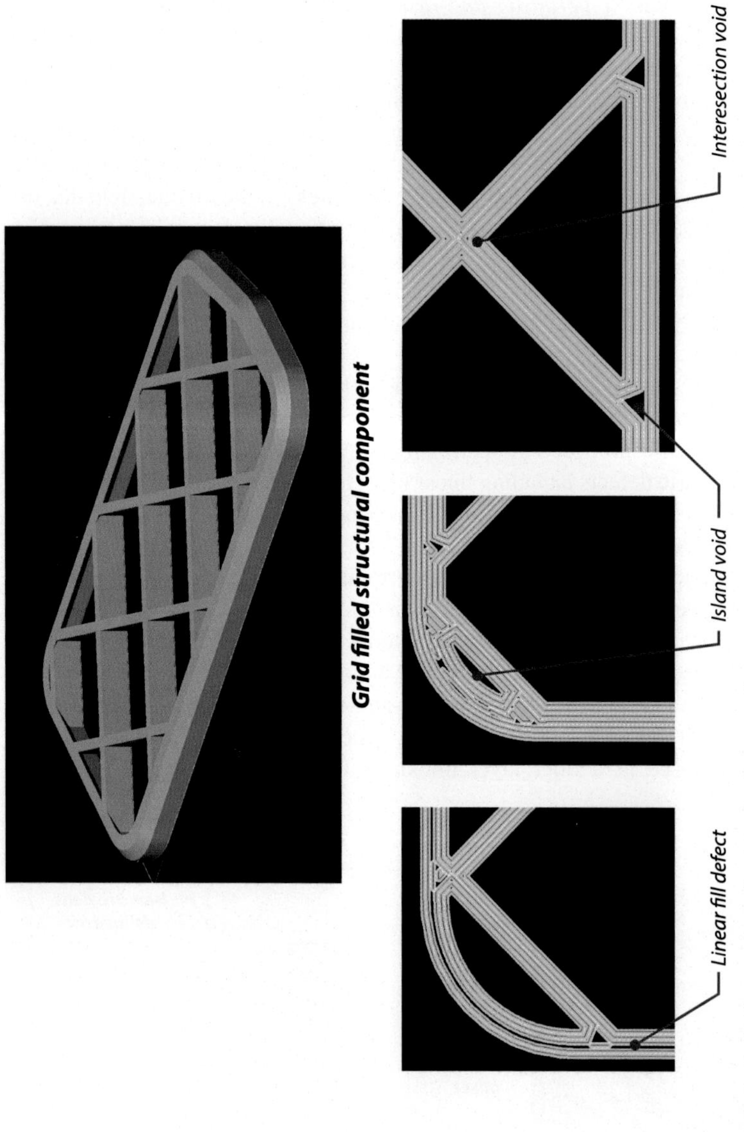

Fig. 3.32 Technical challenges for robust toolpath generation.

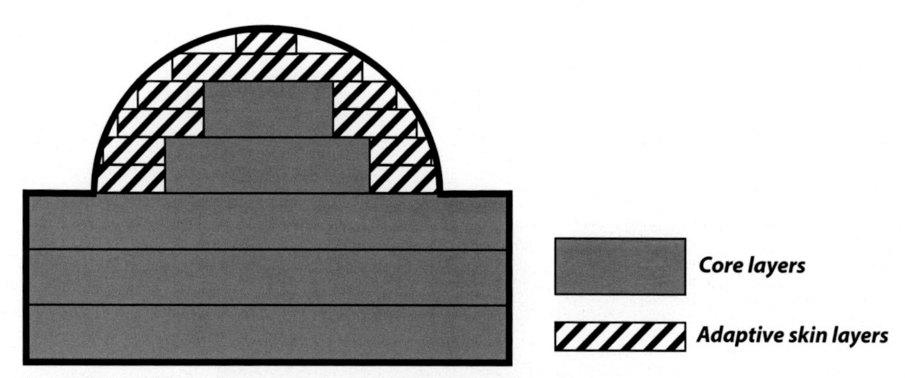

Fig. 3.33 Adaptive slicing with variable thickness skin-and-core.

increased processing speed without compromising surface roughness. These methods can potentially also reduce slice data file size; however, it is typically required that these adaptive layers be an integer multiple of the intended layer thickness (Fig. 3.33).

One overarching challenge associated with the generation of advanced DFAM tools to specify optimized AM toolpaths is a lack of formal standards for digital toolpath formats. Furthermore, many commercial AM systems employ commercially encrypted toolpath data formats, further stymying their generalized application within custom digital DFAM tools.

3.3.7 Manufacture

In response to the complexity of AM methods and the requirements for certification of high-value products, AM systems are increasingly fitted with process data acquisition and feedback control. Process feedback systems include position sensors, pyrometers and thermal cameras. These systems have the potential to generate large data sets that must be managed to enable robust manufacture. In particular, imaging systems provide an exceptional opportunity for process certification during manufacture; however, thermal images can generate inordinate volumes of data, for example, the in-situ thermal data of Fig. 3.34. Practical analysis of this data is typically completed at a *meta-data* level, whereby the raw data is distilled into a manageable subset, for example peak temperature or cooling duration, and this meta-data is used as a proxy to indicate robust manufacture (Fig. 3.34). Although the data is acquired during manufacture, it is also utilized as inspection data and as such is discussed further in the following section.

3.3.8 Inspection and certification

AM enables the fabrication of high complexity geometry that provides significant commercial value; for example, in the fabrication of highly complex engineering structures with inherently high technical efficiency. These high-value structures include

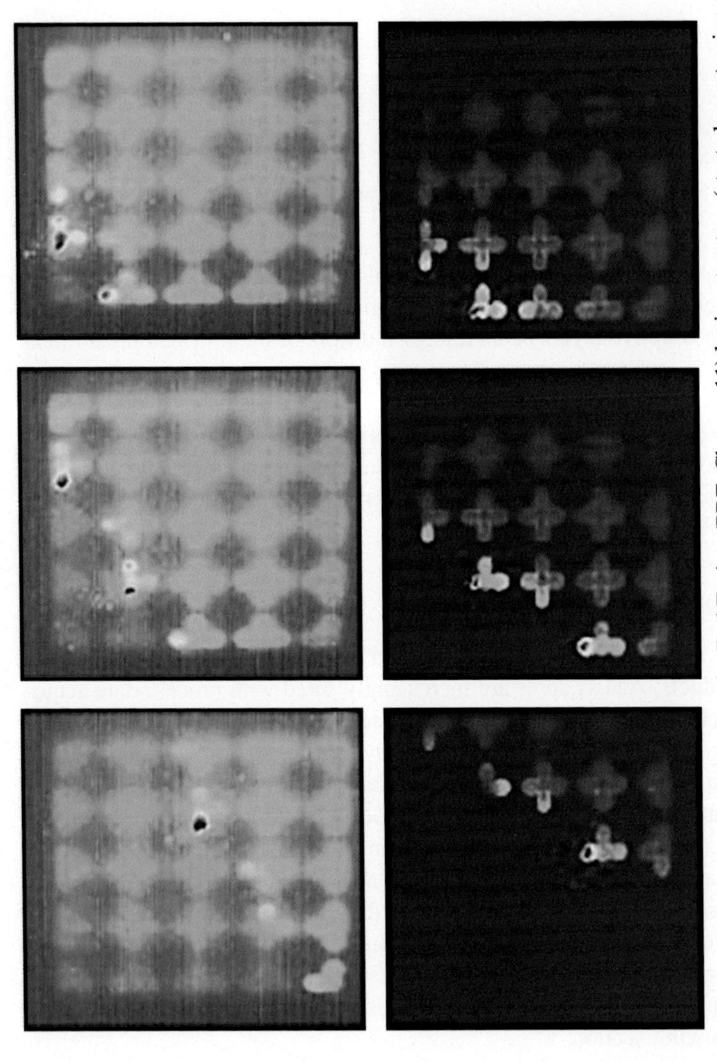

Fig. 3.34 In-situ thermal camera data acquired for a Powder Bed Fusion (PBF, Chapter 11) lattice structure (strut elements viewed in cross-section). Image sequence (left to right) shows the sequential progression of the laser toolpath within a single manufactured layer. Upper images are raw camera data. Lower images are digitally processed meta-data to extract peak local temperature data.

safety-critical aerospace applications and medical implants that must be certified prior to commercial application. The *inside-out* nature of AM (Chapter 4) that enables the generation of high complexity geometry also enables the generation of internal defects that may not be visually apparent in the fabricated specimen. As a result, a lack of certainty in defect inspection increases the safety factors that must be applied to safety-critical AM applications, thereby eroding the opportunity for commercial advantage. Furthermore, AM systems are subject to fundamental challenges in terms of achievable

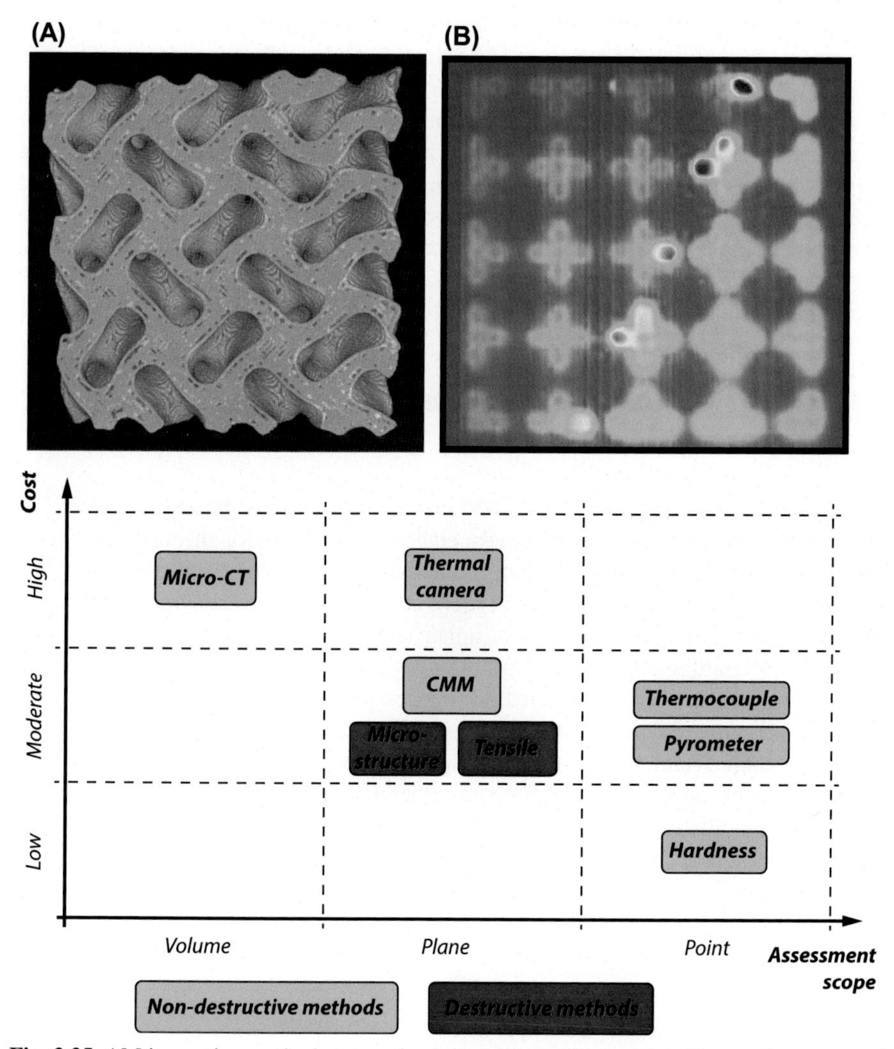

Fig. 3.35 AM inspection methods categorized according to the associated inspection costs and assessment scope. Legend: Coordinate Measurement Machine (CMM). Inset identifies high-value inspection and certification outcomes including (A) micro-CT inspection highlighting internal flaws that exceed an identified critical threshold, (B) in-situ thermal camera data digitally processed to identify regions of high thermal intensity.

geometric precision, and each unique AM system has inherent potential for unique manufacturing defects; for example, thermal systems introduce the potential challenges of thermal distortion, and cured polymer systems introduce challenges associated with partially cured resins.

In response to the challenge of AM certification, numerous DFAM strategies for AM inspection have been developed. Each of these inspection methods varies in terms of associated inspection costs and technical attributes (Fig. 3.35). Particularly, these methods are either destructive or non-destructive, and, of relevance to AM, either allow the inspection of internal features, or are associated with external features only. A paradox exists: specifically, that non-destructive testing of physically large components to the scale of relevant AM defects is technically possible (for example micro-CT scanning) but is prohibitively expensive for all but the highest-value production applications. For typical AM applications, engineering risk management procedures must be applied to achieve a balance between cost and certainty of specimen integrity. For example, high frequency inspection with low-cost methods combined with low-frequency inspection with high-resolution methods; and, destructive testing of witness coupons and systematic definitions of critical allowable defects. Formal DFAM protocols for AM risk management remain an active and open commercial research opportunity.

3.4 Opportunities for advanced digital DFAM opportunities

A summary of the traditional digital AM process workflow is presented in Fig. 3.5. These workflow phases are sequentially elaborated in terms of the inherent digital DFAM data, commercial best practice and associated research opportunities (Section 3.3). Further to these, and ultimately of more interest to AM process optimization, are DFAM opportunities that allow the designer to truncate the traditional AM process workflow by omitting the need for specific phases altogether. These advanced digital DFAM opportunities allow increased processing efficiency concurrently with increased certainty and documentation rigour. Furthermore, advanced digital DFAM tools can enable seamless transfer of digital data along the entire process workflow, allowing entirely algorithmic AM design whereby design decisions can be implemented at computational speeds and data bandwidths, rather than being stymied by specific human intervention and *ad hoc* decision-making. If this algorithmic design can be implemented such that design decisions can be autonomously made to optimize function, the design process is generative (Chapter 7), whereby the designer is responsible for specifying the functional objectives rather than the specific AM inputs required to achieve these functional requirements.[17]

[17] Generative design provides significant opportunity for commercial benefit, but is predicated on a deep understanding of the influence of design variables on the associated design function; resulting in significant restrictions on the scope of design scenarios for which generative design is commercially viable (Chapter 7).

These opportunities are significant for those attempting to commercialize AM and to develop research programs that support these commercial activities. The following points summarize key benefits of *advanced* digital DFAM tools.

- The ability to truncate design steps, reducing overall design time and allowing greater insight into specific design problems early in the design phase; thereby offsetting the paradox of understanding and flexibility by utilizing the present value of knowledge.
- Allowing consistent documentation of embedded design processes, as is required for high-risk commercial design including medical, automotive and aerospace applications.
- Allowing ultra-high complexity design to be implemented by generative methods implemented on high performance computers.
- Allowing novel design methods and algorithms to be implemented within proprietary *commercial-in-confidence* software.

These identified key benefits are discussed below in the context of the advanced digital DFAM tools identified in Fig. 3.36:

- ***CAD to slice data:*** An opportunity to convert CAD data directly to representative slices such that the potential data loss and costs incurred by the Volumetric Data phase can be avoided (loop A, Fig. 3.36).
- ***CAD to toolpath data:*** Such that toolpath data can be directly prescribed and generated from the CAD data phase (loop B, Fig. 3.36).

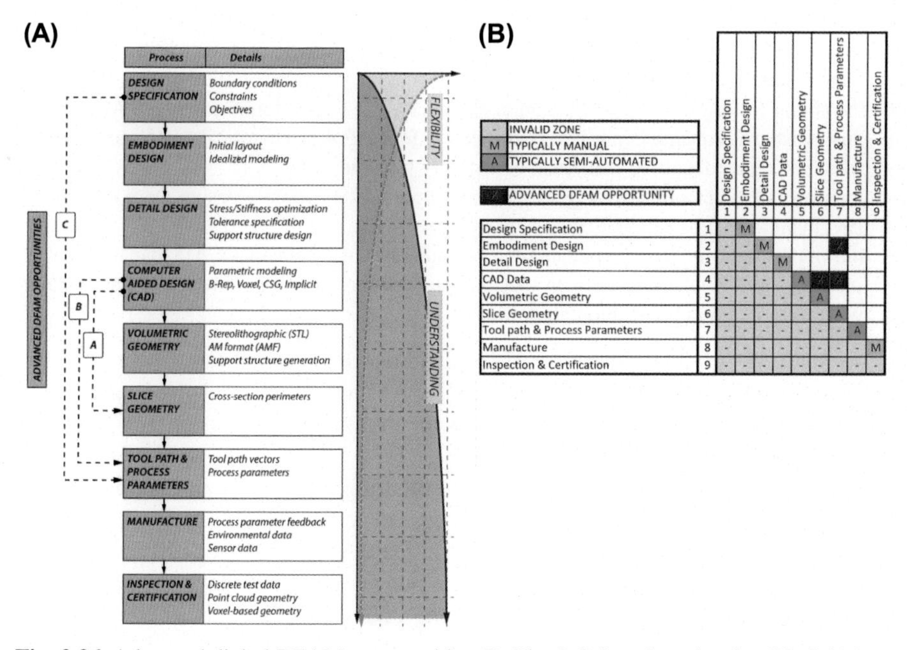

		1 Design Specification	2 Embodiment Design	3 Detail Design	4 CAD Data	5 Volumetric Geometry	6 Slice Geometry	7 Tool path & Process Parameters	8 Manufacture	9 Inspection & Certification
Design Specification	1	-	M							
Embodiment Design	2	-	-	M						
Detail Design	3	-	-	-	M					
CAD Data	4	-	-	-	-	A				
Volumetric Geometry	5	-	-	-	-	-	A			
Slice Geometry	6	-	-	-	-	-	-	A		
Tool path & Process Parameters	7	-	-	-	-	-	-	-	A	
Manufacture	8	-	-	-	-	-	-	-	-	M
Inspection & Certification	9	-	-	-	-	-	-	-	-	-

Fig. 3.36 Advanced digital DFAM opportunities (B, Fig. 3.6) based on the simplified AM design workflow (A, Fig. 3.5).

- ***Design specification to toolpath data:*** The opportunity to generate robust toolpath data directly from the design specification data is highly challenging, and potentially feasible only for simplified design scenarios (loop C, Fig. 3.36). However, this opportunity is highly attractive as it allows algorithmic integration of the entire design process as is required for generative design (Chapter 7).

3.4.1 CAD to slice data

The typical AM workflow involves the conversion of CAD data to a volumetric data representation. This data is typically provided in the STL format but is increasingly presented in second-generation volumetric data formats such as AMF and 3MF (Section 3.3.4). This volumetric data phase may be beneficial as it provides a common data input to the slice data phase; however, it does introduce additional data handling and processing costs, as well as the potential for geometric error incurred during conversion from the native CAD data.

An emerging DFAM opportunity exists to disrupt this typical AM workflow. Advancing directly from the CAD phase to the slice data phase allows the cost and resolution challenges of the volumetric data phase to be bypassed (loop A, Fig. 3.36). Furthermore, embedding the slice data within the native CAD format provides significant opportunities for enhanced AM design control from directly within the CAD system, including:

- the potential to adaptively modify slice thickness and associated resolution according to existing CAD design variables, including required geometric tolerance and material of manufacture
- the ability to formalize build orientation and slice layer data within the native CAD format; this assists with quality control and certification for performance-critical component design, where this manufacturing data must be robustly documented
- reduced risk of incorrect slicing or orientation parameters when a design team outsources AM manufacture to an external third-party
- the opportunity for build orientation to be integrated with existing CAD and CAE tools, including structural analysis, such that material anisotropy can be directly accommodated within orientation optimization algorithms.

At the time of writing, these DFAM capabilities are not well implemented within existing commercial CAD tools. However, as AM technologies evolve and become increasingly commercially relevant, so does the motivation for commercial CAD systems to implement such DFAM capabilities.

Despite the opportunities associated with omitting the volumetric data phase, this strategy requires that the manufacturing process expertize required for optimal orientation and process optimization be available at the CAD data phase. This requirement is compatible with commercial best practice but may be incompatible with existing AM workflows where the component nesting and slicing is completed later in the workflow.

3.4.2 CAD to toolpath data

An extension of the previously stated opportunity is to move directly from CAD data to the toolpath data (loop B, Fig. 3.36). This advanced digital DFAM opportunity bypasses both the volumetric and slice data phases and therefore allows for significant reduction in computational time while reducing the opportunities for geometric error. Furthermore, the direct connection of the CAD and toolpath data phases allows for the integration of CAD design with advanced AM process simulation methods that predict the influence of toolpath on AM manufacturability,[18] as well as being compatible with the requirements for generative design (Chapter 7). This design opportunity allows high-level AM manufacturing data to be defined early in the design process and formally embedded within the associated CAD data; thereby providing data security for designers who intend to outsource manufacture to independent third-parties.

This opportunity requires that the CAD data be converted directly to a toolpath format that is technically robust and readable by the intended AM technology. Consequently, it is necessary that the intended AM process be well understood. As presented in Section 3.3.6, this requirement is technically challenging, especially for AM systems that employ commercially encrypted toolpath data formats. Furthermore, this opportunity requires that the input CAD data be compatible with direct conversion to AM toolpath data.

Although the near-term opportunities for direct CAD to toolpath data are somewhat limited, they are compatible with commercially important design scenarios, especially those associated with lattice and Triply Periodic Minimal Surfaces (TPMS). These complex, but efficient structural elements are eminently compatible with AM technologies, and their geometry can be algorithmically defined, thereby enabling direct conversion from the CAD design phase (or even the detail design phase) directly to slice and toolpath data. This strategy is gaining momentum within the DFAM research community (Fig. 3.37) but remains an open opportunity for commercial and research contribution.

3.4.3 Design specification to toolpath data

The output of the design specification phase includes the formal specification of the boundary conditions, design objectives and constraints associated with a specific design program. The opportunity to generate robust toolpath data directly from this design specification is highly attractive from a commercial and technical perspective. This opportunity is technically challenging and is potentially only feasible for particular design scenarios. It is presented here as an aspirational research opportunity for the development of highly advanced digital DFAM tools.

The simplified AM design process represents a series of design phases that are typically associated with AM design. These phases are either fully or partially automated or are manually implemented (Fig. 3.36). Manual processes allow design flexibility, allowing the expertize and experience of the design team to positively influence the designed solution. Manual methods are therefore compatible with the requirements

[18] AM process simulation is complex, especially for thermal systems; however, these DFAM tools are becoming an increasingly important part of the AM design process (Chapter 4).

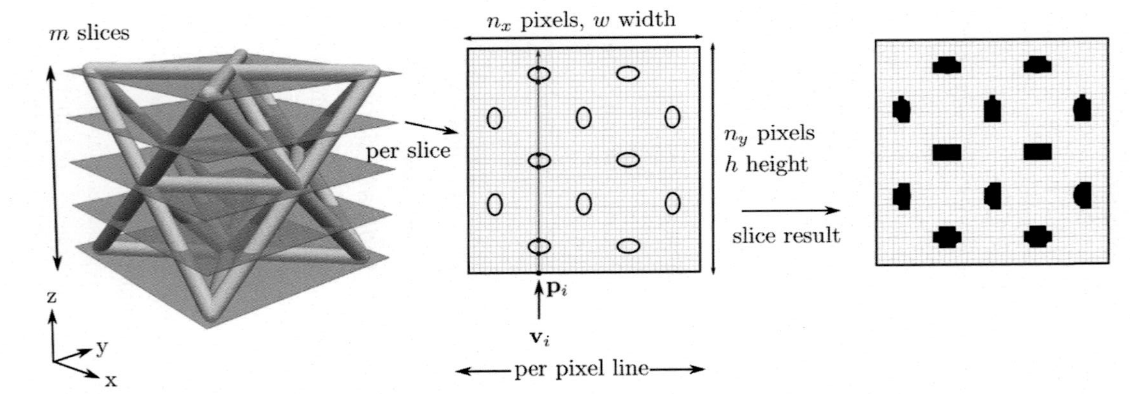

Fig. 3.37 Advanced DFAM tool enabling parametric CAD representation to be converted directly to planar slice data represented either by explicit slice perimeters, or by an implicit voxel field [22]. This slice data can then be algorithmically converted to optimal toolpath data.

of revolutionary design, or for complex or highly dimensional scenarios where no formal DFAM tools exist. However, manual design processes do incur challenges, including relatively large design time, possibility for error, and variability in response between individual designers.

An advanced digital DFAM opportunity exists whereby the manual phases of the design process are fully automated, allowing toolpath data to be generated directly from the formal design specification phase. These scenarios enable fully algorithmic implementation of the design process, a strategy that is formally known as *generative design* and enables highly complex design outcomes to be implemented with high efficiency and repeatability and with robust documentation. These strategies require that all design phases be compatible with automated methods and are currently restricted to a limited subset of AM design scenarios. These generative design opportunities are more fully presented in Chapter 7 and include, for example:

- design of lattice and TPMS structures
- conversion of medical imaging data directly into technically useful AM products and systems
- automated design of permutations of well understood structures or systems.

These opportunities may be described as Boundary Condition to Additive Manufacture (BC2AM) methods and present an important emerging commercial research opportunity. An example of a clinically implemented BC2AM system is presented in Fig. 3.38.

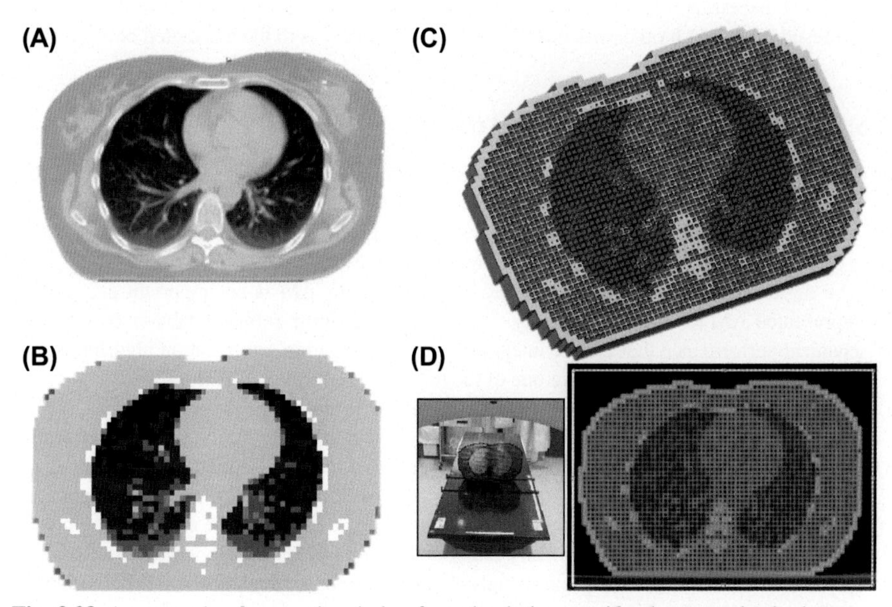

Fig. 3.38 An example of generative design from the design specification to toolpath phases. In this scenario medical imaging data is defined at the design specification phase. Based on custom DFAM tools, the subsequent design phases are digitally implemented (Chapter 7 and 9).

3.5 Summary of chapter outcomes

The AM process is inherently digital; from the initial specification of the design requirements through the generation of candidate solutions and their definition in slice and toolpath data, as required for the specific AM technologies. Increasingly, these AM technologies include in-situ monitoring systems that provide digital data for process control and monitoring. Consequently, the commercialization of AM is entirely predicated on the robust, interconnected and efficient digital design DFAM capabilities. Despite the criticality of digital DFAM tools, they have arguably received limited research attention. In particular, advanced digital DFAM tools that propose holistic strategies for digital DFAM are relatively sparsely reported. In response to the limited availability of digital DFAM tools, this chapter integrates robust engineering design strategies with observations of commercial best practice in AM design in order to identify strategic research opportunities for the development of commercially relevant digital DFAM tools. These opportunities are identified with respect to the identified engineering design phases below.

3.5.1 Engineering design

- Digital methods that actively enable the robust integration of engineering design philosophies. For example, to utilize the present value of knowledge (Section 3.1.1), parallel set narrowing (Section 3.1.2) and the selection of appropriate DFAM strategies and data handling techniques for revolutionary and evolutionary design projects.
- It is not necessary that these contributions be digitally implemented; they can also be achieved by research outcomes that educate AM designers as to the associated best practice.

3.5.2 Embodiment and detail design

- DFAM tools that can accommodate topological optimization within the embodiment and detail design phase are critically important for commercial AM design, especially for high-value systems that necessitate high efficiency solutions. AM-integrated topological optimization provides a core opportunity for generative AM design, and is presented in Chapters 6 and 7.
- The digital integration of material selection methods for AM is an important and under-represented AM research opportunity. These methods require robust databases of AM mechanical performance that can be integrated within orientation optimization algorithms.
- Build orientation algorithms were one of the first areas of formal DFAM research, and a significant range of DFAM tools have been presented to automate build orientation. These methods are often based on simplified proxies for AM manufacturability, typically allowable inclination angle; and are optimized for the objective of minimal support material consumption. The integration of advanced metrics of AM manufacturability, notably temperature fields for thermal AM systems, will allow a new generation of DFAM orientation tools.
- Support generation tools are typically either automated implementations of pre-set support geometries based on allowable build angle. In particular for thermal systems, the algorithmic deployment of optimal support structures remains an open research opportunity.

3.5.3 Volumetric data

- The current generation of volumetric meshing algorithms is based on the triangulation of arbitrarily selected seeding nodes. These methods generate watertight, but not necessarily optimal, volumetric data in terms of the functionally relevant data size.
- Particularly, in the attempt to avoid faceting error, designers often reduce the facet size well below the resolution of the associated AM technology, thereby generating data structures that are significantly larger than necessary and causing critical bottlenecks in the AM process. The optimization of volumetric data generation is an open research question and requires that expert systems be developed that enable the formal definition of the required facet resolution for a specific AM technology and associated geometry.
- These expert systems require that experimental data be generated on the effect of volumetric data resolution on specific geometric inputs for AM systems of interest. This data should preferentially be generated for high-value structural elements, in particular, lattice structures and TPMS (Chapter 5).
- A concurrent research opportunity is the generation of DFAM tools that generate volumetric data with resolution based on the functional requirements of various AM features. Such digital DFAM tools will be particularly useful for commercial applications that require the generative design of product with functional requirements such as form and fit of interacting surfaces.
- Regarding research strategies for volumetric data generation, it is potentially helpful to reference the analogous design problem of FEM mesh generation, whereby finite-elements are generated automatically (rather than manually) to achieve a numerically robust mesh. Although this research has received significant effort, the generation of a universal meshing algorithm remains elusive, and the current best practice remains to include some user input to clarify regions of criticality and interest, and thereby restrict the dimensionality of the mesh generation such that the intended numerical outcomes are robustly achieved. This outcome will likely be mirrored in AM research for surface mesh generation DFAM tools.
- Despite the inherent limitations of first-generation volumetric data formats, the STL format remains the default AM industry standard. Digital DFAM tools that can enhance the interoperability and functionality of the STL format are commercially useful; for example, tools that formalize the addition of surface data or accommodate multiple processing parameters.
- Second-generation volumetric data formats such as AMF and 3MF enable high efficiency outcomes including three-dimensional nesting; however, few digital DFAM tools exist to optimize for these novel features.

3.5.4 Slice and toolpath data

- Second-generation volumetric data formats provide an opportunity to embed data that assists in enhanced efficiency slicing and toolpath generation. DFAM tools that can accommodate slicing errors, for example adaptive slicing, are much required.
- Adaptive slicing with variable skin-and-core thickness provides an opportunity to achieve increased processing speed without compromising surface roughness. These methods require robust experimental data on the effect of these slice parameters on the as-manufactured structure.
- An overarching digital DFAM challenge is a lack of formal standards for digital toolpath formats. In particular, commercial AM systems often employ encrypted toolpath data formats.
- AM toolpath directly influences the mechanical and functional properties of AM components. Sub-optimal toolpath specification introduces defects that act as structural

discontinuities that compromise mechanical performance. DFAM tools that actively minimize these defects, or provide predictive estimates of their structural influence, are necessary for structurally-critical AM applications.

3.5.5 Manufacture, inspection and certification

- Emerging AM inspection methods include in-situ data monitoring, for example the use of thermographic cameras to identify critical defects and to provide process feedback control. To provide useful feedback and to avoid inordinate volumes of data, *meta-data* measures are required that can summarize build overall quality within a reasonable data budget. Much experimental work is required to formalize appropriate meta-data measures and their integration within process control and certification systems.

3.5.6 Advanced DFAM opportunities

The simplified AM design process provides a reflection of the current AM design practice but can also restrain innovation by presuming that these current practices are optimal and inflexible. Advanced DFAM opportunities potentially allow the designer to truncate current AM process workflows by omitting the need for specific phases altogether. Such advanced digital DFAM opportunities provide the potential to increase design efficiency and repeatability while providing the documentation rigour required for performance-critical applications.

Advanced digital DFAM tools can allow the seamless transfer of digital data throughout the entire design process, allowing entirely algorithmic AM design. These generative design methods allow the design to be implemented at computational speeds and without *ad hoc* decision making; thereby enabling the mass-customization of AM design (Chapter 7). Advanced digital DFAM tools include the following.

- ***CAD to Slice data:*** The volumetric data phase provides a useful, common, intermediary phase between CAD data and the slice data phases required for AM. Despite the benefits of this data phase, it is technically unnecessary and does introduce potential for geometric error, processing costs and potentially large file size. An emerging digital DFAM strategy is to omit the volumetric data phase and generate slice data directly from CAD data (loop A, Fig. 3.36). This DFAM strategy also provides an opportunity for AM designers to enforce specific slicing requirements early in the design phase; and furthermore, provides a commercial benefit to CAD software developers intent on increasing the commercial benefit of their product.
- ***CAD to toolpath data*****:** A natural extension of the *CAD to slice data* opportunity is to move directly from the CAD representation to the toolpath data required as input for the AM manufacturing phase (loop B, Fig. 3.36). This digital DFAM strategy provides a significant opportunity to prescribe high-level AM manufacturing data directly within the associated CAD data fill, a significant advantage for designers who wish to ensure that subsequent manufacturing stages are implemented as intended. A particular challenge to these opportunities is a lack of formal methods for defining slice and toolpath data, as well as the tendency for AM hardware manufacturers to encrypt this data.
- ***Design specification to tool path data:*** The opportunity to generate robust toolpath data directly from the design specification data is highly challenging but is potentially feasible for simplified design scenarios (loop C, Fig. 3.36). This generative design opportunity is

commercially significant and may be described as enabling Boundary Condition to Additive Manufacture (BC2AM) outcomes (Chapter 7).

References

Engineering design philosophy

[1] Eppinger SD, Ulrich KT. Product design and development. 5th ed. McGraw Hill Education; 2011.
[2] Samuel A, Weir J. Introduction to engineering design. Elsevier; 1999.
[3] Budynas RG, Nisbett JK. Shigley's mechanical engineering design, vol. 8. New York: McGraw-Hill; 2008.
[4] Dym CL, Little P, Orwin EJ, Spjut E. Engineering design: a project-based introduction. John Wiley and Sons; 2009.

Commercial best practice AM applications

[5] Krznar N, Pilipović A, Šercer M. Additive manufacturing of fixture for automated 3D scanning—case study. Procedia Engineering 2016;149:197—202.
[6] Mendible GA, Rulander JA, Johnston SP. Comparative study of rapid and conventional tooling for plastics injection molding. Rapid Prototyping Journal 2017;23(2):344—52.

Topology optimization

[7] Leary M, Merli L, Torti F, Mazur M, Brandt M. Optimal topology for additive manufacture: a method for enabling additive manufacture of support-free optimal structures. Materials & Design 2014;63:678—90.
[8] Panesar A, Abdi M, Hickman D, Ashcroft I. Strategies for functionally graded lattice structures derived using topology optimisation for additive manufacturing. Additive Manufacturing 2018;19:81—94.
[9] Guo X, Zhou J, Zhang W, Du Z, Liu C, Liu Y. Self-supporting structure design in additive manufacturing through explicit topology optimization. Computer Methods in Applied Mechanics and Engineering 2017;323:27—63.
[10] Primo T, Calabrese M, Del Prete A, Anglani A. Additive manufacturing integration with topology optimisation methodology for innovative product design. International Journal of Advanced Manufacturing Technology 2017;93(1—4):467—79.
[11] Mirzendehdel AM, Suresh K. Support structure constrained topology optimisation for additive manufacturing. Computer-Aided Design 2016;81:1—13.

Systematic material selection

[12] Ashby M. Materials selection in mechanical design. 5th ed. Elsevier; 2017.
[13] Ashby MF, Brechet YJM, Cebon D, Salvo L. Selection strategies for materials and processes. Materials & Design 2004;25(1):51—67.

Integrated topology optimisation and material selection for AM systems

[14] Emmelmann C, Sander P, Kranz J, Wycisk E. Laser additive manufacturing and bionics: redefining lightweight design. Physics Procedia 2011;12:364—8.

Build orientation optimisation and support structure generation

[15] Das P, Chandran R, Samant R, Anand S. Optimum part build orientation in additive manufacturing for minimizing part errors and support structures. Procedia Manufacturing 2015;1:343–54.

[16] Calignano F. Design optimisation of supports for overhanging structures in aluminum and titanium alloys by selective laser melting. Materials & Design 2014;64:203–13.

[17] Vaidya R, Anand S. Optimum support structure generation for additive manufacturing using unit cell structures and support removal constraint. Procedia Manufacturing 2016;5:1043–59.

[18] Hussein A, Hao L, Yan C, Everson R, Young P. Advanced lattice support structures for metal additive manufacturing. Journal of Materials Processing Technology 2013;213(7): 1019–26.

[19] Aremu AO, Brennan-Craddock JPJ, Panesar A, Ashcroft IA, Hague RJ, Wildman RD, Tuck C. A voxel-based method of constructing and skinning conformal and functionally graded lattice structures suitable for additive manufacturing. Additive Manufacturing 2017;13:1–13.

[20] Chen Y. 3D texture mapping for rapid manufacturing. Computer-Aided Design and Applications 2007;4:761–71.

Slice data generation

[21] Haipeng P, Tianrui Z. Generation and optimisation of slice profile data in rapid prototyping and manufacturing. Journal of Materials Processing Technology 2007;187:623–6.

[22] Messner MC. A fast, efficient direct slicing method for slender member structures. Additive Manufacturing 2017;18:213–20.

Detail DFAM

4

Previous chapters have developed the strategic requirements for engaging in a program of AM: specifically, the economic opportunities (Chapter 2) and associated digital design philosophies (Chapter 3). Once these high-level Design for Additive Manufacture (DFAM) considerations have been satisfied, it is necessary to define the particulars of the physical design such that materials, geometry and processing conditions are optimized for AM. This detail design activity is nominally associated with embodiment and detail design phases as defined in Chapter 3.

AM technologies share common attributes associated with the fundamental method of manufacture, which essentially involves the sequential addition of common source materials based on a digitally specified design. These fundamental attributes dictate a series of generalizable DFAM strategies that are common to all AM technologies (Section 4.1). Further to these generalizable strategies, each AM classification is based on a specific manufacturing process and technology implementation, resulting in specific design strategies of merit for each AM classification (Section 4.2).

This chapter begins with a definition of the distinguishing attributes inherent to all AM systems. From this definition, generalizable DFAM strategies are introduced and illustrated with reference to specific case studies. Following these generalizable strategies, a series of design metrics are introduced to assist in the decision-making process to determine the preferred AM classification for a given set of techno-economic design requirements (Section 4.3). The chapter then concludes with a summary of specific DFAM strategies that can be implemented to enhance production outcomes for a specific AM application (Section 4.4). To promote deep understanding, these DFAM strategies are presented with a fundamental analysis of the underlying failure modes and are demonstrated with reference to commercial best practice examples.

4.1 Generalizable DFAM strategies

This text takes an holistic approach to DFAM and considers the strategic considerations of production economics (Chapter 2) to be equally relevant to the technical engineering requirements and detail-oriented decisions associated with component design for a specific AM technology. These detail-oriented activities are the focus of this chapter, and when required for clarity are referred to as *detail-DFAM* in this text.

By referring to the fundamental definition of AM technologies, it is possible to identify distinguishing attributes that are generally typical of all AM technologies. These attributes are introduced and developed below, and are used to systematically define generalizable DFAM strategies with reference to case studies based on commercial best practice.

Design for Additive Manufacturing. https://doi.org/10.1016/B978-0-12-816721-2.00004-X

4.1.1 Distinguishing attributes of AM

ISO/ASTM[1] defines Additive Manufacture as the '*process of joining materials to make parts from 3D model data, usually layer upon layer, as opposed to subtractive manufacturing and formative manufacturing methodologies*' [1]. Based on these observations, Additive Manufacturing (AM) can be distinguished from Traditional Manufacture (TM) by a series of techno-economic design attributes (Fig. 4.1):

Sequential material addition

AM is fundamentally distinguished from traditional manufacture by the sequential addition of materials to incrementally generate the component of interest. This incremental addition may be made to a supporting platen which is subsequently removed, or to bulk material that is then integrated within the final manufactured component.[2]

Common source material

Whereby a standard common source material is utilized for the manufacture of various geometries (this material form may be granulated, filament, sheet or other). This form is typically fixed for a specific AM technology classification.

Distinct hardware implementation

Specific AM technologies are implemented with a distinct set of hardware that directly effects the manufacturability of specific geometries and materials. AM hardware is often based on a Cartesian geometry, resulting in a planar or layerwise fabrication; however, alternative layouts are feasible, including polar and series-robotic.

Digital dataflow

AM is implemented by a digital data workflow. For example, a digital definition of the intended production geometry is digitally processed to provide toolpath specifications for the AM process of interest. Optimization of the digital DFAM attributes of AM systems is critical to their commercial application. Digital DFAM strategies are developed in Chapter 3 and, of particular importance, include algorithmic methods for generative design (Chapter 7).

4.1.2 Case Study: DFAM strategies applied to high-value aerospace structure

As a consequence of the above-listed distinguishing attributes, a series of generalizable DFAM design strategies can be defined that are broadly applicable to manufacture within the AM discipline. These strategies are here presented sequentially, in the context of the

[1] The cooperation between the International Organization for Standardization (ISO) and American Society for Testing and Materials (ASTM) is governed by a Partner Standards Developing Organization (PSDO) agreement (Chapter 3).

[2] Including, for example, an input geometry manufactured by TM to be value-added by AM.

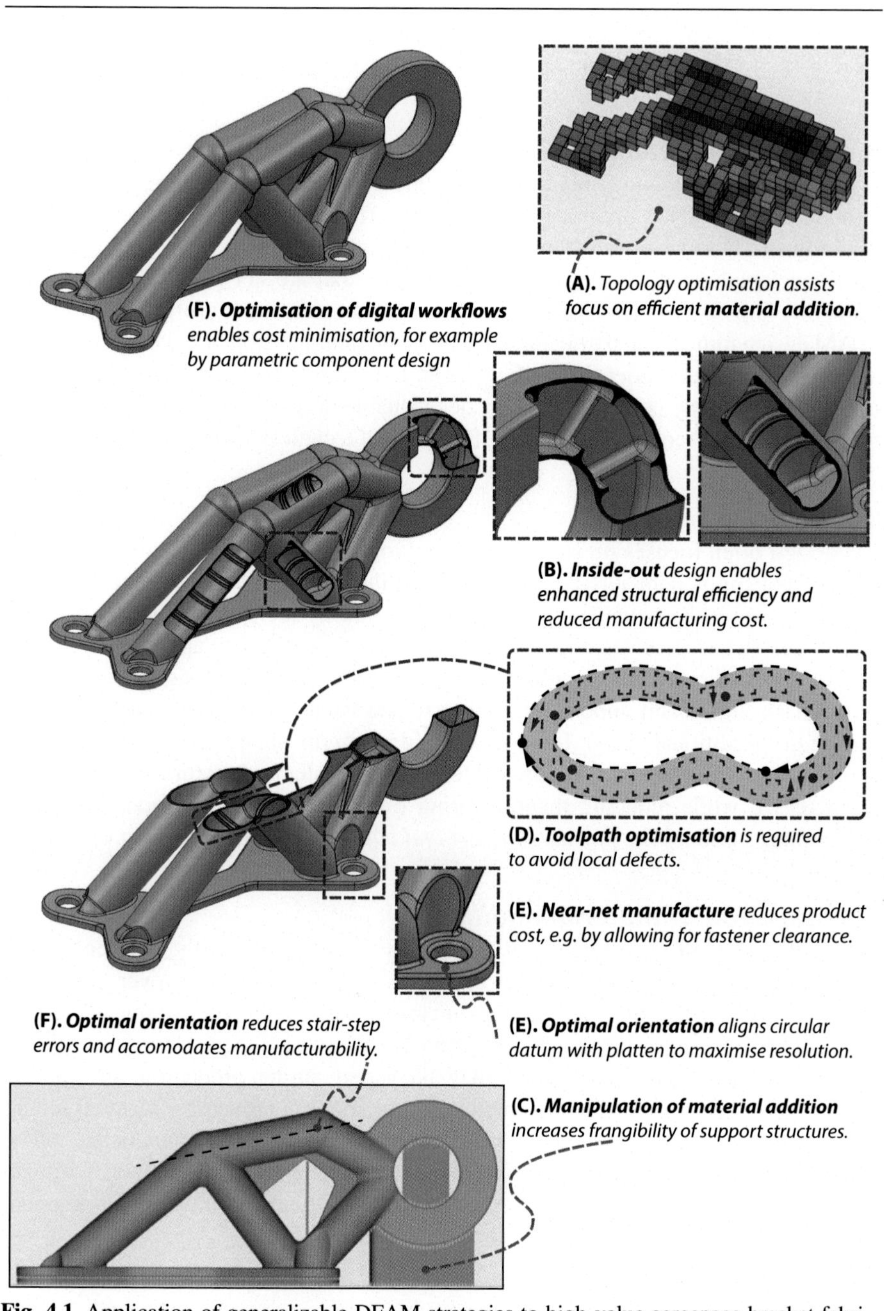

Fig. 4.1 Application of generalizable DFAM strategies to high-value aerospace bracket fabricated with Powder Bed Fusion (PBF): (A) Focus on material addition, especially enabled by topology optimization. (B) Inside-out design to maximize structural efficiency while reducing manufacturing cost. (C) Manipulation of material addition to enhance support structure removal. (D) Toolpath optimization to avoid local defects. (E) Near-net manufacture focus to reduce holistic component cost. (F) Orientation design to improve manufacturability and product function, (G) Optimization of digital workflows to allow minimum manual effort in production and promote generative design methods.

design of a high-value aircraft structure (Fig. 4.1) and should be applied concurrently with the best practice economic (Chapter 2) and digital DFAM strategies (Chapter 3).

A focus on material addition

A focus on the *addition of materials*, as opposed to their subtraction, assembly or forming, is necessary to optimize AM outcomes. This *AM mindset* is especially useful for designers who have transitioned from TM manufacturing sectors, and who may have developed design strategies and preferences that are not distinctly compatible with AM. The AM design case studies presented in this text are useful in developing an AM design mindset, in particular, a focus on the minimization of material volume as well as the design and orientation of geometry to minimize the necessity of supporting structures. Topology optimization provides valuable design guidance for the minimization of material required to achieve the intended technical function (Fig. 4.1A).

Inside-out design

TM design often focuses on an *outside-in* approach, whereby the design focus is on the external component surface.[3] For example, the design of a TM forged aerospace component typically achieves its technical function by the specification of external component surfaces. However, AM enables an *inside-out* design approach, whereby the internal features of the manufactured structure are a primary design consideration (Fig. 4.1B). Significant opportunities exist to add value to AM components by redirecting this traditional design focus to an inside-out philosophy. By focussing on the internal structural geometry, as is feasible with AM systems, structural efficiency can be increased while reducing manufacturing time and associated production costs. This strategy utilizes the *paradox of cost and complexity* as discussed in Chapter 2.

Manipulation of material addition

For a specific AM classification and technology implementation, the combination of material input and the mechanism of material addition is fixed. However, it is often possible to achieve unique technical advantage by varying process parameters or material inputs during production. For example, in Directed Energy Deposition (DED) applications, the alloy composition may be varied during production in order to achieve functionally graded microstructure and mechanical properties; Material Jetting (MJT), Binder Jetting (BJT) and Powder Bed Fusion (PBF) process parameters can be modified such that structural materials become more frangible for support removal; and Material Extrusion (MEX) systems can be intermittently entrained with filament and particulate additives for technical advantage (Fig. 4.1C).

[3] This focus on external component surfaces can occur with a TM mindset, as for machining, forming and fabrication; the toolpath, forming surface and input sub-components are all defined with reference to the external surfaces of the manufactured product.

Toolpath optimization

The set of geometric vectors that describes the instantaneous addition of material are defined as the *toolpath*. Sub-optimal toolpath specifications can result in local geometric defects that compromise the structural robustness of the AM product. The efficient design of AM systems requires that DFAM tools be available to optimize toolpaths for the intended product function in a manner that accommodates the proposed AM technology (Fig. 4.1D).

Net-shape manufacture

AM enables innovation in commercial product design, especially associated with high-complexity and low-batch size scenarios. An inherent challenge to the commercial application of AM is the associated unit-cost, which is potentially high in comparison with TM, especially for large-volume production (Chapter 2). Production cost is a function of multiple variables. One potentially significant contributor to unit-cost is associated with post processing, including heat treatment, final machining and inspection. By focussing design effort on the ability for AM to enable net-shape manufacture, overall production costs can be optimized, thereby increasing the cost-competitiveness of AM. Specific AM opportunities for net-shape AM manufacture include strategic component design to minimize inspection costs and the avoidance of local machining by specifying geometry that inherently allows for press-fit of mating components and fastener assembly (Fig. 4.1E.).

Orientation design

The optimality of a proposed AM design is highly sensitive to the orientation of the component within the build volume, and with respect to the hardware kinematics, i.e. Cartesian, polar, robotic or other. This consideration includes component nesting within the build volume to optimize production time, local geometry, thermal stresses and material solidification. For example, by modifying global and local feature orientation allows thermal stresses, stair-step effects and near-neat enabling features to be optimized (Fig. 4.1F).

As a note of caution, these strategies attempt to provide a comprehensive set of generalizable DFAM strategies for the commercially optimal application of AM technologies. The extent to which these guidelines are applicable depends on the specifics of the AM classification, and exceptions exist for the generalizable guidelines presented here. To provide specific insight, AM system classifications are presented below to enable discussion of DFAM strategies of merit for specific AM classifications.

4.2 Review of AM classifications and inherent design implications

ISO/ASTM categorises AM systems according to the fundamental process by which they enable the addition of source material to generate the manufactured product.

Table 4.1 Summary of inherent attributes of ASTM AM process classifications.

ISO/ASTM classification	Materials		Typical architecture	Typical size	Typical resolution	Overhang support	Thermal process?	Typical material addition geometry	Materials
Vat Polymerization (VPP)	Single	• Curable liquid monomer • Liquid provides some buoyancy to fused solid • Materials may degrade and may damage cells	• Planar	• Mod-Large	• Very high	• Low	• May be thermal	• Point • Plane	• Polymer
Material Jetting (MJT)	Multiple	• Curable liquid • Materials can be combined within jet • Materials may degrade and may damage cells	• Planar	• Mod	• High	• Nil	• Non-thermal binding • Thermal curing	• Linear	• Polymer
Binder Jetting (BJT)	Single	• Input material in powder form • Material consolidated post-binding	• Planar	• Mod	• High	• Very high	• Non-thermal	• Linear	• Metal • Polymer • Ceramic
Material Extrusion (MEX)	Multiple	• Material input ther-mosetting polymer • Filament or pelletized form	• Planar • Cylindrical • Robot	• Small-Large	• Mod-High	• Low-Mod	• Thermal	• Point	• Polymer

Powder Bed Fusion (PBF)	Single	• Input material of powder form • Must be fusible	• Planar	• Small-Mod	• Mod	• Mod-High	• Thermal	• Point	• Metal • Polymer • Ceramic
Sheet Lamination (SHL)	Multiple	• Planar (sheet) input material • Sheet joined by adhesive or fusion	• Planar	• Mod-Large	• Low-Mod	• Very high	• Non-thermal	• Plane	• Metal • Polymer
Directed Energy Deposition (DED)	Multiple	• Material input in powder or wire • Materials can be combined within meltpool	• Planar • Cylindrical • Robot	• Mod-Large	• Very low-Low	• Poor	• Thermal	• Point	• Metal • Polymer • Ceramic

Materials: allows single or multiple materials and are further detailed in terms of specific material attributes. Typical system architectures are identified including planar, cylindrical and robotic. Necessity for support structures are identified as being: Nil overhang allowed to fully supported. Thermal systems are identified. Typical resolution is reported in the z-direction for commercial systems. Typical size reported for commercially robust systems at time of publication.

The technical attributes of these ISO/ASTM classifications are presented in detail in Chapters 8–14, and are briefly summarized below in terms of the *fundamental attributes* of the specific classification and the specific *DFAM considerations* inherent to the classification (Table 4.1):

4.2.1 Vat Polymerization

Fundamental attributes of Vat Polymerization (VPP): Liquid monomer is locally polymerized at point or plane by a UV laser or other curing method (Chapter 10).

DFAM considerations: The implementation of VPP systems involves a single curable polymeric material that is by necessity implemented with a planar architecture. Curing is typically thermal, or by exposure to a specific EM radiation, e.g. UV light. Local curing may be incomplete, resulting in the presence of locally uncured monomers, which may be relevant for the function of medical and biological devices that are in contact with living cells or reactive chemical agents. Although VPP is associated with a planar architecture, material addition may occur at a point or plane depending on the specific technology implementation. Local structure and internal porosity defects are characteristic of this curing type with potential planar defects due to sub-optimal toolpaths and inhomogeneous layer exposure. VPP enables a precise curing zone that can result in very high geometric resolution. The liquid polymer vat provides some floatation support for the as-manufactured structure; however, allowance for unsupported overhang is very low. VPP can be a thermal or non-thermal system depending on the polymerization method, although thermal impact is modest in comparison to other thermal AM systems and there is little no requirement for heat transfer through the manufactured structure. The removal of uncured polymer from the manufactured component must be accommodated.

4.2.2 Material Jetting

Fundamental attributes of Material Jetting (MJT): Liquid material is deposited by a mobile jetting system and subsequently solidified (Chapter 9).

DFAM considerations: As solidification occurs subsequent to liquid material deposition, the accommodation of overhanging structures is effectively nil, and additional support material must be used to accommodate any overhanging geometry. To enable removal post-manufacture, support material may be of a specific composition, for example to allow chemical breakdown for removal, alternately build material may be deployed in an intentionally frangible manner, for example by the intentional introduction of porosity defects. Typical architecture is planar, and observed defects are of this planar type – for example, streaking due to failed jetting elements or material impact with jetting assembly. MJT technologies allow accommodation of an extensive range of materials, as well as allowing these materials to be integrated in-situ, allowing the *digital* design of materials with extreme flexibility in material properties such as strength, stiffness, and colour.

4.2.3 Binder Jetting

Fundamental attributes of Binder Jetting (BJT): Powdered material is deposited within the build space and locally bound with a deposited binding agent (Chapter 13).

DFAM considerations: BJT systems are non-thermal planar powder-bed systems. Consequently, the support for overhanging structures is very high, and should be utilized to maximize the value of BJT components. Specific opportunities include vertical nesting and excellent support by the powder bed for overhanging structures. BJT systems are not directly compatible with multiple materials. Multiple layer thickness implementations may be a challenge for these systems as binder penetration into the powder may be limited. Powder removal is to be accommodated, as are the technical requirements for the curing of the bound powder material.

4.2.4 Material Extrusion

Fundamental attributes of Material Extrusion (MEX): Input material is drawn through an extrusion nozzle and deposited as required (Chapter 8).

DFAM considerations: Material input is typically a thermosetting polymer, and may either be of a filament or pelletized form; the latter providing economic opportunities due to the low cost of polymer pellets available for injection moulding applications. Commercial MEX systems are also becoming popular for the extrusion of biological and chemically reactive materials. Material extrusion systems are typically planar, but can also be implemented by either robotic or cylindrical architectures. Multiple materials can be accommodated, but require multiple extrusion heads, or material change-systems that practically limit the feasible number of discrete materials. Extensive opportunities exist for innovation in extrudate delivery, including particle and filament inclusions. Thermoplastic MEX systems are fundamentally thermal; however, cooling is to the local environment, and structural conductive paths are not required. Accommodation of overhanging surfaces is moderate and is dependent on the polymer thermal setting characteristics and extrudate geometry.

4.2.5 Powder Bed Fusion

Fundamental *attributes of Powder Bed Fusion (PBF):* A power source scans a planar powder bed, resulting in layerwise fusion (Chapter 11).

DFAM considerations: PBF systems are fundamentally planar and utilize metallic, ceramic or polymeric materials in powdered form. Variable layer thickness can be accommodated, but is limited by the available system power, and the propensity of the heat source to penetrate the powdered bed material. High local temperatures must be accommodated by heat transfer paths – both through the manufactured structure and associated supports and into the powder bed. Powder bed heat transfer results in the physical enlargement of manufactured geometry due to the partial adhesion of powder particles and the lateral conduction of powder source into neighbouring powder. This phenomenon is a function of the toolpath, process parameters, local geometry, and powder properties. Thermal distortion is typically accommodated by support structure design, part orientation and system preheating. Local oxidization can be an issue and is typically mitigated by vacuum or shielding gas, the former being necessary for systems that utilize an electron beam as the fusing energy source. Heat transfer rates limit the achievable inclination of unsupported overhangs, although partial fusing of the powder can be implemented to increase overhang support and allow three-dimensional part nesting. Powder removal must be accommodated. Potential defects are of the layer type and include local porosity and lack-of-fusion. These defects can be mitigated by process optimization and toolpath design. Potential damage to recoating mechanism must be accommodated.

4.2.6 Sheet Lamination

Fundamental attributes of Sheet Lamination (SHL): Sheet material is cut as required and sequentially joined by lamination.

DFAM considerations: Planar (sheet) material is a highly economical input that can enable efficient fabrication of complex functional structures. For connected layer elements, the support of overhanging geometry is excellent, so long as layer material and thickness are sufficiently strong. Individual layer geometry must be sufficiently robust to accommodate the layer transfer and lamination techniques. Isolated material regions, or filigree structures may be infeasible. Mechanical properties in the lamination direction may be poor and must be accommodated.

4.2.7 Directed Energy Deposition

Fundamental attributes of Directed Energy Deposition (DED): Input material is directly combined with a moving energy source and is locally fused to previously fused geometry of the component under manufacture (Chapter 12).

DFAM considerations: DED systems can accommodate polymers and ceramics, but are typically implemented for metals in either powder or wire form. Commercial DED implementations are typically of robotic or planar architecture. The absence of a supporting material results in a relatively low accommodation of overhanging geometry. DED is highly thermal and significant heat transfer from the heat source must be accommodated for robust manufacturing outcomes.

4.3 Selection of AM technologies

The practical selection of a preferred AM classification requires the identification of synergies between the product design requirements and the techno-economic attributes of the AM technology under consideration.[4] For example, the specific design requirements may necessitate colour materials to represent the final product aesthetics;

[4] In simple terms, *objectives* define system performance, while *constraints* define feasible solutions. The role of constraints and objectives in formally defining a specific design are presented in Chapter 3.

multiple materials to allow complex functional outcomes; mechanically robust materials to allow efficient load bearing; and, physical size and tolerance requirements for form-and-fit requirements. Furthermore, the intended AM component will have a series of economic requirements associated with the required production volume and allowable unit-cost.

These attributes can be qualitatively assessed with reference to the DFAM considerations inherent to specific AM classifications (Table 4.1). Furthermore, the following charts provide explicit guidance to allow the designer to move directly from the fundamental design requirements to the identification of a set of feasible AM classifications (Figs. 4.2 − 4.5). These charts are based on plausible published data from established commercial suppliers made at the time of publication and are intended to be used for initial material and AM classification selection, as demonstrated in the following summaries and associated case studies.

Once candidate AM classifications are identified, further testing can be completed to identify the specific attributes of the technology platform and materials for the intended AM production system. Furthermore, the charts can be directly applied by AM researchers and technology developers, to identify low Manufacturing Readiness Level (MRL)[5] regions that represent commercial AM opportunities that warrant further research attention.

Part volume versus layer thickness

This data of Fig. 4.2 assists in compromise between the achievable physical component volume and typical layer size.[6] For AM technology classifications that span a large range of part volumes, layer thickness can be typically seen to increase with part volume. This observation represents a pragmatic approach by AM technology manufacturers to enable reasonable production rates for very large components, as well as a technical limitation that achieving fine layer resolution becomes more expensive as system size increases.

Material deposition rate versus physical part volume

Fig. 4.3 assists in compromise between the rapidity of AM manufacture and achievable component volume. Material deposition rates are reported for the maximum material flow rate (for MEX and DED systems) and the maximum volume scan rate (for VPP, MJT, BJT and PBF). Actual production rates are also a function of production time overheads and machine yield.[7] In general, this chart indicates that deposition rates increase with increasing part volume.

[5] Manufacturing Readiness Level (MRL) is highly relevant to the development of innovative product with emerging manufacturing technologies such as AM.

[6] Layer size provides insight into the stair-step effect inherent to AM technologies (Chapter 3), but does not directly indicate minimum feature size, i.e. the smallest physical feature that can be achieved with a specific AM technology. Minimum feature size is a function of process parameters and material as is further discussed within the specific AM technology descriptions (Chapters 8−13).

[7] Machine yield refers to the utilization efficiency of the available production resource, including packing efficiency within available production volume.

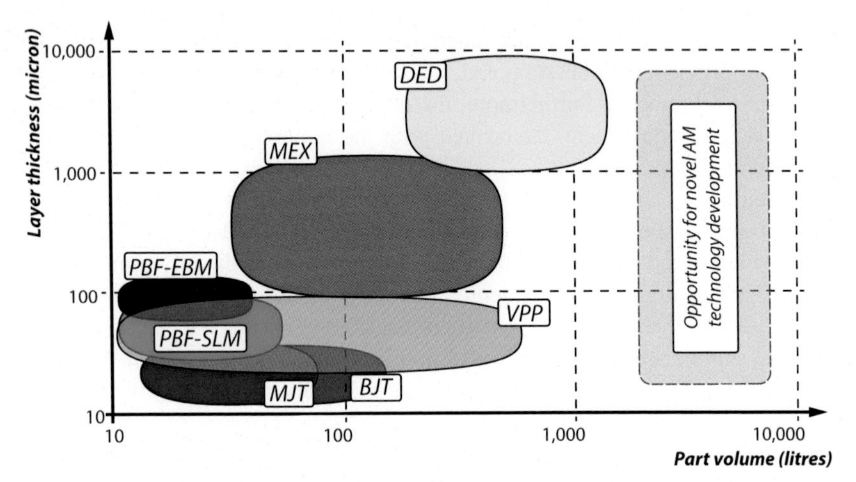

Fig. 4.2 Physical part volume versus layer thickness based on reported data for robust commercial implementations of ISO/ASTM AM classifications at time of writing: Directed Energy Deposition (DED), Material Extrusion (MEX), Vat Polymerisation (VPP), Powder Bed Fusion - Electron Beam Melting (PBF-EBM), Powder Bed Fusion – Selective Laser Melting (PBF-SLM), Material Jetting (MJT), Binder Jetting (BJT).

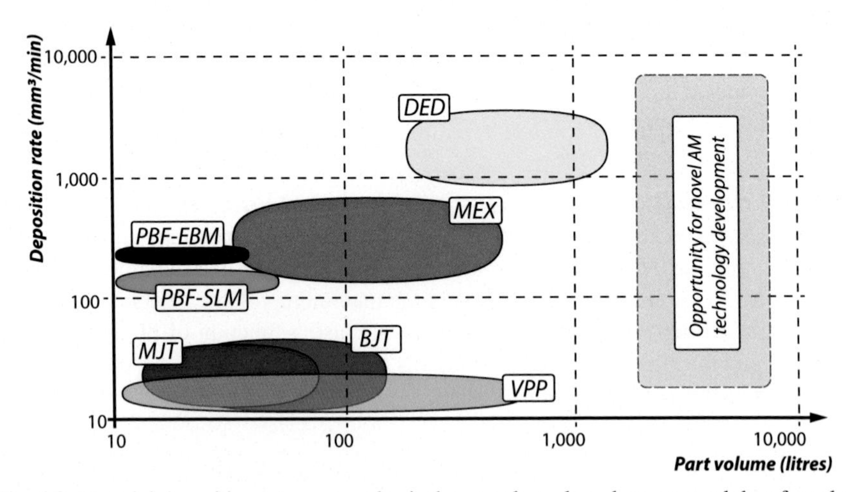

Fig. 4.3 Material deposition rate versus physical part volume based on reported data for robust commercial implementations of ISO/ASTM AM classifications at time of writing: Directed Energy Deposition (DED), Material Extrusion (MEX), Vat Polymerisation (VPP), Powder Bed Fusion - Electron Beam Melting (PBF-EBM), Powder Bed Fusion – Selective Laser Melting (PBF-SLM), Material Jetting (MJT), Binder Jetting (BJT).

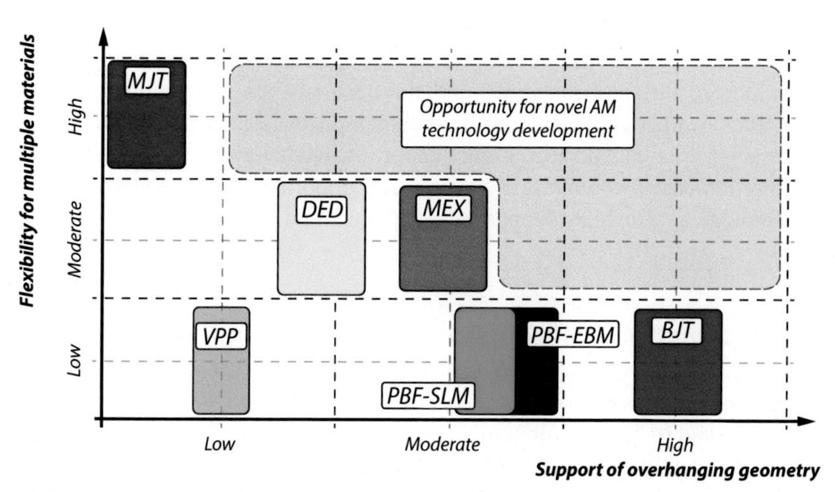

Fig. 4.4 Material flexibility versus support requirements based on reported data for robust commercial implementations of ISO/ASTM AM classifications at time of writing: Directed Energy Deposition (DED), Material Extrusion (MEX), Vat Polymerisation (VPP), Powder Bed Fusion - Electron Beam Melting (PBF-EBM), Powder Bed Fusion – Selective Laser Melting (PBF-SLM), Material Jetting (MJT), Binder Jetting (BJT).

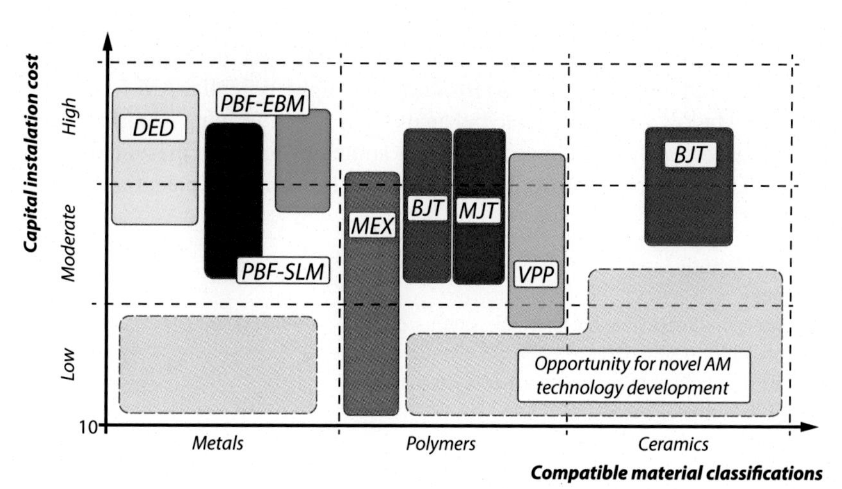

Fig. 4.5 Approximate capital installation cost versus available material types based on reported data for robust commercial implementations of ISO/ASTM AM classifications at time of writing: Direct Energy Deposition (DED), Material Extrusion (MEX), Vat Polymerization (VPP), Powder Bed Fusion - Electron Beam Melting (PBF-EBM), Powder Bed Fusion – Selective Laser Melting (PBF-SLM), Material Jetting (MJT), Binder Jetting (BJT).

Material flexibility versus support requirements

Fig. 4.4 provides insight into the compromise between the flexibility of an AM system to accommodate different materials, and the necessity for supporting structures to accommodate overhanging geometry. This chart indicates that in general the flexibility for materials decreases with increasing support of overhanging geometry; this observation may indicate a commercial and research opportunity for the development of AM technologies with high support of overhangs and accommodation of multiple materials.

Capital installation costs versus available material classification

The capital installation cost varies significantly between AM technologies, and is highly relevant to decisions regarding in-house or outsourced manufacture. The approximate data of Fig. 4.5 attempts to capture the range of costs incurred for the installation of commercially available AM hardware, this data aids in selecting AM technologies that are appropriate for specific product development scenarios, as well as identifying potential commercial opportunities for new technology development.

4.4 Specific AM design strategies

The generalizable AM design considerations identified above are applicable for all AM scenarios. Once a specific AM technology has been selected for a proposed manufacturing exercise, a number of specific DFAM guidelines exist to enable optimization of technical and economic attributes for the specific requirements of the selected AM technology. In order to foster deep insight, potential AM failure modes are presented below with a fundamental analysis of the underlying manufacturing basis for the observed failure. Specific DFAM strategies are then presented based on observed industrial best practice for the following categories of potential AM failure modes.

1. AM architecture
2. Toolpaths
3. Heat transfer (applicable for thermal systems only)
4. Support structures
5. Recoating (applicable for powder-bed systems only)
6. Digital data management
7. Net-shape AM manufacture
8. Miscellaneous

These categories are presented in detail in the following sections to identify the fundamental failure modes and attributes of importance. Illustrative examples are then provided to demonstrate specific DFAM requirements, and discussion is introduced for enhanced DFAM practices and opportunities for commercial technology innovation and DFAM research contributions.

4.4.1 AM architecture

Each of the ISO/ASTM AM classifications dictates a unique system architecture that fundamentally influences the morphology of material addition, achievable rate of material deposition, and associated manufacturability constraints. AM architecture refers to the kinematics of tool path movement, as either cylindrical, planar or robotic; the mechanism by which material addition is enabled, including non-thermal phenomena such as polymerization, and thermal phenomena such as fusion and solidification; as well as the material delivery method, either powder-bed or non powder-bed (Table 4.1). AM system architecture introduces a series of specific DFAM considerations, including:

- AM architecture is commonly planar, resulting in layerwise material addition. This architecture results in continuous geometry being discretized as a series of prismatic sliced geometries (as described in Chapter 3). This discretization results in distinct geometric effects depending on whether the geometry of interest is within the plane of manufacture or has some perpendicular component. For example:
  - For geometry that is within the plane of manufacture, the accuracy of manufacture is dependent on the positional accuracy of the end-effector (e.g. galvanometer directed laser or extrusion stepper-motor), and the accuracy of the material addition method such as melt pool or extrudate solidification. Geometries that require high positional accuracy are preferably aligned within this fabrication plane (Fig. 4.6A).
  - For geometry that has a component perpendicular to the plane of manufacture, the layerwise addition of material introduces a geometric discontinuity known as the *stair-step* effect to the intended geometry. The stair-step phenomenon is a function of the geometry inclination to the normal of the plane of manufacture. Geometry that is subject to stair-step effects may require specific DFAM strategies to be functionally useful (Fig. 4.6B).
- Stair-step effects are often simplified according to the geometric representation of the layerwise input data. This representation does not correctly characterize the effect of the local material addition on the fabricated geometry, which is also a function of the material delivery method, for example:
  - For PBF systems, the manufactured geometry is a function of the local heat transfer from the thermal zone, as well as interaction with the powder bed. Downward facing surfaces potentially exhibit partially adhered particles as they are in contact with the powder bed during local fusion, whereas upward facing surfaces include faces that correspond to the upper extent of the powder bed, and therefore exhibit a reduced propensity to acquire partially adhered particles (Fig. 4.6C).
  - Various scenarios including high energy input and acute local inclination of local geometry can result in excessive local temperatures that can compromise manufacturability. These scenarios can be accommodated by managing process parameters, optimizing build orientation and the use of active support structures.
  - For non powder-bed systems such as MEX, the manufactured geometry is a function of the extrusion head geometry, tool path trajectory, process parameters and the supporting effects of local geometry and active support structures (Chapter 8). Excessively acute orientation to the build platform results in unsupportable overhang that can be addressed by the use of active supporting structures. Phenomenologically similar behaviour can be observed for directed-energy deposition (DED) systems.

(A). *Fabrication plane alignment*
The positional accuracy in the fabrication plane is typically higher than in the perpendicular direction.

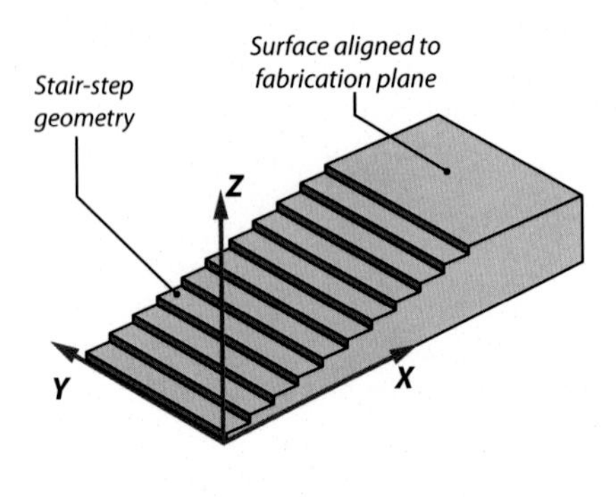

(B). *Stairstepping*
Geometry that is perpendicular to the fabrication plane incurs stair-step effect.

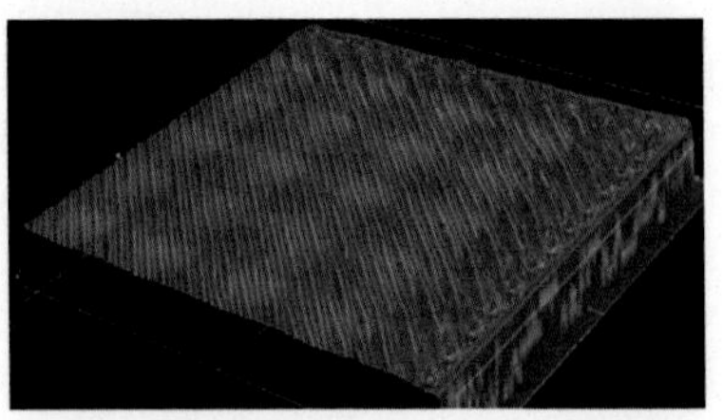

Aligned to fabrication plane

Stair-step geometry

(C). *Partially adhered particles*
Partially adhered particles are not completely fused and are especially common on downward facing surfaces for PBF systems (Chapter11).

Upward facing PBF surface

Downward facing PBF surface

Fig. 4.6 The influence of AM system architecture on manufacturability and associated DFAM guidelines: (A) Gometries that require high positional accuracy are preferably aligned within the fabrication plane. (B) stair-step effects may compromise functionality. (C) For powder-bed systems, up and down-facing surfaces can be fundamentally different due to their contact with powder bed during local fusion.

- Emerging DFAM tools are being developed to accommodate more detailed understanding of the phenomena of stair-step effects; however, there remain significant research opportunities in the development of empirically and phenomenologically correct DFAM representations of stair-step geometry.
- AM architecture plays a significant role in the characteristics of the observed defects for a specific AM technology. For example, planar architectures are more likely to introduce defects within the plane of manufacture, including lack of fusion defects for Powder Bed Fusion systems (PBF, Chapter 11), and geometric stress concentrations for Material Extrusion systems (MEX, Chapter 8).

4.4.2 Toolpaths

The specific AM architecture is used to position an end effector such that material addition is achieved. The trajectory of this end effector is referred to as the *toolpath*. Optimization of the toolpath is a challenging design problem inherent to AM, which results in numerous DFAM strategies to avoid unnecessary fabrication time and to avoid unnecessary toolpath generated defects. Resolution of toolpath DFAM issues requires awareness of material deposition aspects of the particular AM process relevant to allowable toolpath geometry, as well as awareness of mathematical concepts of graph theory, such that toolpath deposition traverses the structural elements of the AM geometry without incurring unnecessary discontinuities or stress concentrations due to sub-optimal toolpath selection.

Toolpaths are limited only by the AM system architecture and end effector design; however, typical toolpath algorithms take a pragmatic approach to toolpath planning that combines a perimeter *skin* structure, which provides structural rigidity and aesthetic surface finish, with an internal raster *core* which allows efficient space filling and accommodation of complex geometry[8] (Fig. 4.7A).

- Toolpath optimization is especially relevant for load bearing AM applications as sub-optimal toolpath generation can result in local porosity and compromised structural performance (Fig. 4.7B).
- A specific challenge associated with toolpath optimization is associated with the feasible nesting of perimeter toolpaths within the layer cross-section of interest. When cross-section thickness is not equal to an integer multiple of the toolpath width, a void may be introduced. Such flawed geometry readily occurs, and is especially challenging to control when cross-section width varies continuously (Fig. 4.7C). Elimination of these defects requires logical toolpath deployment algorithms that, for example, can modify local toolpaths to avoid unnecessary fill-defects.
- This challenge of incomplete filling also applies to thermal energy toolpath systems such as PBF and DED, where porosity may be induced if the toolpaths cannot achieve full coverage of the geometry cross-section (Fig. 4.7J); this challenge potentially results in internal porosity in high-value structures and is discussed in Chapter 11.

[8] This typical AM toolpath methodology is described by disparate terminology, for example *perimeter-and-raster* in fused-filament fabrication (FFF) systems and *skin-and-core* in powder-bed-fusion (PBF) systems.

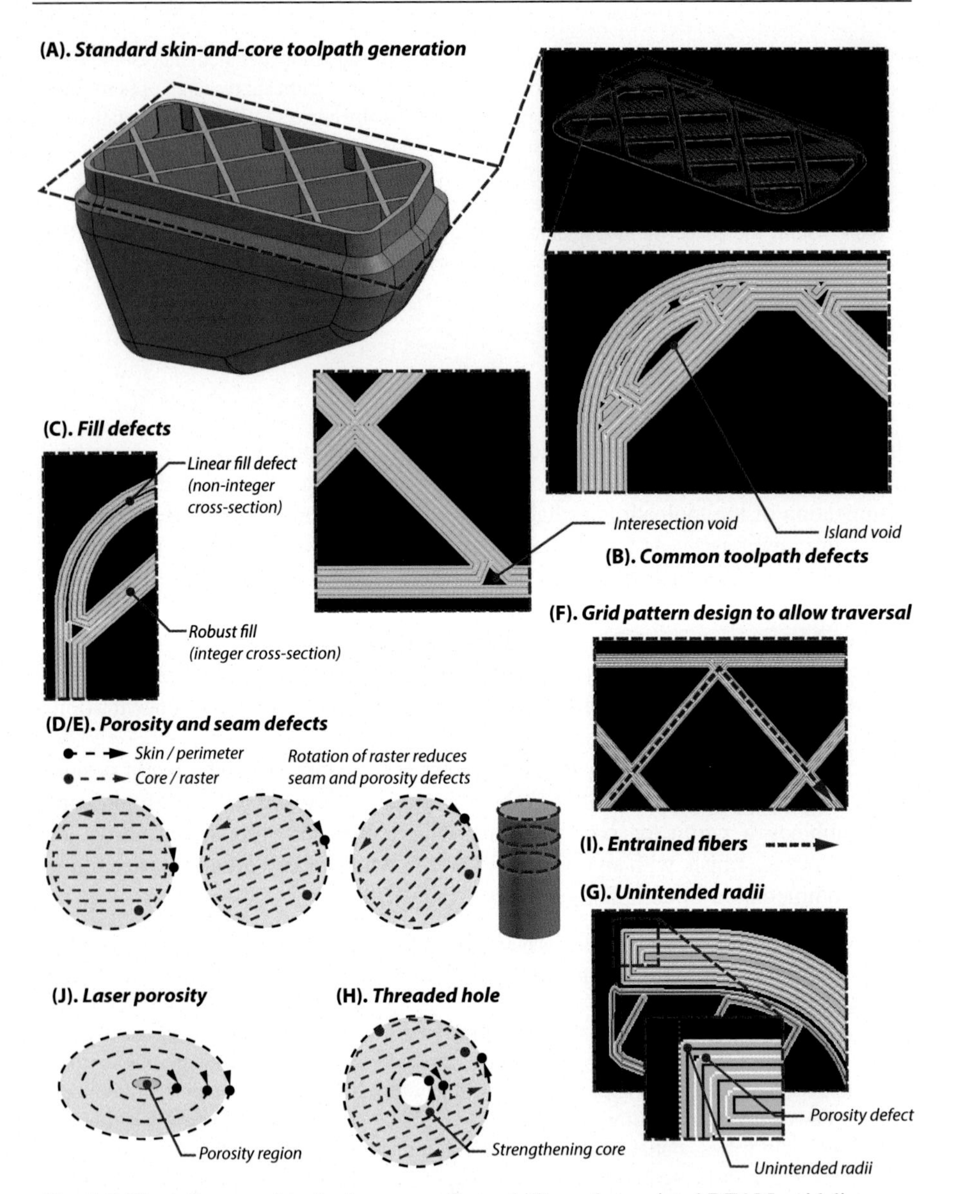

Fig. 4.7 The influence of toolpath on manufacturability and associated DFAM guidelines.

- The core (raster) trajectory is typically based on a grid pattern, which can introduce porosity defects where the toolpath changes direction, and in a linear pattern along toolpath intersections. If core orientation is aligned between sequential layers these defects can accumulate. A typical response is to allow the core pattern to vary in orientation between sequential layers (Fig. 4.7D).
- Similarly, if the skin perimeter is initiated at the same position for adjacent slices, it is possible for a *seam* to be introduced in the manufactured product. This seam is a mechanical and aesthetic defect and can be avoided by alternating the skin perimeter initiation position (Fig. 4.7E).
- Grid support structures are highly efficient, and are readily manufacturable with AM. These structures can prove challenging to optimize for toolpath deployment; and toolpath errors are often associated with a failure to allow continuous traversal of the grid network. By reference to graph theory, network structures can be identified that allow continuous traversal of the structural grid network and thereby avoid unnecessary toolpath defects (Fig. 4.7F). Process parameters (including laser dwell time and material feed rate) can also be optimized to mitigate potential defects within challenging grid locations to enhance manufacturability.
- Geometries must accommodate the minimum toolpath radii at corner features. When corners are excessively sharp, unintended radii are introduced. For internal corners, sub-optimal toolpath planning can result in unintended radii and porosity defects (Fig. 4.7G).
- The objective of net manufacture is important for economically optimized AM design. The flexibility and geometric complexity of AM can be utilized in the direct manufacture of threaded holes, bearing and pin mounts as an effective net-shape design strategy. To enable these features, the toolpath must be appropriately designed by specifying geometries that accommodate net-shape manufacturing outcomes, for example by allowing multiple internal perimeter to provide strengthening for the tapping of a threaded hole (Fig. 4.7H).
- Entrained filament MEX provides an exceptional opportunity to combine the low cost and large volume fabrication achievable with MEX with the high strength of entrained fibres. These entrained fibre applications must understand and respect the toolpath in order to optimize layout and avoid unnecessary interruption of critical fibre layouts (Fig. 4.7I).

4.4.3 Heat transfer (applicable for thermal AM systems)

Thermal AM systems, specifically DED and PBF, enable unique and useful performance attributes for the additive manufacture of high-value systems. Thermal energy must be managed in order to allow robust manufacturability, including the design of thermal load paths; understanding of local temperature field and its impact on manufactured geometry and microstructure; the thermal influence of the powder-bed; and, the effect of geometry and process parameters on thermal gradients and subsequent thermally induced distortion. Specific heat transfer failure modes and associated DFAM tools and strategies include:

- Thermal AM systems generate high local temperatures, which must be dissipated by radiant, conductive and convective heat transfer. These heat transfer modes are of different relative magnitude for different AM systems. These differences are illustrated with respect to three similar, yet distinct AM systems (Fig. 4.8A).
  - Direct Energy Deposition (DED) systems such as LMD typically occur in a locally gas-shielded environment. This environment allows convection to the shielding gas, radiation to atmosphere and conduction through previously solidified geometric features.

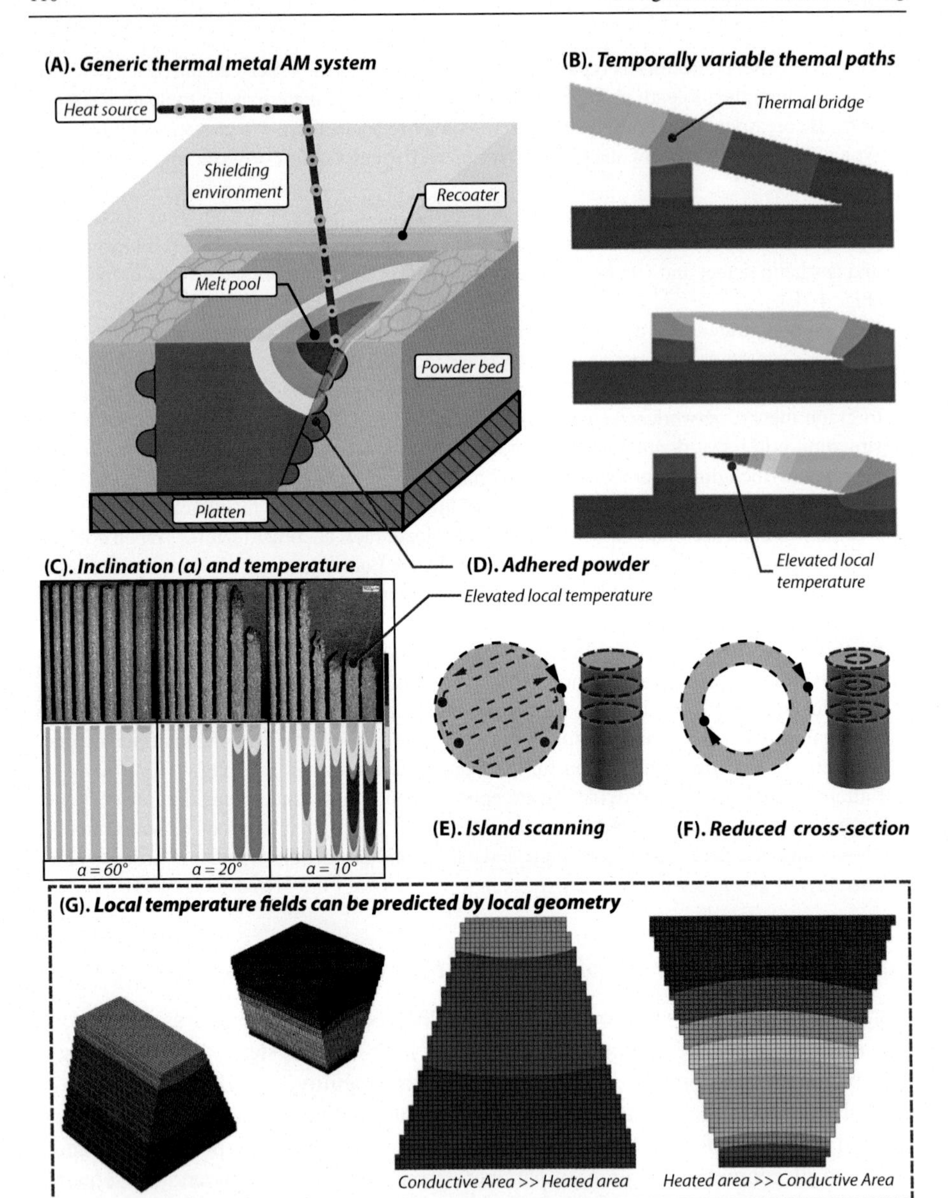

Fig. 4.8 The influence of heat transfer on manufacturability and associated DFAM guidelines. Thermal simulation data based on [2].

  - Powder Bed Fusion (PBF) laser-systems such as Selective-Laser Melting (SLM) are similar to the DED-LMD system described above. However, conduction also occurs via the non-fused elements of the powder bed.
  - Electron Beam Melting (EBM) systems are architecturally similar to the SLM system described above, however with electron beam (rather than laser) energy source. The electron beam necessitates a vacuum environment; consequently, for PBF-EBM systems the convective heat transfer aspect is essentially nil.
  - AM temperature-fields are a function of process parameters, relevant material properties, component geometry and orientation within the build envelope (Fig. 4.8C).
- AM component geometry varies with time, resulting in complex heat transfer paths that are transient in time and space (Fig. 4.8B). The associated thermal fields are highly complex and time consuming to predict, and commercial best practice involves some thermal prediction, as well as operator experience and trial-and error to generate thermally robust outcomes for high-complexity AM geometry. However, computationally efficient DFAM tools for first-order thermal field prediction are emerging, although such tools remain an open research opportunity.
- Inclination within the build envelope can induce variation in temperature fields for otherwise similar geometry. For example, the local temperature field for cylindrical geometry varies with the inclination angle, α. This effect is a function of the increase in exposed cross-section versus the effective conductive path from the heat source to the platen (Fig. 4.8C).
- Platen heating is a useful mechanism for reducing thermal gradients and associated thermal stresses. High platen heating is technically challenging and may be limited in commercial AM systems; especially for systems that utilize a continuous flow shielding gas to isolate the meltpool from the ambient environment, as this gas provides a cooling effect that counters the effect of platen heating.
- Thermal interactions between neighbouring manufactured components can be significant. For example, lattice structures include internal strut elements that can thermally interact with neighbouring struts. Strut elements at the perimeter of a lattice array may see lower resistance to heat transfer and are typically of lower temperature than core strut elements that are surrounded by perimeter elements.
- Processing sequence within the powder-bed is influential to the peak local temperature. For example, *island* scan strategies can be applied such that specific component cross-sections is processed as a series of discrete sub-areas of each larger cross-section, rather than each cross-section sequentially. These strategies can be used to influence peak local temperatures by reducing local thermal intensity (Fig. 4.8E). Concurrently, thermal loads can be reduced by reducing associated cross-section area (Fig. 4.8F).
- MEX systems require that the extrudate be allowed to solidify sufficiently quickly that it can become self-supporting and avoid *die-swell* such that it be sufficiently well formed to provide support to succeeding layers (Chapter 8).
- Temperature fields vary also within local geometric features. Specific phenomena are associated with downward facing surfaces; these features include acute geometry that increases resistance to heat transfer, resulting in increased local temperatures. Powder-bed systems provide some support to this high temperature downward facing surface, thereby increasing manufacturability. However, high-temperature interaction with powder can result in unintended powder adhesion and increased local roughness (Fig. 4.8D). Upward facing surfaces are not in direct contact with powder until subsequent material recoating and therefore exhibit lower roughness.
- DED systems do not include a supporting powder-bed and therefore have a reduced prevalence to the adhesion of partially fused particles, although some particle adhesion does occur.

Furthermore, increased local temperatures on downward facing surfaces compromises manufacturability and can result in increased roughness.
- Local temperature fields can be simply predicted by the effects of local geometry; specifically, that larger geometry cross-sections allow a relatively higher heat input to be provided to the component and relatively small cross-sections provide a relatively higher resistance to heat transfer. This relationship provides a simple basis for the qualitative prediction of local temperatures and can, for example, be used to predict that the cone geometry will induce significantly lower temperature fields than for the inverted cone geometry (Fig. 4.8G).

4.4.4 Support structures

In the context of AM, support structures refer to geometry that is fabricated exclusively for the function of enhancing the manufacturability of candidate geometry by providing physical restraint and heat transfer paths (for thermal systems) during manufacture. Specific DFAM considerations associated with support generation include:

- Enabling near-net manufacture by the positioning of support geometry such that functional surfaces, such as the functional sealing faces of AM geometry, are not compromised by frangible support structures (Fig. 4.9A).
- Specification of support structures such that overall support material required is minimized. This can be achieved by part orientation such that overhanging geometries are minimized as well as by the specification of efficient support structure embodiments (Fig. 4.9B).
- Use of functional support structures that enable manufacture of challenging geometry as well as adding value to the final product. For example, in the thermal system of Fig. 4.9C, the supporting lattice is non-frangible and provides structural rigidity while enabling fluid flow for efficient cooling.

4.4.5 Recoating

AM powder-bed systems, specifically Powder-Bed Fusion (PBF) and Binder Jetting (BJT) utilize a recoating mechanism to regulate the delivery of the unfused powder to the powder bed increment. Various commercial implementations exist, including wiper and roller systems (Chapter 11). These technical recoating systems can introduce challenges to robust AM manufacture and must be accommodated. Specific considerations include:

- The potential for damage to the recoating system, resulting in compromised distribution of subsequent powder layers. This challenge is especially relevant for soft polymer recoater blades that can be damaged by sharp local geometry.
- The potential for manufactured geometry to influence powder distribution. Specifically, it is possible for geometries that are either at high temperature, or are physically oversized, to influence the flow of powder downstream of the problematic geometry.

These challenges can be avoided or mitigated by strategic orientation of parts with respect to the recoater system. Specifically:

- Bulk geometries with orthogonal cross-sections should be oriented at some inclination, β, to the recoater trajectory. This inclination mitigates damage to the recoater system and fabrication defects due to inhomogeneous powder distribution (Fig. 4.10A).
- AM lattice topologies are often specified to maximize structural efficiency; these high efficiency structures are potentially more susceptible to the existence of local flaws due to imperfect recoating. Inclination of the lattice geometry to the recoater trajectory avoids defects that may occur due to multiple recoater intersections (Fig. 4.10B).

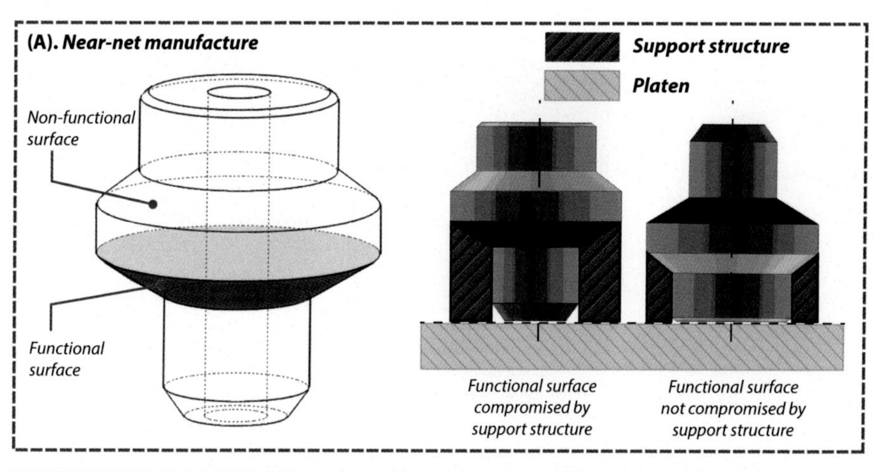

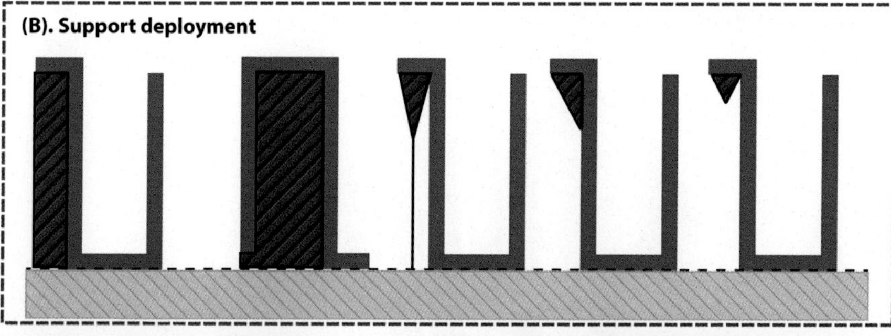

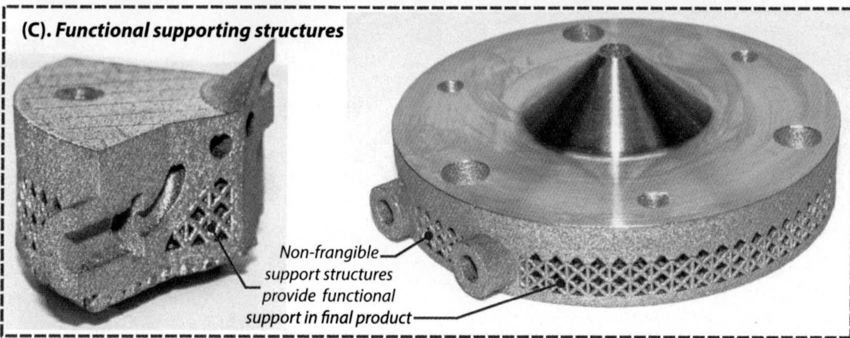

Fig. 4.9 The influence of support structures on manufacturability and associated DFAM guidelines.

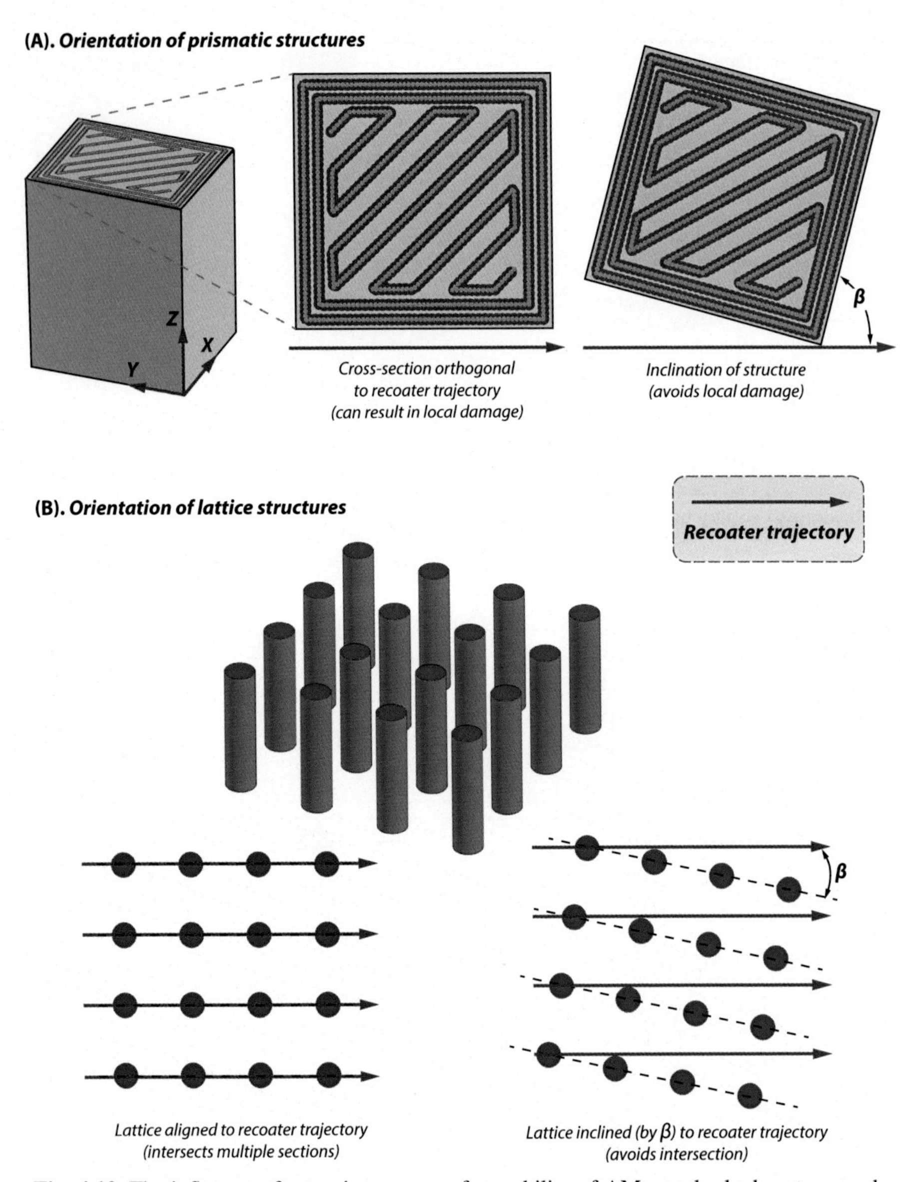

Fig. 4.10 The influence of recoating on manufacturability of AM powder bed systems and associated DFAM guidelines.

4.4.6 Digital data management

AM geometry and production parameters are digitally managed, and production occurs with a common system architecture and source material. Consequently, AM is eminently compatible with digital data management. Robust and efficient digital data management is a critical DFAM attribute that is often overlooked in both commercial practice and research activities. Digital DFAM attributes are presented in detail in Chapter 3, and in summary, include the following considerations:

- Geometric resolution of the AM toolpath must be sufficiently high such that the intended geometric resolution is achieved in the manufactured component. The required resolution is a function of the local curvature of the intended geometry, as well as the AM production parameters and the intended orientation and position within the manufacturing envelope (Chapter 3). Furthermore, resolution is strongly impacted by the selection of the specific digital format used to represent component geometry, and without appropriate DFAM strategies, can rapidly incur file sizes that are computationally excessive (Fig. 4.11).
- Data file size must be appropriate for the intended AM production system. In particular, processing time scales exponentially with file size, and any finite computational system will therefore reach an upper limit to data size beyond which production time is significantly compromised. The requirement for reduced file size is contradictory for the requirements for acceptable resolution, as demonstrated in Fig. 4.12.

4.4.7 Net-shape manufacture

The economic opportunities of AM can be enhanced by minimizing the costs associated with subsequent processing stages: specifically, post-processing and subsequent assembly. Specific net-shape and near net-shape manufacture opportunities and associated DFAM strategies include the following (Fig. 4.13).

Parts consolidation

The integration of product functions within a single manufactured component enables significant economic benefit and enhancement of technical function. Cost reduction is achieved by reduced component count, reduced material consumption and reduced assembly costs. Enhanced technical function is achieved by reduced mass, reduced component volume, enhanced assembly access and the elimination of potential failure modes that exist due to component interaction (Fig. 4.13A). Examples include the integration of hinge, detent and slider components within rapidly fabricated functional AM prototypes (Chapter 8), as well as the high-value rocket manifold presented in Chapter 11.

Design for nesting

AM is typically a batch-process and may result in production throughputs that are lower than are usually associated with traditional mass-production. Part nesting provides a strategic opportunity to increase the throughput of AM systems, either though the 2D tessellation of candidate components within the build volume, or, increasingly, three-dimensional (3D) nesting. Specific DFAM strategies depend on the particular attributes of AM system under consideration, for example (Fig. 4.13B):

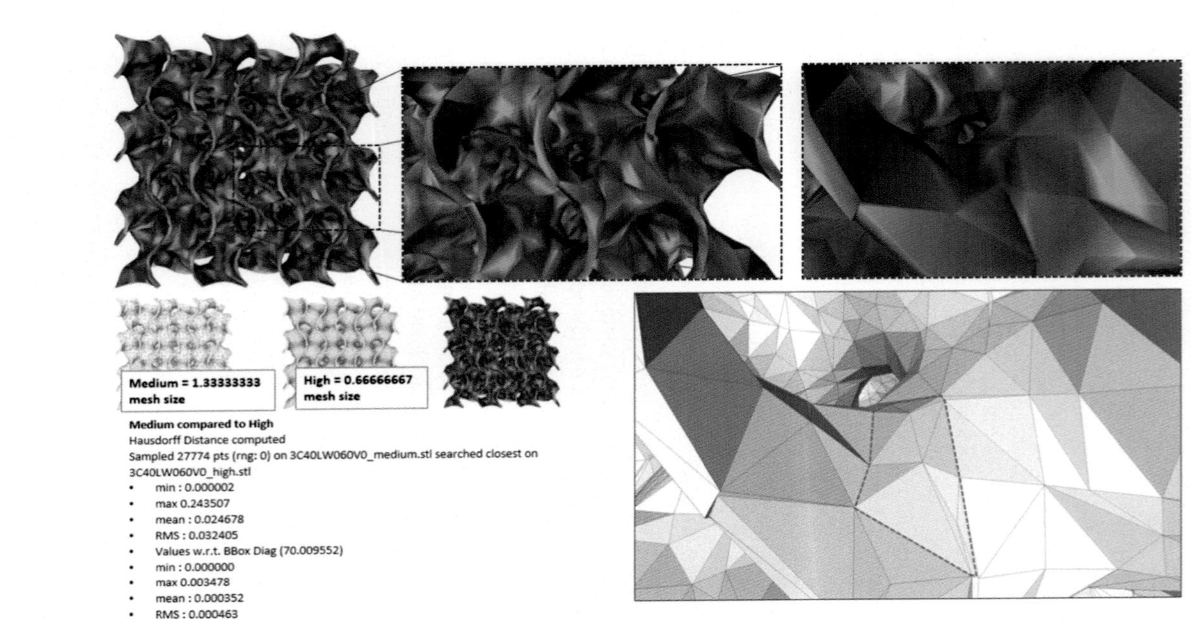

Fig. 4.11 Effect of tessellated data file resolution on manufactured quality.

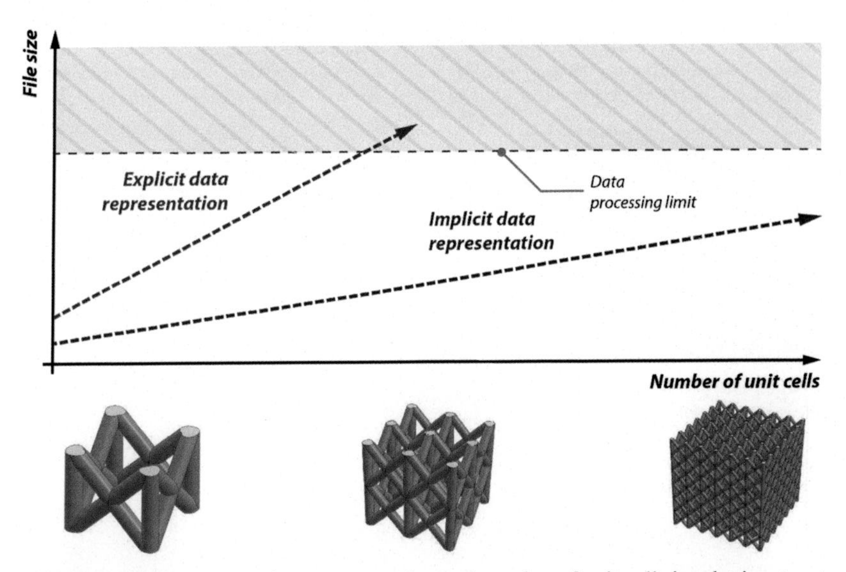

Fig. 4.12 Data file size versus data representation and number of unit cells in a lattice structure. The data processing limit represents the maximum processable file size. This challenge is considered further in Chapter 3.

- Non-thermal powder-bed systems, specifically Binder Jetting (BJT), provide robust support at the powder bed and do not require the accommodation of thermal loads, consequently these systems accommodate unlimited 3D nesting – an attribute that is as yet under-utilized and presents a commercially relevant research opportunity.
- Thermal powder-bed systems, specifically PBF, enable 3D nesting, but require accommodation of thermal loads paths. This opportunity is especially suited to Electron Beam Melting (EBM) systems, where the application of isolated support structures has been technically demonstrated (Chapter 11).
- Vat Polymerization (VPP), Material Jetting (MJT) and Material Extrusion (MEX) systems accommodate 3D nesting, but due to the limited inherent ability of these systems to accommodate overhang or unsupported features, extensive support material may be required to achieve vertical nesting.
- Directed Energy Deposition (DED) systems enable 2D within-plane nesting, however 3D nesting is technically challenging due to the absence of a supporting powder bed, and difficulty in fabricating support structures. However, 3D nesting of DED systems provides a potential research opportunity.

Design for functional geometry

By identifying functionally relevant manufactured surfaces, appropriate orientations and detail designs can be identified and implemented such that net-shape outcomes are achieved (Fig. 4.13C). Functional geometry enables high-value product design, for example the fabrication of patient-specific radiation dosimetry phantoms with continuously variable radiation properties (Chapter 9).

(A). *Parts consolidation*
AM technologies enable complex geometry and functional integration, thereby allowing reduced part-count (integrated hinge and detent, Chapter 8).

(B). *Design for nesting*
AM production efficiency is improved by allowing efficient nesting of components within the build-volume (Chapter 8).

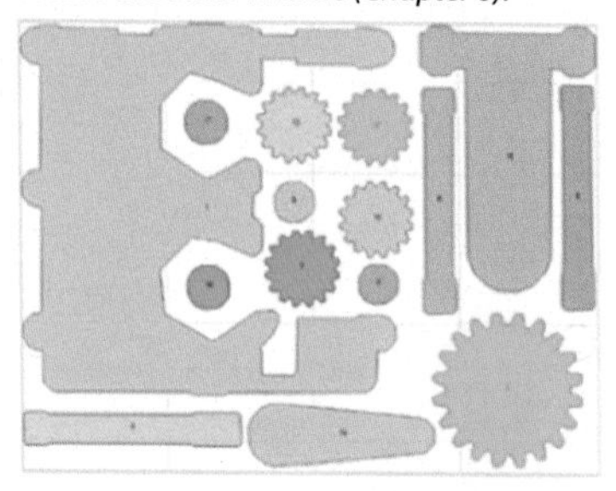

(C). *Functional geometry*
Value is added by integrating functions within manufactured geometry (customised radiation properties, Chapter 9).

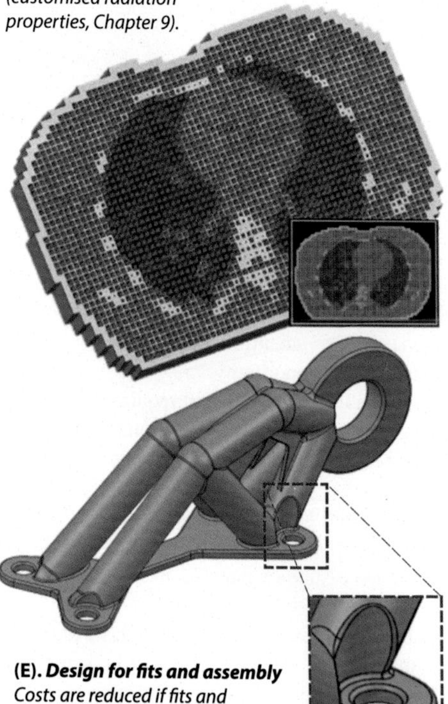

(D). *Elimination of post-processing*
The flexibility of AM technologies eliminates post-manufacture processing, especially machining (integrated cooling channels, Chapter 11).

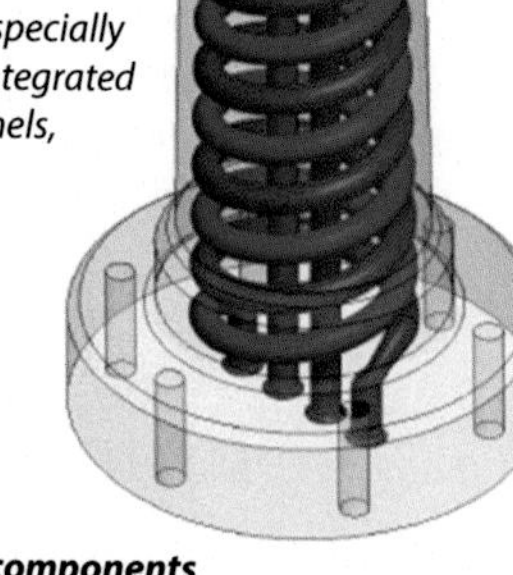

(E). *Design for fits and assembly*
Costs are reduced if fits and assembly features are integrated within the AM geometry (integrated fastener clearance, Chapter 11).

(F). *Hybrid components*
The integration of AM with existing technologies allows the fabrication of hybrid systems with enhanced manufacturing outcomes (rolled plate plus DED, Chapter 12).

Fig. 4.13 DFAM guidelines for net and near-net manufacture.

Elimination of post-manufacture machining

Technical components that are designed with no requirement for post manufacture machining provide an opportunity to significantly reduce overall part cost. Important examples of post-manufacture machining include thread cutting and datum machining. DFAM strategies that can be applied include: the elimination of thread cutting by selection of thread pitch and thread axis to satisfy self-supporting manufacturability requirements; integral cooling channels (Chapter 11); in-situ insertion of threaded inserts; elimination of the need for datum machining by following 3-2-1 methods or similar[9] (Fig. 4.13D).

Design for fits and assembly

Fits and assembly are critical technical requirements for integrated engineering systems. By accommodating the opportunities and design requirements of AM, valuable DFAM strategies can be defined such that AM manufactured componentry requires minimal post-processing to satisfy the requirements for assembly. For example, net-shape AM geometry can be achieved by aligning critical-fit geometries with appropriate manufacturing orientations; as well as by specifying geometry that achieve the intended function without the need for post-processing. Examples of net-shape manufacture include geometry that can directly accommodate press-fit of interacting components, for example rolling element bearings and dowel pins; interference fit geometry that accommodates the fundamental manufacturing capabilities of the associated AM technology, and geometry with integrated fastener clearance (Fig. 4.13E).

Hybrid components

AM technologies provide a unique techno-economic profile that complements the capabilities of Traditional Manufacture (TM). Design scenarios exist that directly combine the advantages of both TM and AM technologies such that high-value design outcomes can be achieved at low unit-cost. Examples include the hybrid manufacture of injection moulding tooling and high-value aerospace structures (where TM processes are used where possible to reduce costs and AM processes are used when geometric complexity exceeds TM capability – for example conformal cooling channels manufactured by PBF on billet material) or the addition of out-of-plane geometry by DMD on rolled plate (Chapter, 12, Fig. 4.13F).

4.4.8 Miscellaneous effects

The previously defined DFAM categories attempt to address the majority of typical challenges to robust component design with AM. However, there exist miscellaneous DFAM challenges of note, especially for thermal AM systems.

[9] The fixation of a body in 3D space requires the constraint of 6 degrees-of-freedom. The *3-2-1 method* is a technique for fixing the movement of a body in 3D space by 3 datum points in a plane, followed by 2 datum points on a perpendicular plane, followed by a single datum point on the third orthogonal plane. By strategically defining these datum with as-manufactured AM geometry enables significant reduction in post-processing costs. This commercially critical DFAM opportunity is under-represented in the current research literature.

- PBF systems that fuse powdered material with a high local energy density can result in the energized ejection of material from the meltpool. This ejected material is potentially significant, as these ejected particles can potentially impact with the powder bed and disturb powder homogeneity as well as introduce particles of modified composition and geometry to the powder bed. This phenomenon can result in progressive disturbance of the powder-bed, resulting in potential variation of metallurgical structure and density between the geometry that is initially fused within the build plane and the final geometry to be fused.
- Systems that minimize material oxidization by the controlled flow of inert atmosphere, for example PBF-SLM, can introduce a contaminated plume to the local meltpool. To mitigate this issue, shielding gas flow should be predictable laminar (with no turbulent eddies to entrain the contaminated plume) and the feature processing order should be selected such that the coincidence of contaminated plume with local meltpool be minimized.
- High-energy fusion can result in the generation of soot that can occlude the laser or electron beam, potentially compromising manufacture.

4.5 Chapter summary

Once the high-level DFAM considerations of production economics (Chapter 3) have been established and a program of AM design has been initiated, it is then necessary to systematically select specific AM technologies and physical design attributes including materials, geometry, and processing conditions that are optimized for the preferred AM technology.

This chapter provides a systematic basis for implementing this detail-DFAM activity: specifically by presenting: a series of AM design strategies that are generalizable to any specific AM technology, and assist in fostering a DFAM mindset; a series of technical design charts that enable selection of the preferred AM technology according to the fundamental design constraints and objectives; and, a fundamental review of the underlying failure modes of specific AM technologies, demonstrated with reference to commercially relevant geometries such that commercial best practice design can be adopted, and fertile research activities can be identified.

AM technologies are fundamentally associated with the *addition* of *common* source materials to manufacture a three-dimensional object based on a *digital* representation of the intended geometry. These fundamental attributes imbue the following generalizable design strategies that are economically and technically relevant to all AM technologies.

- ***A focus on material addition:*** Especially the use of topology optimization to systematically identify regions of efficient material deployment.
- ***Inside-out design***: Where the design focus utilizes the opportunity for AM technology to add internal features and complexity that allows for cost-efficient manufacture of mass-efficient products. This opportunity is defined as the *paradox of increased complexity and reduced cost* (Chapter 2).
- ***Manipulation of material addition:*** There exist opportunities to achieve unique technical advantage by varying process parameters and material inputs during AM production. These opportunities vary with the specific AM technology, and provide significant research opportunities for the development of novel commercial products.
- ***Toolpath optimization***: Where the toolpath is considered to be an independent variable that can be optimized for technical and economic benefit. The effective design of AM toolpath remains an open DFAM research opportunity.

- ***Net-shape manufacture focus:*** A recurring economic challenge for AM is the effect of post processing on unit-cost. By utilizing the design flexibility of AM to achieve net-shape manufacture, overall system costs can be optimized to be competitive with TM systems that may require extensive post-processing to achieve the required product function.
- ***Orientation design:*** Component orientation is of significance to the cost competitiveness of a proposed AM design. DFAM considerations, including component nesting, functional surfaces and thermal stresses can assist in utilizing orientation for the design of high-value AM product.

ISO/ASTM standards categorize seven specific AM technologies according to the fundamental process by which the source material is joined to generate the manufactured product [1]: Vat Polymerization (VPP), Material Jetting (MJT), Binder Jetting (BJT), Material Extrusion (MEX), Powder Bed Fusion (PBF), Sheet Lamination (SHL) and Directed Energy Deposition (DED). Key techno-economic attributes of these AM technologies are briefly presented in Table 4.1; Selection charts (Figs 4.2–4.5) are provided that summarize the AM classifications according to: part volume versus layer thickness, material deposition rate versus production volume, material flexibility versus support requirements, and material deposition rate versus available material types. This data enables the AM product designer to systematically move from the design requirements of the product under consideration to the identification of a set of feasible AM classifications. Once candidate AM classifications are identified, detailed analysis can be completed to select a preferred AM classification and the specific technology platform and materials for the intended AM production system. These charts also provide insight for AM researchers and technology developers to identify commercially relevant opportunities for the development of novel AM technology.

Once the preferred AM technology has been selected, specific component geometry, orientation and process parameters must be defined. Robust AM outcomes require that potential AM failure modes be predicted and avoided. Potential AM failure modes have been identified within the following categories:

- ***AM architecture:*** AM architecture introduces a series of specific and important DFAM considerations that must be accommodated to achieve robust production outcomes.
- ***Toolpaths:*** Resolution of toolpath DFAM issues requires awareness of material deposition aspects of the particular AM process relevant to allowable toolpath geometry, as well as awareness of mathematical concepts of graph theory.
- ***Heat transfer (applicable for thermal systems only):*** Thermal energy must be managed in order to allow robust manufacturability.
- ***Support structures:*** Specific DFAM considerations associated with support generation need to be considered to enhance manufacturability of candidate geometries.
- ***Recoating (applicable for powder-bed systems only):*** Technical recoating systems can introduce challenges to robust AM manufacture and must be accommodated.
- ***Digital data management***: Robust and efficient digital data management is a critical DFAM attribute that is often overlooked in both commercial practice and research activities.
- ***Net-shape manufacture:*** The economic opportunities of AM can be enhanced by minimizing the costs associated with subsequent processing stages: specifically, post-processing and subsequent assembly.
- ***Miscellaneous effects:*** There exist miscellaneous DFAM challenges of note that need to be addressed, especially for thermal AM systems.

To provide deep (rather than rote) DFAM insight, failure modes within these categories have been presented with specific reference to the underlying manufacturing basis for the observed failure, as well as with illustrative examples of observed industrial best practice for risk mitigation. Based on these examples, enhanced DFAM practices and opportunities for commercial technology innovation and DFAM research contributions have been identified.

References

International standard reference for principles and terminology of AM

[1] ISO/ASTM 52900:2015(en). Additive manufacturing — general principles — terminology. Geneva, Switzerland: International Organization for Standardization (ISO); 2015.

PBF thermal field prediction

[2] McMillan M, Leary M, Brandt M. Computationally efficient finite difference method for metal additive manufacturing: A reduced-order DFAM tool applied to SLM. Materials & Design 2017;132:226–43.

Design of lattice and zero-mean curvature structures

Naturally occurring structural geometries provide profound insight into fundamentally efficient topologies for structural optimization. Such structures are typically incompatible with traditional manufacturing methods but can be readily achieved with AM if fundamental manufacturing requirements are accommodated (Chapter 4). This chapter identifies naturally occurring structures with highly efficient topologies and engineered equivalents that are of particular interest to AM are presented, specifically lattice structures (Section 5.1) and zero-mean curvature structures (Section 5.2). These structures ground external loads with high efficiency by tension and compression (and potentially bending) of lattice elements, or by stresses within the curved membrane structure. Design tools available to deploy and optimize these structures are explicitly defined, in particular, the use of Maxwell's stability criterion to characterize lattice structural response and the accommodation of edge-effects when designing practical zero-mean curvature structures. The available experimental data is briefly summarized in order to provide insight into the commercial opportunities enabled by these high-value AM structures. The available DFAM data (experimental, theoretical and applied) are then evaluated in the context of high Technical Readiness Level (TRL)[1] commercial imperatives such that industrial best practice outcomes are presented and a roadmap to commercially relevant research outcomes is generated.

5.1 Lattice structures

Cellular structures refer to a broad classification of naturally occurring and engineered structures that tesselate to fill space (Fig. 5.1). These cellular structures may either be *closed cell*, for example, as first observed in the structure of cork by Robert Hooke [1], or *open cell*, for example in naturally occurring honeycomb structures.

Lattice structures refer to the open-celled arrangement of *strut elements* with defined connectivity at specified *nodes* (Fig. 5.2). These lattice arrangements are readily tessellated to fill space and allow highly efficient paths for the grounding of external and internal loads. Consequently, lattice structures are often observed in both naturally occurring biological systems such as the cellular structures of trabecular bone, and in engineered structural systems, for example three-dimensional truss structures used to efficiently ground loads in civil engineering structures. Lattice structures may have either stochastic or periodic arrangements. Biological lattice systems

[1] The Technical Readiness Level (TRL) is a formal measure of the readiness of a specific technology to be considered for a commercial application of interest. High TRL technologies are more proven for commercial system applications than are low TRL activities.

Design for Additive Manufacturing. https://doi.org/10.1016/B978-0-12-816721-2.00005-1

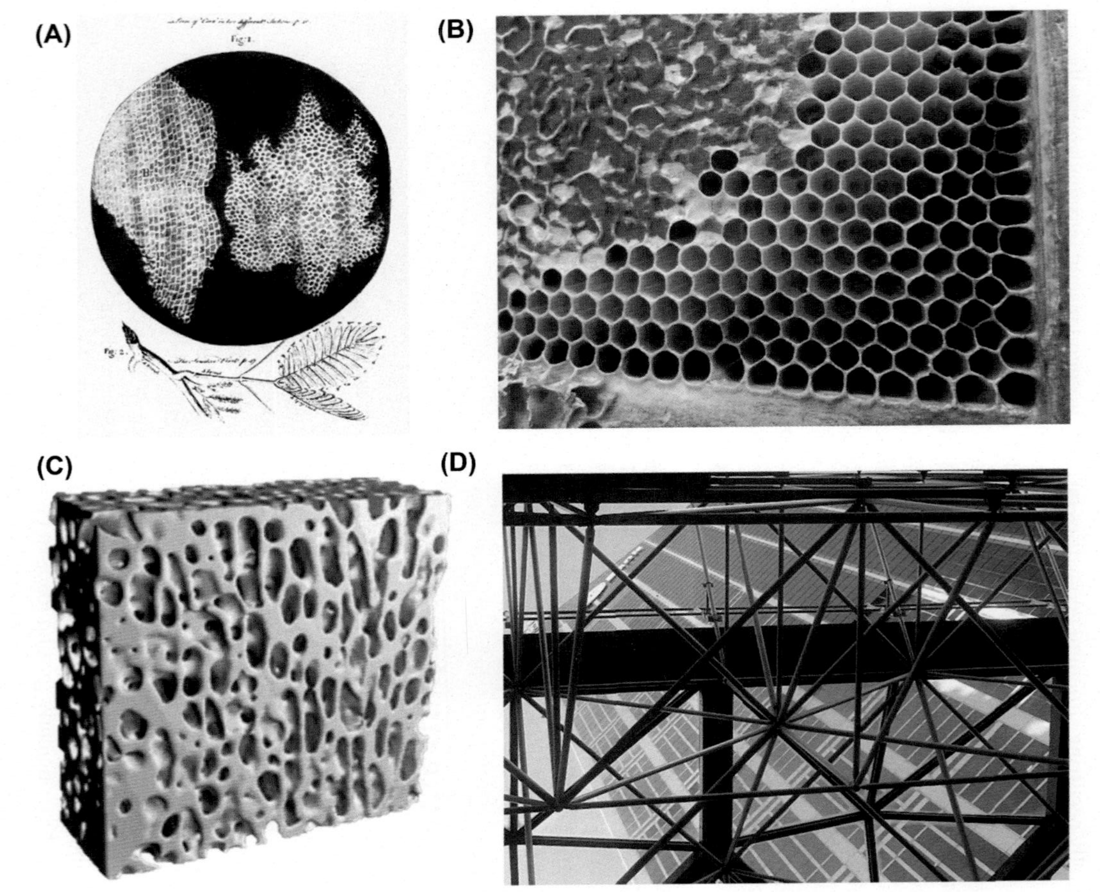

Fig. 5.1 Cellular and lattice structures: (A) First observation of a cellular structure, reported for cork by Robert Hooke in 1665 [1], (B) open hexagonal cellular structure of honeycomb, partially filled with honey, (C) naturally occurring lattice structure observed in trabecular bone [2], (D) civil engineering lattice used to efficiently span space (note pin-jointed connections).

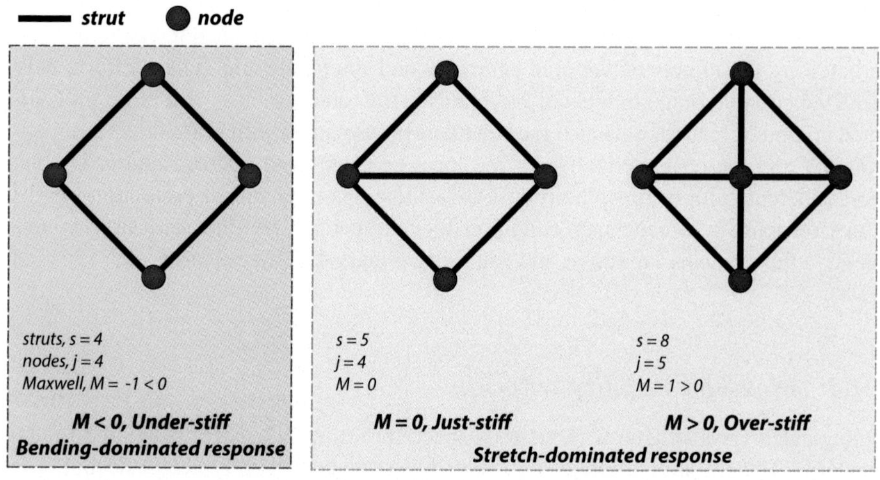

Fig. 5.2 Schematic representation of potential planar lattice configurations, Maxwell number, *M* (Eq. 5.1) and associated response type, either bending or stretch-dominated.

generally feature stochastic arrangements, whereas engineered lattice structures are typically periodic; however engineered structures with stochastic and hierarchical attributes are emerging within the research literature and in commercial applications.

The mechanical response of lattice systems is qualitatively predictable according to the associated *Maxwell stability criterion* (Section 5.1.1). This established relationship provides robust insight into the structural loading modes and anticipated mechanical response associated with a specific lattice topology. Despite the usefulness of the Maxwell criterion, there remain nuances in its application to engineered AM systems that are potentially misunderstood, resulting in the potential for flawed design decision-making and missed opportunities for design optimization. These nuances are identified in order to minimize the risk of design error and to motivate DFAM research contributions in this space (Section 5.1.2).

Depending on the Maxwell criterion for a specific lattice structure of interest, internal loading is either dominated by bending or by axial stretching. These classifications allow insight into the observed modes of deformation and failure (Section 5.1.3). Based on simplified cellular geometry, the *Gibson-Ashby* model allows prediction of the lattice mechanical response as a function of the lattice density, internal loading type and properties of the parent material (Section 5.1.4). The accuracy of this prediction may be enhanced by reference to experimentally assessed data. Alternatively, summary data assessed by the Gibson-Ashby model provides a valuable DFAM tool by allowing prediction of the feasible range of mechanical response achievable for a specific AM technology and associated lattice geometry, as well as currently overlooked research opportunities. A summary of such data is provided for metallic AM lattice structures in Section 5.3.

The quantitative response of AM lattice and zero-mean curvature structures is highly challenging to predict, and experimental data reveals nuanced complexity in mechanical response, especially when subject to plastic deformation, local buckling, cyclic loading or impact. Complexity in the prediction of mechanical response is exascerbated by the effects of variable geometry and microstructure. These effects defy idealized computational models of mechanical structural response; and more sophisticated computational models also face challenges associated with scale-effects, dimensionality and the characterization of the local geometry and microstructure. Despite these inherent complexities, the robust and efficient computational prediction of AM lattice response is a necessary condition for commercial AM lattice design; consequently, this remains an active and multidisciplinary DFAM research area (Section 5.1.8).

5.1.1 Maxwell stability criterion

Triangulated truss structures (known as space-frames in 3D space) provide a fundamentally efficient structural system. Such truss structures are typically designed to ground internal and external loads by a combination of tensile and compressive axial loading within the associated strut elements.[2] For external loads to be robustly equilibrated by internal forces within the truss structure, a sufficient number of strut elements with appropriate nodal connectivity must exist. Maxwell proposed a set of mathematical conditions that must be satisfied[3] for the loads be grounded in a mechanically robust manner [3]. The *Maxwell stability criterion* is stated such that there are 2 free equilibrium-equations[4] (3 for 3D space) associated with each node, j, and that each strut, s, represents an unknown equilibrating force; furthermore, there are 3 external forces (6 for 3D space), resulting in the inequalities of Eqs. (5.1) and (5.2).

$$M = s - 2j + 3, \text{ for planar (2D) truss systems} \tag{5.1}$$

$$M = \text{s} - 3j + 6, \text{ for 3D space-frames} \tag{5.2}$$

For scenarios where $M < 0$, there are insufficient struts to equilibrate the external forces, and the *under-stiff* truss system becomes a mechanism (Fig. 5.2). For $M = 0$, the strut elements are arranged such that the strut loading is determinant for any external loading; such structures are defined as *just-stiff* and contain no structurally redundant strut elements. Additional strut elements can further increase the Maxwell number, $M > 0$, making the truss *over-stiff*.

[2] Of the potential loading modes, axial tensile loading is highly efficient, whereas bending provides low efficiency as it *can* introduce the buckling failure mode, which compromises structural efficiency.

[3] The Maxwell stability criterion is a necessary, but not sufficient, condition to determine the response of a truss structure. It is also necessary that redundant connections and pre-stressed elements be considered [18].

[4] Three equations of equilibrium exist in a 2D plane (6 in 3D space), however rotational equations are trivially solved at a point and are not considered further.

5.1.2 Observed mechanical response (unit cell)

For AM lattice structures, the analogy of truss structures breaks down, as the nodes are not pin-jointed (as in a typical truss system) but are of a continuous structure that can accommodate the transfer of bending across node elements, i.e. the end-fixity is *encastre* or *built-in*. Despite the imperfect analogy, the Maxwell stability criterion enables a profound insight into the mechanical response of proposed lattice topologies. For AM lattice structures that are under-stiff, lattice deformation is resisted by bending at the nodes. Due to the large strains induced by the associated bending stresses, these structures are highly compliant, and are known as *bending-dominated* ($M < 0$). Conversely, for structures that are either just-stiff or over-stiff, loads are grounded by tension and compression of the strut elements only, and there is no bending applied across the node.[5] These structures are known as *stretch-dominated* ($M \geq 0$) and display relatively high stiffness.

The mechanical response of a proposed lattice system is highly dependent on the associated loading mode; a nuance that is somewhat misunderstood in the general engineering community. The simplified numerical analyses of Figs 5.3 and 5.4 serve to illustrate this point for *constant force* and *constant displacement* scenarios, respectively. In response to a constant external force, an under-stiff structure has insufficient strut elements to resist external loads without bending, and is therefore bending-dominated, and is highly compliant with large induced stresses (Fig. 5.3). Conversely, the just-stiff structure resists a similar external load by internal tension and compression; this stretch-dominated structure resists the constant force with little deflection and low associated stresses.

The scenario is reversed for constant displacement scenarios (Fig. 5.4). For these scenarios, the high stiffness of the stretch-dominated structures ($M \geq 0$) results in inherently large forces for a given deflection, as well as substantial induced stresses. Conversely, the bending-dominated structure ($M < 0$) is highly compliant to the applied deflection and accommodates the imposed deflection with very little induced external force or associated stress.

The significance of both loading type (either constant deflection or constant load) and structural response (either bending or stretch dominated) on lattice deflection and induced stresses is of critical importance for the design of lattice structures that are structurally sound and optimized for their intended function. The flexibility of AM systems in allowing the fabrication of customized lattice structures allows for design optimization according to the specific requirements of the engineered system; including functionally-graded lattice and lattice systems with tuned structural response in order to provide variable mechanical response (Section 5.1.5).

[5] Upon bulk deformation, these strut elements may deform significantly and introduce node bending.

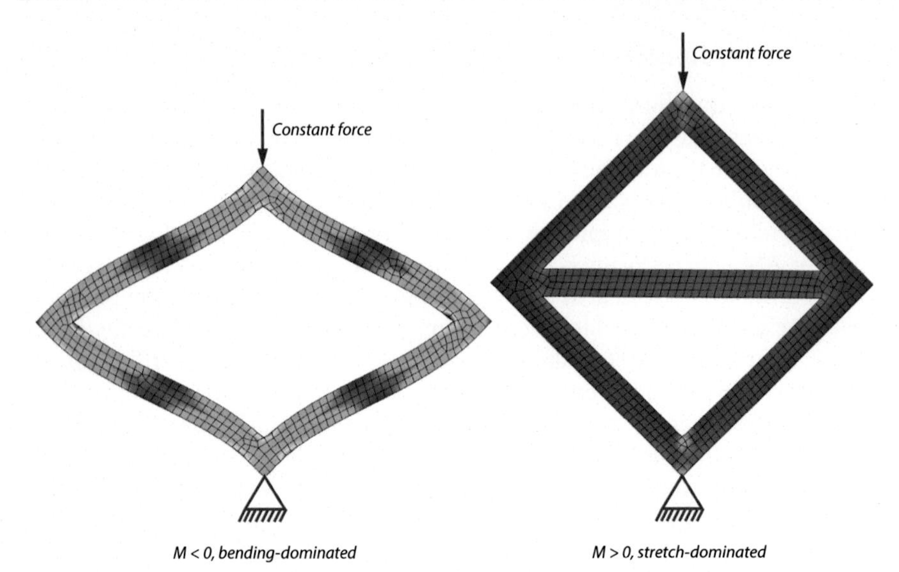

Fig. 5.3 Mechanical response to a *constant force* in bending-dominated and stretch-dominated structures (Fig. 5.2). Colour indicates local stress, which is relatively low for the stretch-dominated structure.

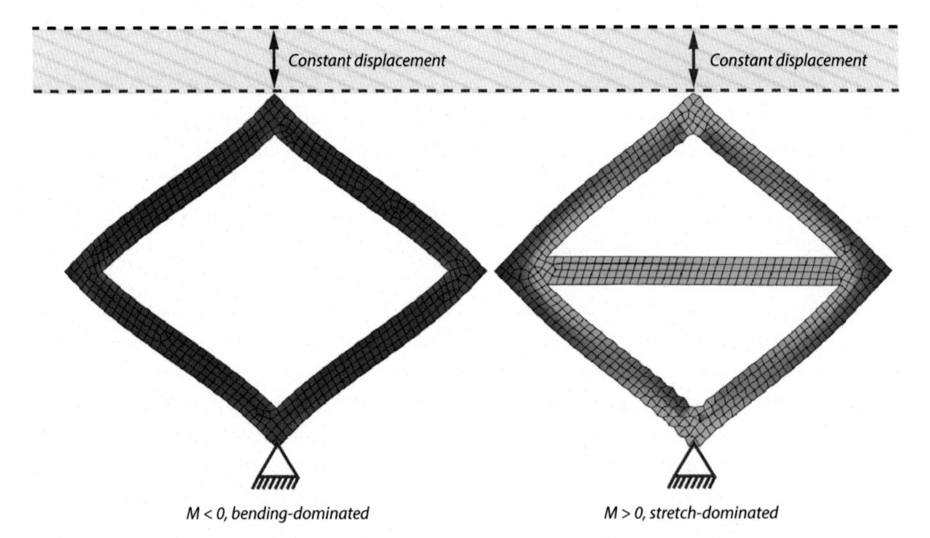

Fig. 5.4 Mechanical response to a *constant displacement* in bending-dominated and stretch-dominated structures (Fig. 5.2). Colour indicates local stress, which is relatively low for the bending-dominated structure.

5.1.3 Observed mechanical response (lattice system)

A lattice system typically consists of an assemblage of individual unit cells. The mechanical response of such a *lattice system* is dependent on the structural response according to the Maxwell stability criterion. The overall behaviour of AM lattice systems has been documented both experimentally and mechanistically and may involve the following observed responses (Fig. 5.5).

Initial plastic consolidation

Whereby local stress concentrations due to AM geometric artefacts and non-uniform loading result in local stresses that exceed the elastic limit of the parent material and result in unrecoverable plastic deformation at relatively low load magnitude.

Linear elastic response

Occurs post consolidation where the response is linear and elastic, with a well-defined lattice Young's Modulus, E^*.

Non-linear elastic response

Depending on specific geometry and topology, a lattice structure *may* display elastic (i.e. recoverable) non-linear buckling of the lattice struts elements. The strain over which this non-linear elastic behaviour is observed is highly dependent on specific lattice topology and local geometry.

Yield strength, or elastic limit, σ^*_{el}

Signifies the elastic limit of the lattice structure and occurs once lattice buckling, or crushing is no longer elastically recoverable.

Unloading Modulus, E^*_{ul}

Refers to the observed modulus post plastic yield. Limited experimental data is available; however, this value has been observed to be higher than Young's Modulus, E^*. The specific value of the unloading modulus is critically important to the design and manufacture of patient-specific medical implants and other mechanical systems that are subject to a stiffness constraint and are exposed to cyclic loading.

Ultimate compressive strength, σ^*_{pl}

Associated with the peak nominal stress of the lattice structure. This stress (and associated strain) is used to define the absolute safety factor for load-limited applications. For stretch-dominated lattice systems σ^*_{pl} occurs catastrophically as the deformation mode transitions to local strut buckling and fracture and is followed by a significant drop in load bearing capacity. For bending-dominated structures, σ^*_{pl} does not represent a change in deformation mode but does represent the peak bending stresses that can be accommodated before bulk plastic deformation occurs throughout the lattice.

Crushing strength

Occurs with progressive plastic deformation beyond the ultimate compressive strength. Crushing stress characteristics are highly dependent on the specific lattice topology: stretch-dominated systems indicate a cyclic response associated with the sequential catastrophic failure of critical zones within the lattice structure, whereas bending-dominated systems indicate a gradual transition to a plateau stress; perhaps without a distinct

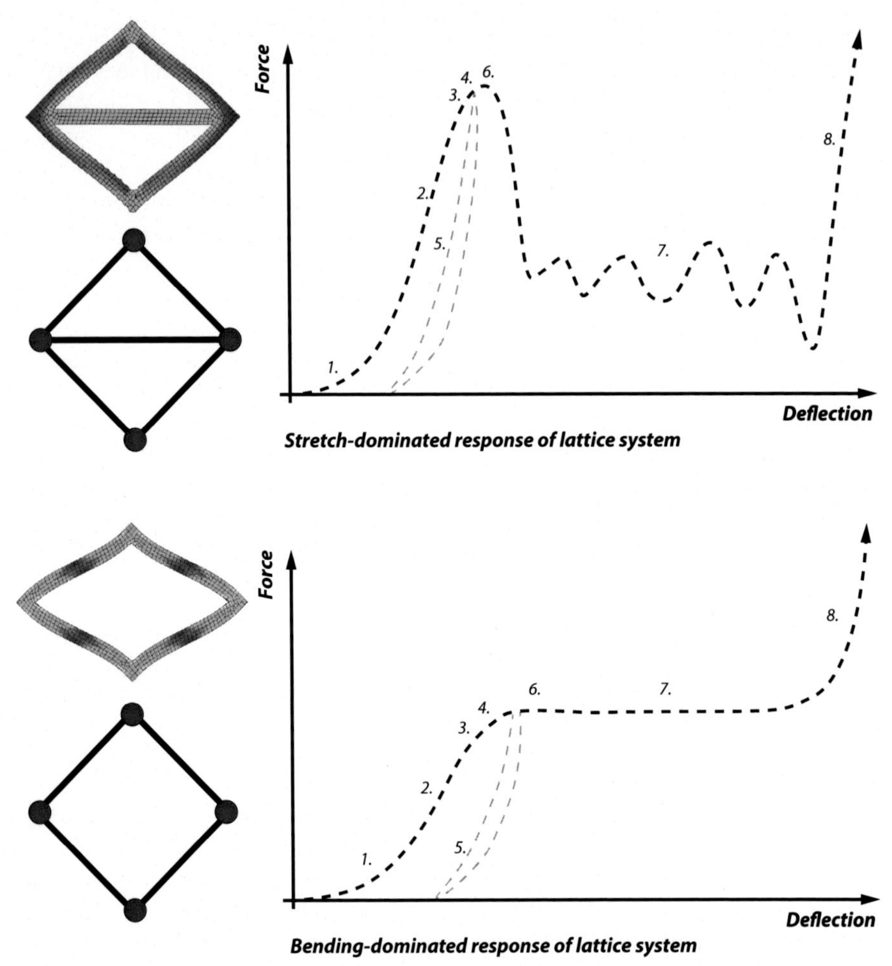

Fig. 5.5 Typical stretch- and bending-dominated mechanical response of a lattice system indicating potentially observed behaviours, including: 1. Initial plastic consolidation, 2. Linear elastic response, 3. Non-linear elastic response, 4. Yield, 5. Unloading Modulus, 6. Ultimate compressive strength, 7. Crushing strength, 8. Densification.

observable ultimate compressive strength, σ^*_{pl}. The opportunity to engineer this observed crushing response is a significant commercial advantage of engineered AM lattice.

Densification

Occurs when plastic deformation of lattice elements is sufficient such that contact is made between normally separate lattice elements. During densification, the lattice behaves increasingly like a solid structure and the associated stress increases exponentially. This densification is only observed in materials with sufficient ductility to

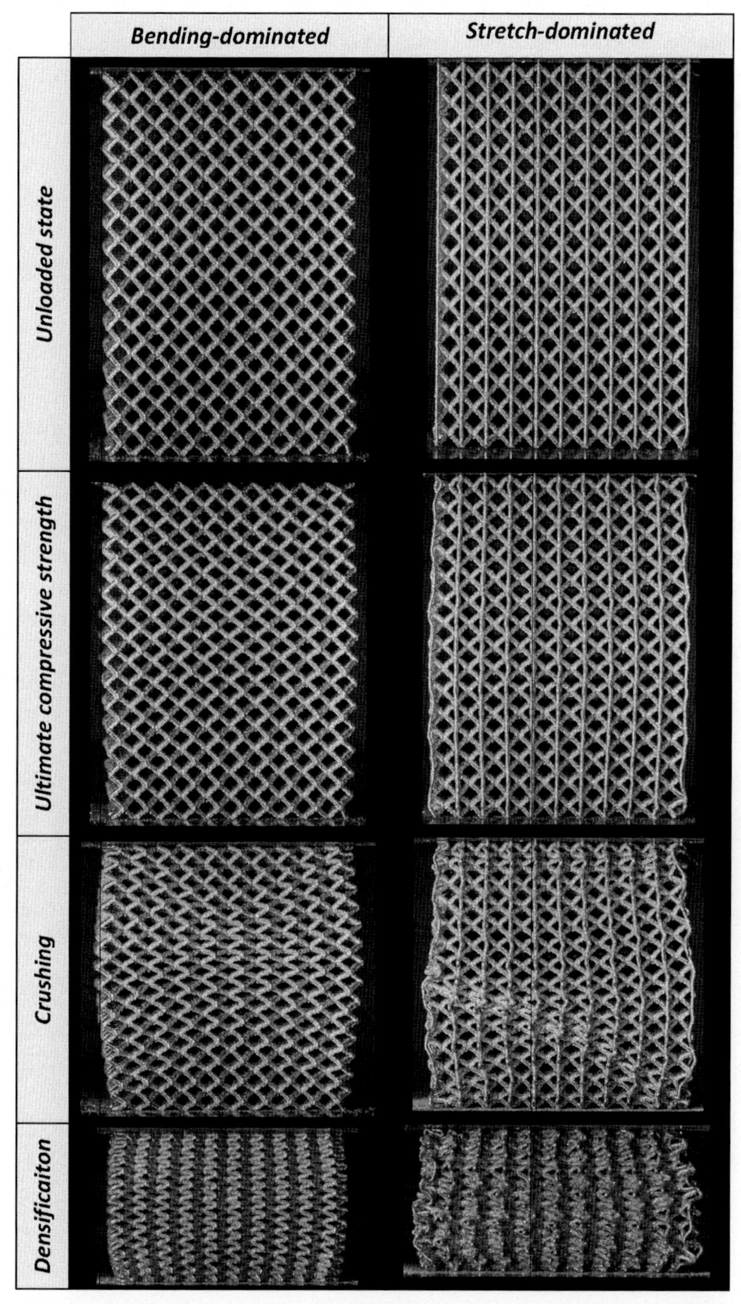

Fig. 5.6 Observed mechanical response of lattice structure displaying bending-dominated (left) and stretch-dominated response. Specimens fabricated by Selective Laser Melting (SLM) with Inconel 625; a highly ductile material that provides insight into the full range of potential lattice deformation response. Specimens fabricated according to *[5]*.

allow this large deformation without fracture. Increasingly brittle materials will fail catastrophically at lower values of observed deflection (Fig. 5.6).

These observations apply to lattice structures that are distinctly within either stretch- or bending-dominated regimes. Additional complexity occurs when structures display transitional behaviour between these distinct regimes. These observations are mostly experimental in nature (Section 5.1.7) but present a commercially valuable research opportunity for the design of AM structures with engineered mechanical response that would be technically infeasible with traditional manufacturing methods. Cost-effective design of such structures requires further understanding within both experimental and numerical research disciplines.

5.1.4 Prediction of AM lattice response

Numerous predictive models have been established to relate the lattice topology and material properties to the observed mechanical properties, most notably the seminal Gibson-Ashby model [4]. This fundamental model is derived from first principles based on insights into cellular geometry and associated mechanical failure modes, resulting in predictive relationships of the type defined in Eq. (5.3). Specifically, these relationships predict the lattice mechanical response, P^*, to be proportional to some known response of the solid material P_S, and the ratio of lattice to solid material density, ρ^*/ρ_S, to the power of some exponent, n, and scaled by some proportionality constant, C. The predicted exponent depends on whether the structure is bending- or stretch-dominated as summarized in Table 5.1 for a number of pertinent scenarios.

Table 5.1 A selection of Gibson-Ashby relationships for bending-dominated and stretch-dominated mechanical response.

Response type	Mechanical property	Relationship	Equation
Bending-dominated	Modulus (E^*)	$E^* = CE_s\left(\frac{\rho^*}{\rho_S}\right)^2$	(5.4)
	Plastic collapse $\left(\sigma^*_{pl}\right)$	$\sigma^*_{pl} = C\sigma_{y,s}\left(\frac{\rho^*}{\rho_S}\right)^{1.5}$	(5.5)
	Elastic collapse $\left(\sigma^*_{el}\right)$	$\sigma^*_{el} = CE_s\left(\frac{\rho^*}{\rho_S}\right)^2$	(5.6)
Stretch-dominated	Modulus (E^*)	$E^* = CE_s\left(\frac{\rho^*}{\rho_S}\right)^1$	(5.7)
	Plastic collapse $\left(\sigma^*_{pl}\right)$	$\sigma^*_{pl} = C\sigma_{y,s}\left(\frac{\rho^*}{\rho_S}\right)^1$	(5.8)
	Elastic collapse $\left(\sigma^*_{el}\right)$	$\sigma^*_{el} = CE_s\left(\frac{\rho^*}{\rho_S}\right)^2$	(5.9)

Associated derivations available in Refs. [4] and [4a].

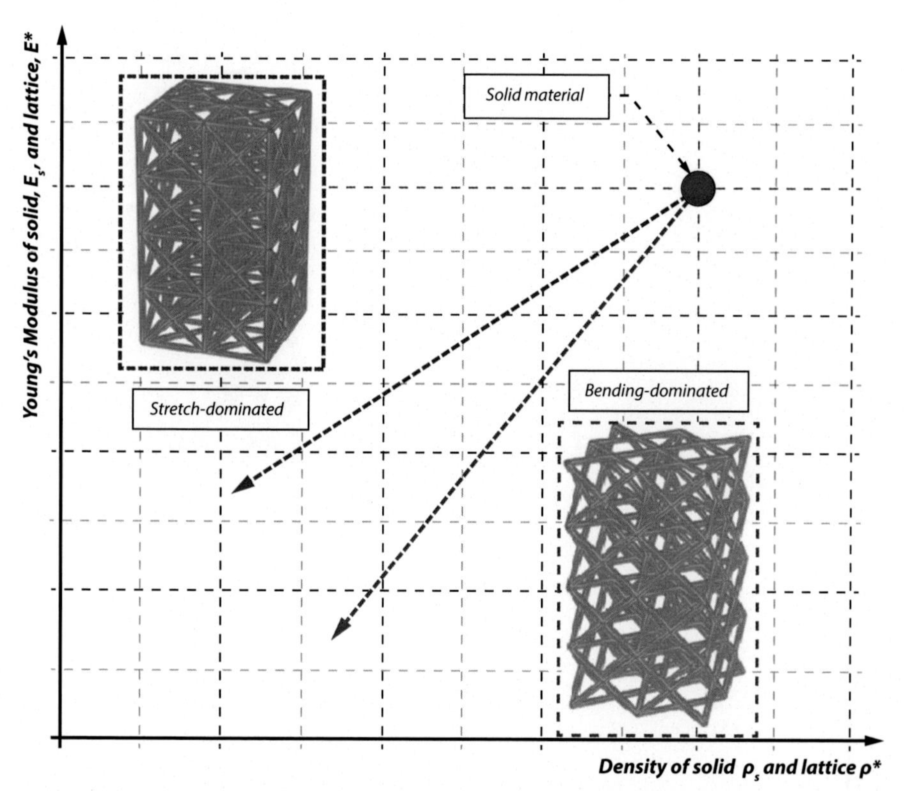

Fig. 5.7 Gibson-Ashby model illustrating Young's Modulus for bending and stretch-dominated structures with reference to the solid material.

$$P^* = C\,P_S\left(\frac{\rho^*}{\rho_S}\right)^n \tag{5.3}$$

Fig. 5.7 provides further insight into the range of mechanical responses to be expected based on Gibson-Ashby relationships of the type of Eq. (5.3). Specifically, the Maxwell stability criterion can be used to directly characterize the lattice response as being either bending- or stretch-dominated. From this understanding, the Gibson-Ashby model can be utilized to predict the specific mechanical response of the associated lattice system. These predictions are compatible with the experimentally observed mechanical response of AM lattice structures (Section 5.1.7); However, variability exists due to simplifications inherent in the Gibson-Ashby model and variation between the idealized and as-manufactured AM lattice geometry. Due to the potential for variability in Gibson-Ashby model predictions, experimental results or validated numerical predictions are required for the design of high-value AM lattice applications.

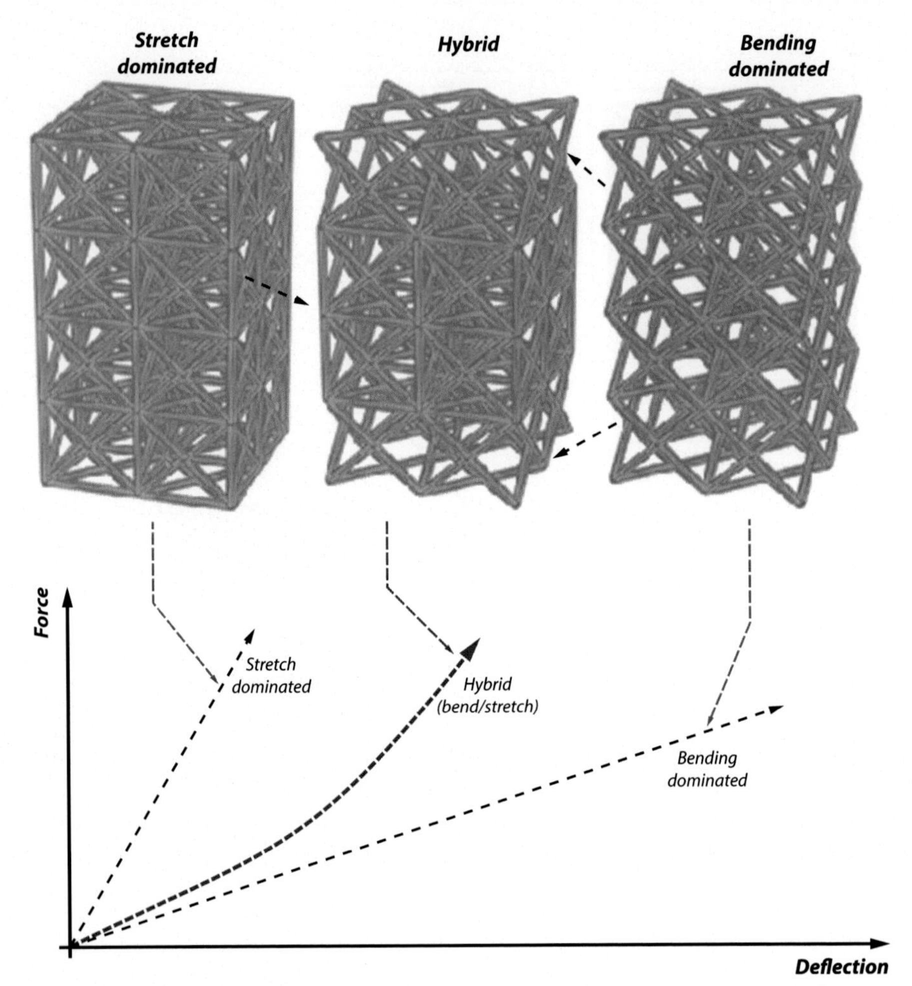

Fig. 5.8 The strategic functionalization of stretch-dominated and bending-dominated lattice structures enables the manufacture of hybrid lattice systems with bespoke mechanical response.

5.1.5 Hybrid lattice design

The nuanced response of practical lattice systems, whereby the local stresses and observed deformation are a function of both the Maxwell number *and* the associated loading type (Section 5.1.2), enables the design team to make informed and robust engineering design decisions in response to the technical requirements of the AM design project. This insight enables, for example, the specification of functional AM lattice structures that include under-stiff elements to avoid inducing excessive forces in response to the initial displacement and include over-stiff elements to avoid undue deflection (Fig. 5.8).

Cell type	Body Centred Cubic	Body Centred Cubic (Z Struts)	Face Centred Cubic	Face Centred Cubic (Z Struts)
	BCC	BCCZ	FCC	FCCZ
Struts, s	8	12	16	20
nodes, j	9	9	12	12
Strut inclination angles, α	35.3°	35.3°, 90°	45°	90°,45°
Maxwell number, M	-13	-9	-14	-10
Strut aligned to load direction	NO	YES	NO	YES

Fig. 5.9 Abbreviated list of candidate lattice unit cells and associated Maxwell number, indicating the existence of strut elements with alignment to loading in the Z-direction, and the inclination, α, of strut elements with respect to the X-Y plane.

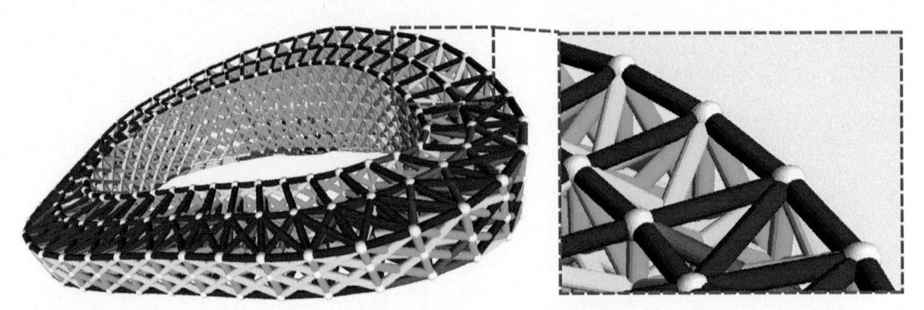

Fig. 5.10 Opportunities to geometrically permute the cubic unit cell to provide conformal geometry, including toroidal and conformally manufactured geometry associated with a bespoke patient-specific medical implant.

5.1.6 Conformal lattice structures

Periodic lattice structures provide a basis for the design of engineered lattice structures that are conformal to the required 3D geometry. There are infinite combinations of unit cells that can tessellate to fill space. However, hexahedral (cubic) unit cells are often specified in commercial engineering practice as they are readily defined, provide reasonable flexibility in the design of various lattice structures, have a reasonably well characterized mechanical response, and can be conformal to fill space. An abbreviated list of cubic lattice cells is presented in Fig. 5.9. As the number of cells increases, so does the associated Maxwell number, implying that additional cells provide a stiffening effect, as does the existence of struts aligned to the loading direction.

These cubic cells can be permuted to alternative geometry that is conformal in space and can be combined in homogeneous or inhomogeneous distributions of lattice cell elements to manipulate mechanical response as required (Fig. 5.10).

5.1.7 Experimental observation of AM lattice response

Fundamental predictive models of lattice mechanical response such as the as the Gibson-Ashby model provide useful insight into the anticipated mechanical response of a proposed AM lattice system (Section 5.1.4). However, the mechanical response of as-manufactured AM lattice systems is complex and recourse to experimental data is required to provide confidence for the design of high-value applications.

For example, the mechanical response of Inconel 625 lattice structures provides a useful basis for experimental investigation. The high ductility of these Inconel structures allows the effect of lattice topology and associated geometry to be characterized for strain up to full-densification (Fig. 5.11). Such data can be used for the validation of proposed numerical models as well as for the identification of unexpected failure modes. For example, a potentially unexpected transition is observed for scenarios that would nominally display stretch-dominated and bending-dominated response and vice versa (Fig. 5.11). This phenomenon appears to be due to a transition between local buckling and local crushing behaviour. This observation provides evidence that

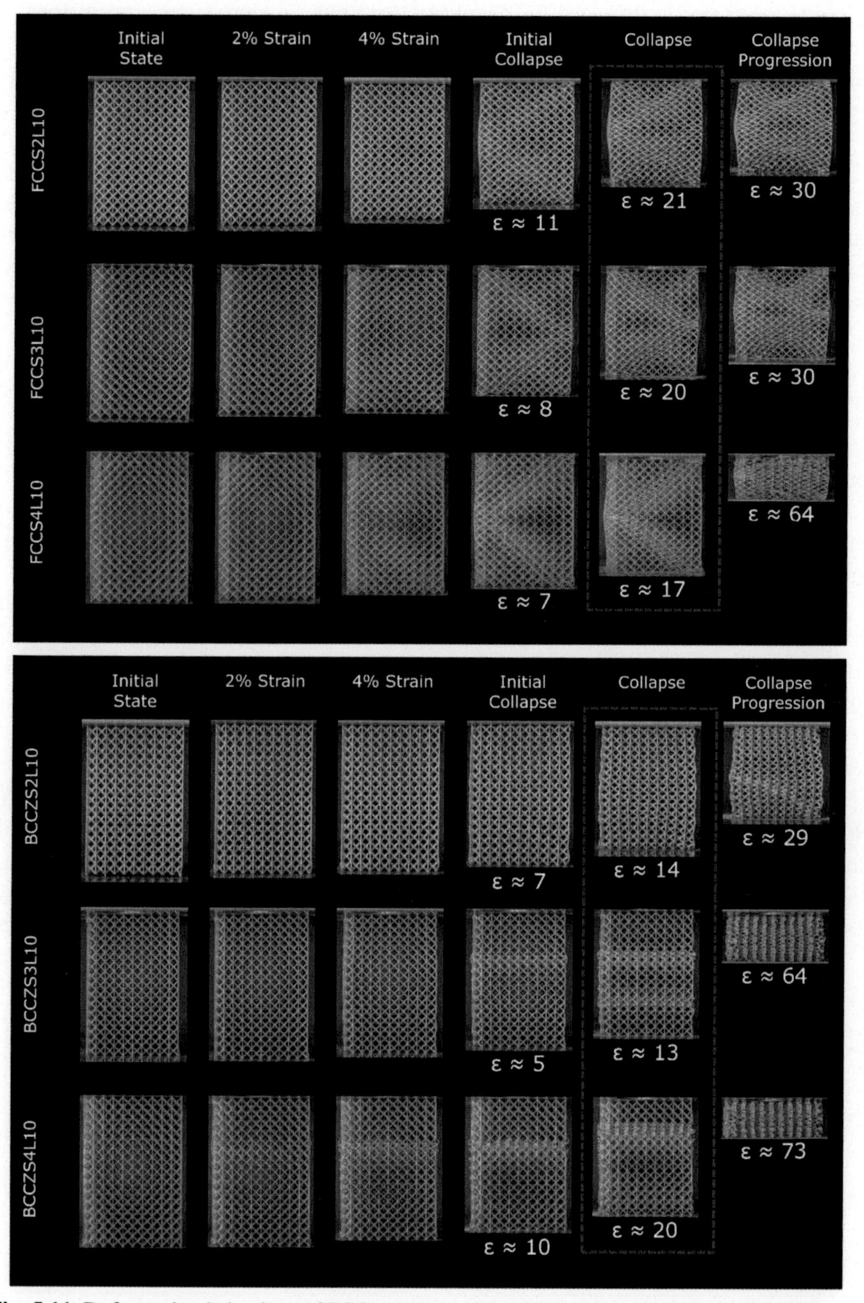

Fig. 5.11 Deformation behaviour of FCC (upper) and BCZ (lower) topologies for cell size of 2, 3 and 4 mm (Fig 5.9). Red highlighted data identifies that cell size can induce a transition between bending-dominated and stretch-dominated failure mechanisms. Specimens fabricated according to Ref. [5].

topology selection (in this case BCZ and FCC) provides a *coarse* mechanism for the tuning of mechanical response, whereas local geometry (in this case cell size) provides a *fine* tuning mechanism. The development of custom DFAM tools to enable these tuning mechanisms for commercial lattice design remains an open research question.

Further research is required to understand the nuanced interactions of topology and local geometry on mechanical response. This DFAM research benefits from numerical analysis methods that enable rigorous insight into the structural mechanics of complex lattice structures. Despite the opportunities for numerical analysis to enable commercially relevant research outcomes, there exist many unresolved challenges for the application of numerical DFAM methods to the prediction of AM lattice response.

5.1.8 Numerical analysis of AM lattice

Robust numerical analysis of AM lattice structures can enable profound insight into the fundamental basis for experimentally observed mechanical response. As an example of this opportunity, the lattice elements of Fig. 5.11 that observed unexpected transitional behavior are analyzed numerically to reveal the effects of local buckling and crushing as being a driving mechanism for this observed transition (Fig. 5.12).

Despite the opportunities for numerical analysis to add value to AM research and design, there exist a number of technical challenges to its robust and timely application. These challenges provide a roadmap to the commercially useful DFAM research contributions and include:

5.1.8.1 The curse of dimensionality

Numerical models of AM lattice structures potentially require a very large number of finite elements to robustly simulate the mechanical response. Computational costs increase exponentially with the number of elements within a specific simulation, rapidly resulting in numerical models that are prohibitively costly. This challenge is

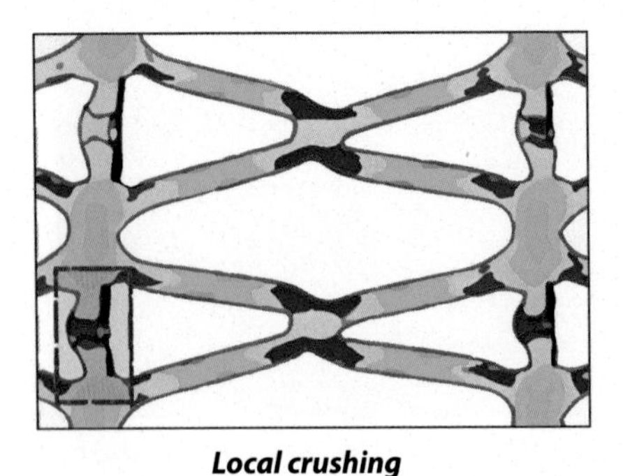

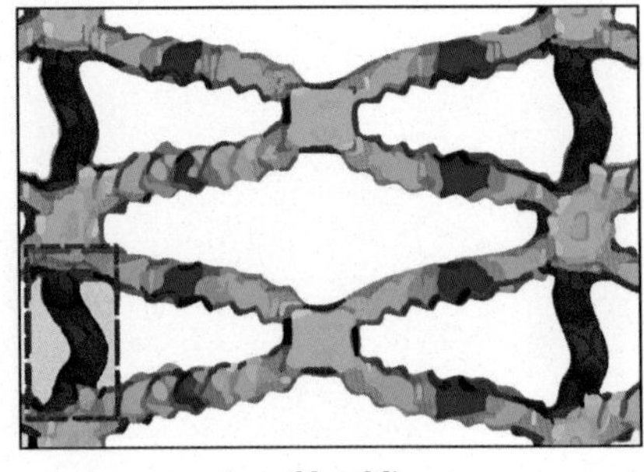

Fig. 5.12 Numerical analysis applied to generate insight into the experimentally observed failure mechanisms of Fig. 5.11. In this case, for a constant topology (BCZ), local geometry initiates a transition between crushing and local buckling.

exacerbated for continuum models especially those that intend to increase simulation validity by the use of geometrically complex micro-CT scan data.

5.1.8.2 Scale effects

An ongoing technical challenge for numerical analyses is associated with the accommodation of phenomena that occur at multiple length scales. Relevant AM phenomena are notoriously multi-scale. For example, the initial plastic deformation observed in AM lattice deformation is associated with local stress concentrations associated with micron size defects; however, the bulk load transfer is dictated by geometric features multiple orders of magnitude larger. Similar scale effects exist for the simulation of AM thermal processes where the physical size of the melt pool is dimensionally insignificant in comparison to the bulk component geometry.

5.1.8.3 Material data availability

The availability of reliable material property data for AM structures is relatively low. Furthermore, the correlation between AM design inputs and the as-manufactured microstructure and associated mechanical response is not well understood and is the focus of much active research effort. The generation of robust AM material property data remains a critically important research activity for the commercial application of AM.

5.1.8.4 Geometric data acquisition

Small-scale geometric artefacts can initiate significant variation in the observed mechanical response; consequently, simplified representations of AM lattice geometry can result in erroneous predictions, as discussed below. To offset this potential error, it is useful to obtain 3D spatial fields by non-destructive methods such as micro CT imaging. These methods offer opportunities for the acquisition of geometrically robust data, but potentially result in very large data sets that are computationally challenging.

It is apparent that the challenges associated with the numerical analysis of AM lattice structures are significant. However, the commercial motivation to overcome these challenges is high, and in response, creative DFAM approaches are emerging. This nascent field of AM research will likely yield unexpected and useful approaches over time. For example, the cost and time required for the explicit manufacture and non-destructive imaging of complex lattice structures is high. A computationally efficient mechanism to overcome this challenge is to statistically analyse micro-CT imaging data in order to allow the statistical generation of a sample of pseudo-random strut elements (Fig. 5.13). From this sample, a plausible full-scale lattice system can be generated that provides insight into mechanical response that is not captured by an idealized analysis of the input CAD data (Fig. 5.14) and furthermore provides a useful DFAM tool by enabling simulation of complex virtual AM lattice systems as part of the early embodiment and design phases (Chapter 3).

Although increasing data is available on the experimental and numerical response of AM to quasi-static loading, there remains both a significant research opportunity

Fig. 5.13 A series of pseudo-random strut elements (left) generated by statistical analysis of micro CT imaging of the AM strut elements of Fig. 5.11. Such methods enable the generation of plausible representations of complex lattice systems that display the statistically variable deformation behavior expected of as-manufactured strut elements (right), but without the measurement costs of explicit manufacture and imaging. Data generated by the method of [5a].

and a commercial necessity for investigations into more complex failure modes, specifically fatigue and impact.

5.1.9 Fatigue failure mode of AM lattice

Cyclic loading is observed to induce catastrophic failure at a significantly lower stress than occurs for monotonic loading. This phenomenon, termed *fatigue*, was first formally investigated by August Wohler in the fatigue investigation of rolling stock axles. Subsequent research has led to a deep understanding of the fatigue failure mode. Fatigue occurs by various sequential phases, namely:

- *crack-initiation* phase, where in the absence of a pre-existing flaw, a crack is initiated by the action of some cyclic strain with this initiation exacerbated by stress concentrating local effects, including local surface roughness and high stiffness microstructural phases
- *crack-propagation* phase that occurs in which the initiated fatigue crack propagates at some constant rate according to the *Paris* crack-growth law
- *fracture* phase, whereby the fracture toughness is insufficient to accommodate loading through the remaining cross-sectional area, and failure occurs.

Fatigue failure is catastrophic[6] and occurs without significant deformation or visible warning and is therefore of significant importance for the design of high-value and safety-critical systems.

[6] The term catastrophic is used to indicate that load bearing function is completely lost without warning in this failure mode.

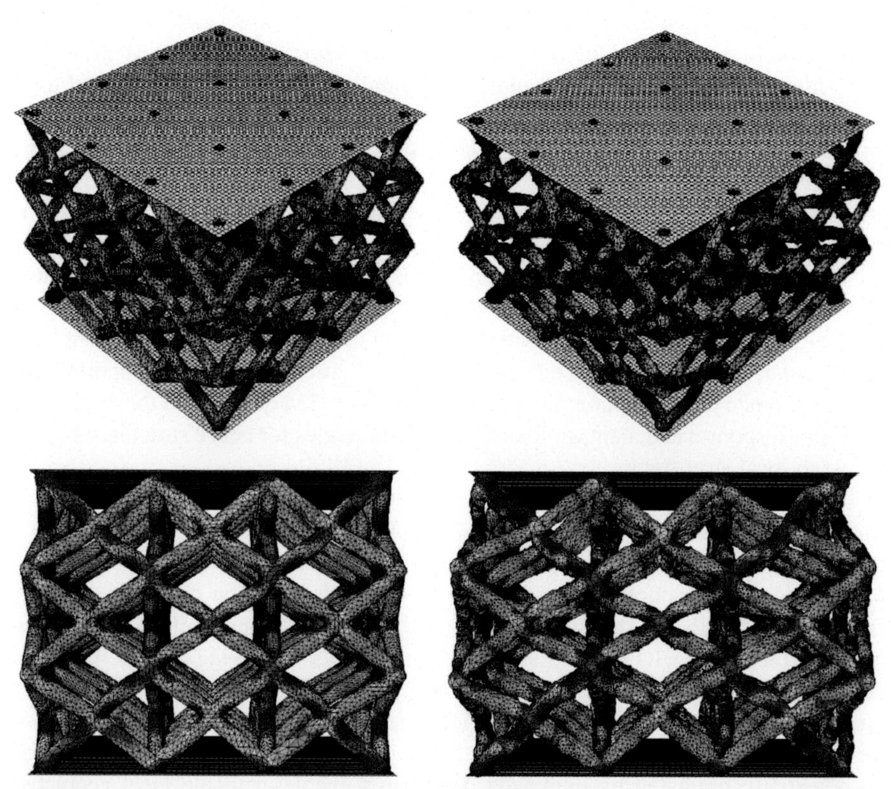

Fig. 5.14 Numerical analysis of idealized CAD geometry (left) and a statistically generated assemblage of pseudo-random strut elements (right). The statistically generated strut elements capture the observed mechanical response with higher fidelity while enabling the simulation of virtual AM lattice as part of the early design phase. Data generated by the method of [5a].

Despite being an intensive research activity for over a century, the prediction of these fatigue phases is complex, and multiple relevant phenomena are observed. When the cyclically applied stress is plotted against the associated number of cycles to failure, an *S-N* curve is generated. Depending on the material type, loading conditions and environment, the S-N curve may display a monotonically decreasing relationship between stress and life, or the curve may display an endurance limit. At stresses below the endurance limit, structures can theoretically accommodate unlimited cycles without failure. This endurance limit is typically associated with ferrous alloys and typically occurs at around 10 million cycles. The fundamental basis for the existence or absence of a fatigue-limit remains an open research question; however, for practical design it is important to note that S-N specimens are typically idealized representations of actual components, and that a theoretical fatigue limit can be absent for specimens with the stress-concentrating geometry associated with as-manufactured structures, or when loading occurs in harsh environments.

Fatigue remains an extremely challenging failure mode to repeatably predict. Fatigue prediction tools include stress, strain and fracture approaches; of these, the stress-life approach is perhaps the most readily understood and will be referred to in this text. Due to the challenges of robust fatigue failure prediction, various design philosophies have evolved, including *safe-life design*, where the useful service-life of the system is defined and the fatigue-life is designed to exceed this value by some appropriate safety-factor; *finite-life* design, where a conservative prediction of fatigue life is used to identify when dynamically loaded components are to be removed from service; and *damage-tolerant design*, where the system is designed such that the overall system function is robust to fatigue failure. The practical utilization of these philosophies in the design of commercial AM systems remains an open DFAM research opportunity; in particular, the opportunities inherent to AM structure design. These opportunities include geometry optimization; prediction of plausible AM defects and their influence of fatigue response; and, damage-tolerant design strategies for lattice structures that are robust to local strut failure.

Fatigue is a weak-link failure-mode, and local stress-concentrating effects can significantly reduce the observed fatigue-life. AM lattice structures include significant opportunities for stress-concentrating effects at both the macro- and micro-scales. Macro-scale stress concentrations are related to the deformation characteristics that are associated with the structural determinacy of the lattice unit cell. The Maxwell number (Section 5.1.1) identifies lattice structures as being either bending-dominated or stretch-dominated. Accordingly, these lattice structures introduce distinct stress concentrations within the local geometry, particularly within the intersecting nodes, and strut mid-span (Fig. 5.12). These local stress concentrations can be potentially mitigated by optimized design that enables variable local geometry;

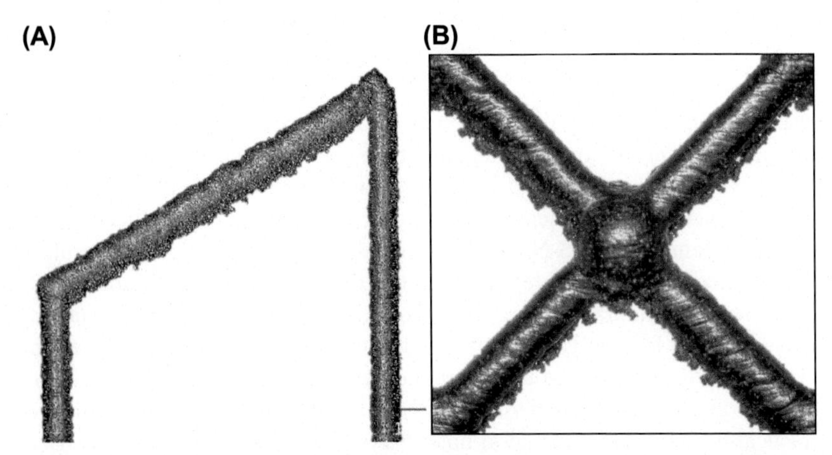

Fig. 5.15 Fatigue-life is strongly affected by artefacts and potential defects of manufacture, including (A) stair-step effects and adhered particles, especially on downward facing surfaces, and (B) asymmetric distortion of intended lattice node geometry.

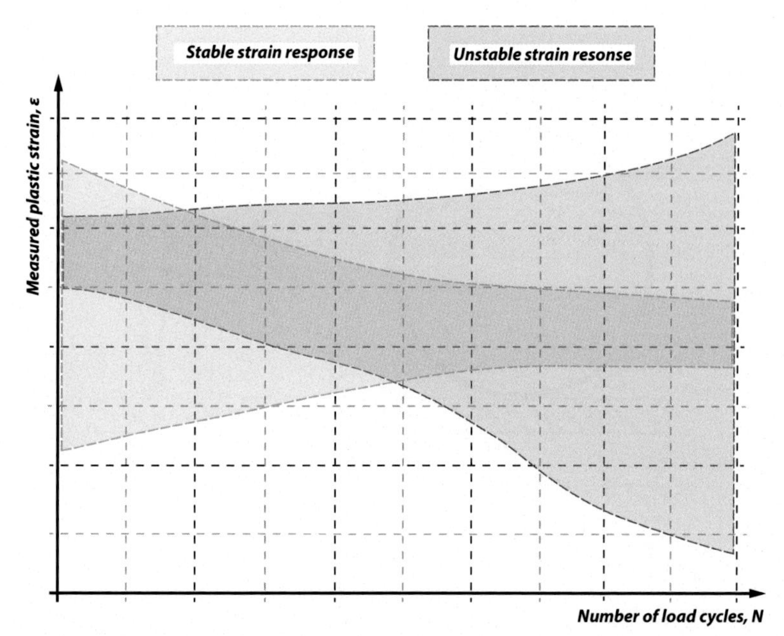

Fig. 5.16 Schematic representation of strain-life response of AM lattice structures indicating plastic strain response that either stabilizes or remains unstable until fracture with increasing number of cycles.

however, relatively sparse research contributions have been made in this area. Even with robust design practices, AM lattices structures include numerous opportunities for geometric and microstructural artefacts of relevance to the fatigue failure mode. Geometric artefacts include partially attached particles, especially on downward-facing surfaces; asymmetric distortion of structural features; stair-stepping effects; and artefacts introduced by digital data representations (Fig. 5.15).

Very little data exists on the fatigue response of AM lattices structures due to the relatively recent nature of AM technologies and the significant technical effort required to acquire robust fatigue data. This omission provides a significant research and commercial opportunity, especially for practically relevant scenarios; for example, the effects of variable amplitude loading, non-axial loading, local geometry, temperature and load ratio. In particular, the effect of *cyclic load ratcheting* is poorly understood, whereby the observed plastic strain varies according to the associated number of cycles. This cumulative plastic strain can either increase in amplitude, resulting in relatively low fatigue-life, or can stabilize, as is indicative of AM lattice structures that are robust to dynamic loading (Fig. 5.16). This phenomena is of critical importance to the design of cyclically loaded AM structures, but is poorly understood and remains a significant DFAM research opportunity.

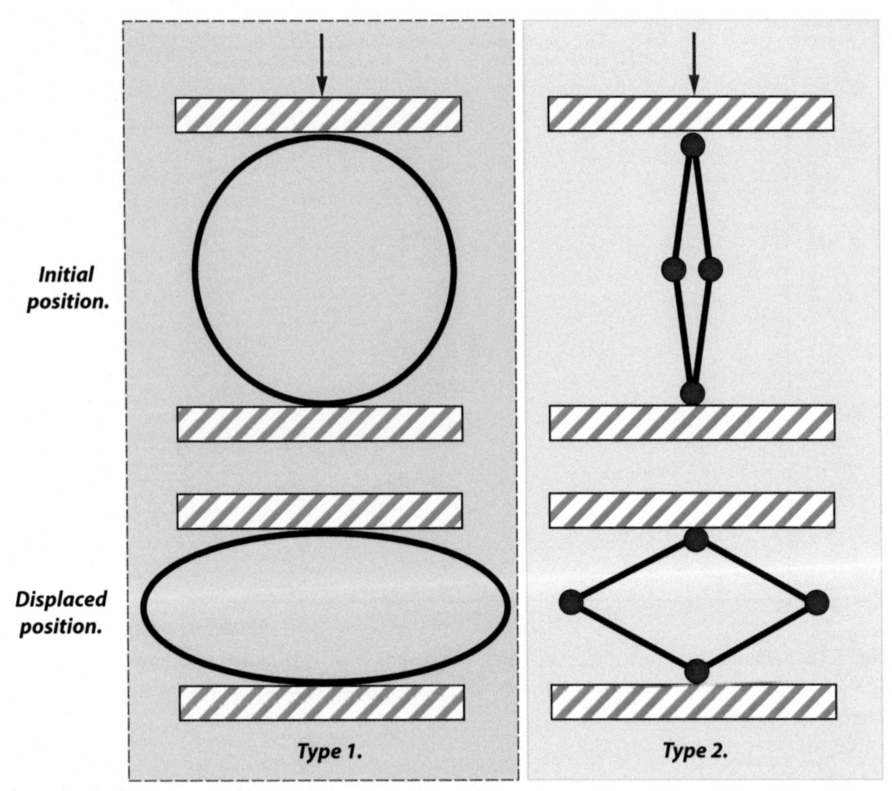

Fig. 5.17 Geometry that does not incur significant lateral acceleration (Type 1) displays a lower stiffness and strength when subject to impact loading when compared to structures that are stiffened by lateral inertial forces (Type 2) [6]. This observation enables the design of AM impact structures with high mass efficiency, although little DFAM research exists in this space.

5.1.10 Impact failure mode

Engineered AM structures can provide otherwise technically infeasible design capabilities in the field of impact engineering. Impact results in high strain-rate deformation with kinematic deformation-wave effects and adiabatic strain that results in a mechanical response that cannot be directly predicted from the observed quasi-static response. There exists very little robust data on the impact response of AM lattice structures; however, fundamental studies provide insightful reference for the design of impact loaded lattice structures, especially associated with the phenomenon of *inertial strengthening* described by Calladine [6]. This phenomenon essentially refers to a potential increase in resistance to impact loading due to the lateral acceleration of specific structures. For structures that primarily deform in the direction of loading, material acceleration is in the direction of the applied force; however, structures that have geometry that accelerate laterally during

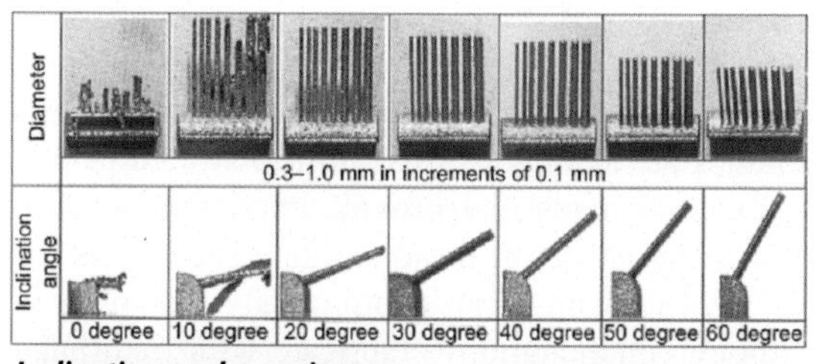

Inclination angle specimens

Horizontal span specimens

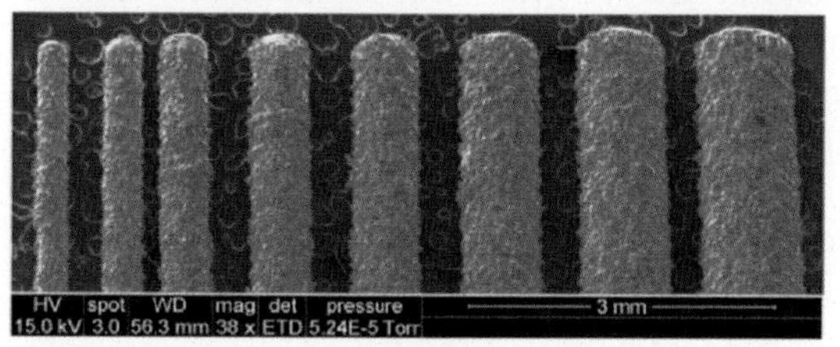

SEM examination of local artefacts

Material		Ti6Al4V									
Diam. (mm)		0.1	0.2	0.3	0.4	0.5	0.6	0.7	0.8	0.9	1.0
Angle (degrees)	0	L	L	L	L	L	L	L	L	L	L
	10	L	L	M	M	M	M	M	L	L	L
	20	L	L	H	H	H	H	H	H	H	H
	30	L	M	H	H	H	H	H	H	H	H
	40	L	H	H	H	H	H	H	H	H	H
	50	L	H	H	H	H	H	H	H	H	H
	90	L	H	H	H	H	H	H	H	H	H

DFAM manuacturability table

Fig. 5.18 Schematic representation of potential definitions of AM lattice manufacturability and the associated DFAM design table.

loading will induce a lateral opposing force (Fig. 5.17). This lateral force induces a stiffening effect that provides an important opportunity for the design of impact bearing AM structures with very high mass efficiency, as well as allowing bespoke impact properties that are technically feasible only with AM methods.

5.1.11 Lattice manufacturability

Struts in the loading direction can ground loads with high efficiency, resulting in a stretch-dominated response for scenarios that are actually under-stiff according to a fundamental Maxwell analysis (Fig. 5.9). This observation can be useful for the design of lattice structural elements that are technically challenging for AM; for example, horizontally-oriented lattice struts can be challenging to manufacture; these challenging struts can potentially be omitted while maintaining stretch-dominated response if the loading direction is dominantly aligned to the remaining strut elements.

Accommodating these technical lattice design constraints requires formal definitions of AM manufacturability. The specific requirements of these definitions vary according to the specific AM hardware and intended application, but typically include assessment of the effect of inclination angle, α, and specimen diameter, d, on local geometry and associated internal defects; effect of horizontal span parallel to the plate; and the local artefacts generated by this geometry for specific process parameters and material input. These effects are typically quantified by a DFAM design table that characterizes the observed AM manufacturability and allows robust design decisions to be confidently made (Fig. 5.18), as well as providing a basis for the expert systems required for generative design (Chapter 7).

5.2 Triply periodic minimal surfaces

A *minimal surface* defines a membrane structure with an average curvature of zero at any surface point. These membranes are analogous to soap-film structures that are highly efficient at grounding external loads and provide exceptional mass-specific mechanical response. A particular subset of these minimal surfaces is the Triply Periodic Minimal Surface (TPMS), which can continuously fill space in three perpendicular directions. Although challenging to fabricate with traditional methods, TPMS are readily manufacturable with AM technologies, and an increasing volume of commercial and research attention is being focussed on the application of AM TPMS for high-value products including porous bio-implants, energy absorbing structures and thermal systems. Of these TPMS, the *Shoen gyroid* structure has attracted particular attention due to its robust manufacturability and exceptional mechanical properties. This gyroid can be fabricated by continuous curvilinear tool paths and is therefore compatible with many AM technologies, including metallic, ceramic and polymer structures.

The following section presents a fundamental introduction to minimal surfaces and TPMS and introduces the application of TPMS to engineering applications, including metallic structures for highly efficient structures as well as polymeric AM technologies in the manufacture of radiation dosimetry phantoms. The intent of this review is to demonstrate the diverse potential commercial application of these interesting and valuable structures as well as to formalize potential DFAM research activities to support commercial applications in this space.

5.2.1 Physical and mathematical definitions

Soap bubble films are a familiar example of a minimal surface. These surfaces take up their complex forms in response to their physical obligation to balance overall surface tensions. In doing so they generate surfaces that minimize local[7] surface area and balance curvature; specifically, that any infinitesimal surface area cut from a minimal surface results in the smallest possible physical area for the given surface boundary. Furthermore, the curvature along principal curvature planes, k_1 and k_2, is equal and opposite at every point; the mean local curvature, H, is zero (Eq. 5.1); and the Gauss curvature, K, is less than zero.

$$\mathrm{H} = (\mathrm{k}_1 + \mathrm{k}_2)/2 \quad \text{Mean curvature, H} = 0 \text{ for minimal surface} \tag{5.10}$$

$$K = k_1 k_2 \quad \text{Gauss curvature, K} < 0 \text{ for minimal surface} \tag{5.11}$$

[7] The reference to local geometry is important. Minimal surfaces minimize the local area for a given local boundary. Therefore, multiple minimal surfaces may exist within the same global boundary. Each of these minimal surfaces may have unique overall surface area.

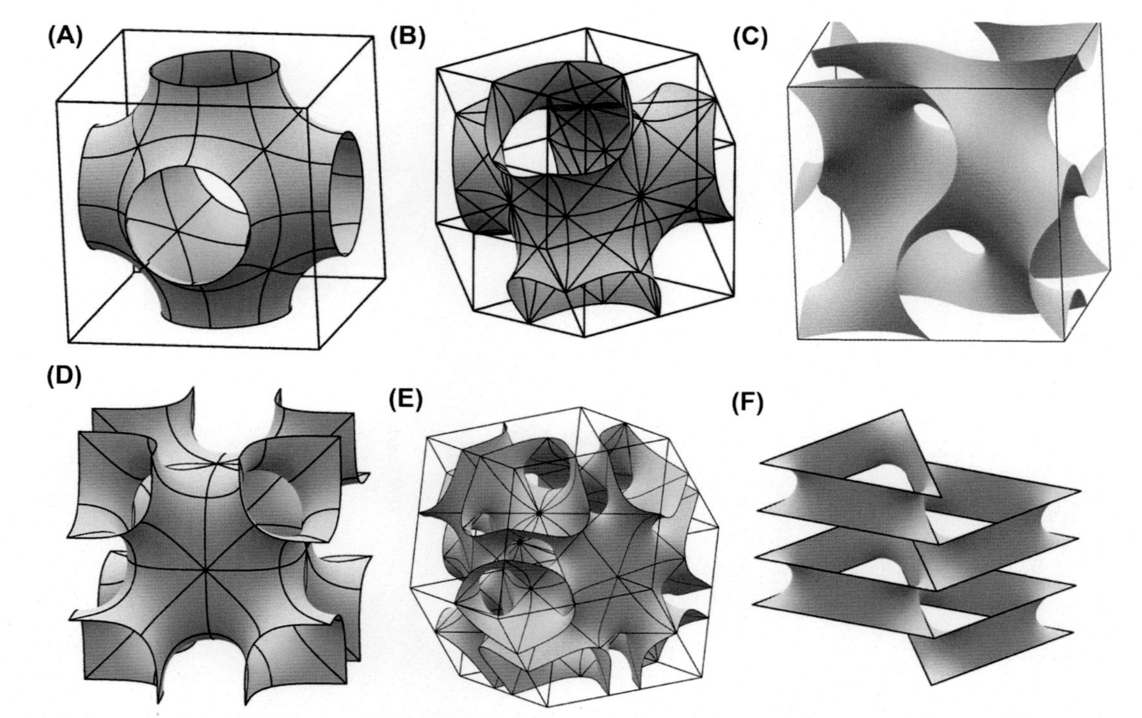

Fig. 5.19 A range of sample unit cells from the range of known TPMS: (A) Schwarz' P Surface cubic unit cell, (B) Schwarz' D Surface rhombic dodecahedron unit cell, (C) Schoen's Gyroid Surface cubic unit cell, (D) Schoen's I-WP Surface, (E) Schoen's Complementary D Surface with rhombic dodecahedral unit cell, (F) Schwarz' H Surface with triangular prism unit cell. TPMS provide the useful engineering quality of filling 3D space with various crystalline symmetries. Labyrinthine domains are colour coded. All images courtesy of Prof. Kenneth Brakke, Susquehanna University, PA [http://facstaff.susqu.edu/brakke/evolver/examples/periodic/periodic.html].

To be useful for engineered applications, it is desirable that the minimal surface be predictable in filling 3D space. Minimal surfaces that fill space according to crystallographic symmetries[8] are defined as Triply Periodic Minimal Surfaces (TPMS); these minimal surfaces include many variants of technical interest (Fig. 5.19) and are finding increased application in the design of high value AM engineering structures.

Of the minimal surfaces shown in Fig. 5.19, the Diamond (D), Gyroid (G) and Primitive (P) surfaces have cubic crystalline symmetry, and have been shown to be approximated by level-set equations in local Cartesian coordinates, X, Y, Z, according to a specified iso-value, t (Eqs. 5.12–5.14). The fundamental D, G, and P surfaces correspond to an iso-value of zero; however, the iso-value may be modified to generate TPMS variants, including structures that are representative of trabecular bone. In this text, these variants are referred to as *TPMS lattice* in order to distinguish these from the lattice structures of Section 5.1. The effect of iso-value variation on mechanical response remains an open research question.

$$\cos(Z)\sin(X+Y)+\sin(Z)\cos(X-Y)=t \quad \text{Diamond (D) surface} \tag{5.12}$$

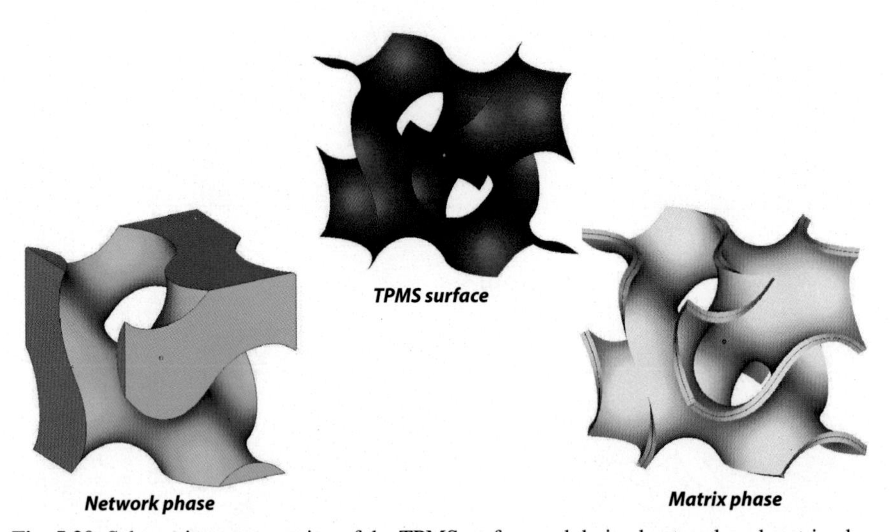

Fig. 5.20 Schematic representation of the TPMS surface and derived network and matrix phase structures.

[8] Crystallographic symmetries allow space filling by the tessellation of various unit cells. Cubic unit cells are considered here as they provide an orthogonal tessellation that is useful for engineering applications. Alternate tessellations may provide technical advantage and are an open research question.

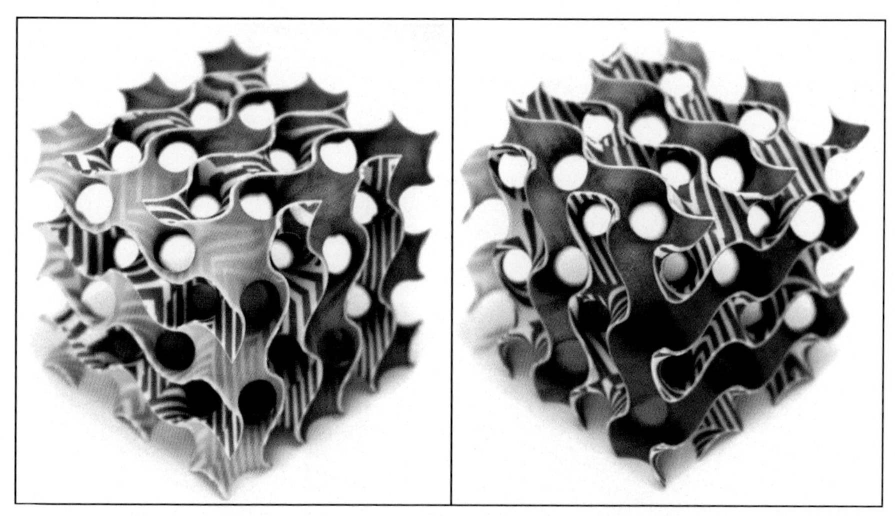

Fig. 5.21 *Material Jetting (MJT, Chapter 9) AM technology applied to visualise the complexity of the TPMS gyroid structure, including colorized labyrinthine domains. Designed and fabricated by Phil Pille at RMIT University Advanced Manufacturing Precinct.*

$$\sin(X)\cos(Y) + \sin(Y)\cos(Z) + \sin(Z)\cos(X) = t \quad \text{Gyroid (G) surface} \tag{5.13}$$

$$\cos(X) + \cos(Y) + \cos(Z) = t \quad \text{Primitive (P) surface} \tag{5.14}$$

The TPMS are defined by a level-set relationship in terms of the iso-value, t. To generate a physical structure with tangible volume, these level-set relationships can be *thickened* by some finite distance about $t = 0$ (Fig. 5.21). Alternatively, these equations isolate the bounding volume into *labyrinthine domains* that can provide a basis for the Boolean selection of a physical volume. In the first case, these domains are in intimate proximity, but remain physically separated without enfolded voids. These structures, are termed *matrix phase*, can be utilized in the fabrication of heat exchangers and osmotic devices, whereby fluids are physically (or functionally) isolated; or in the design of systems with high mechanical efficiency (Fig. 5.20). The second case, known as *network phase* systems, is an active research area that shows promise in generating mechanical systems that efficiently ground external loads [7], especially for fatigue-limited scenarios where geometric stress concentrations are to be avoided (Section 5.1.9).

Of the known TPMS permutations, the Gyroid, identified by Schoen in 1970 has captured much engineering interest due to its remarkable geometric characteristics and mechanical properties (Fig. 5.21). The Gyroid has no reflection symmetry or linear features, enabling the mitigation of stress concentrations when subject to mechanical loading: this outcome is particularly important for weak-link failure modes such as fatigue.

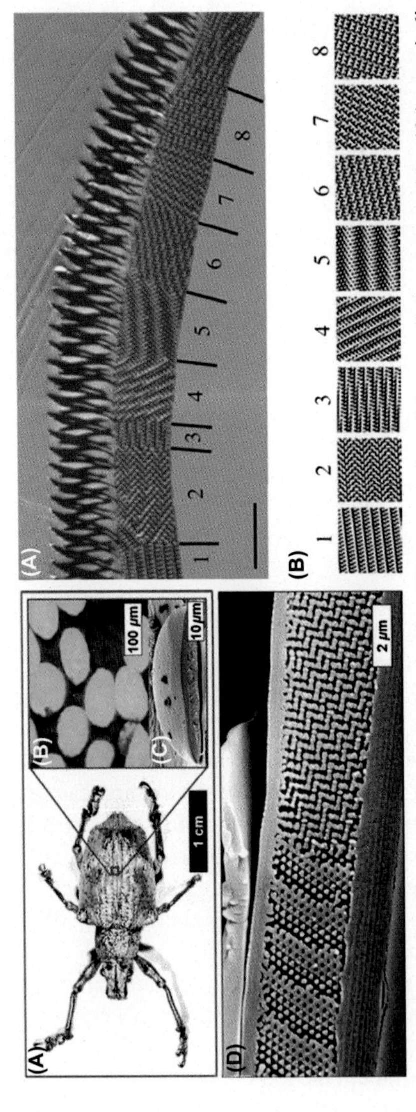

Fig. 5.22 Observation of biological minimal surfaces in nature. Left: Iridescent scales of the weevil *L. augustus*, identifying periodic geometry similar to minimal surfaces [8]. Right: TEM micrograph cross-section of a wing scale from the *P. Sesostris* butterfly. Scale bar is 2.5 μm and black structures are chitin. Inset identifies simulated Gyroid structures in various projections with $t = -0.3$ [9].

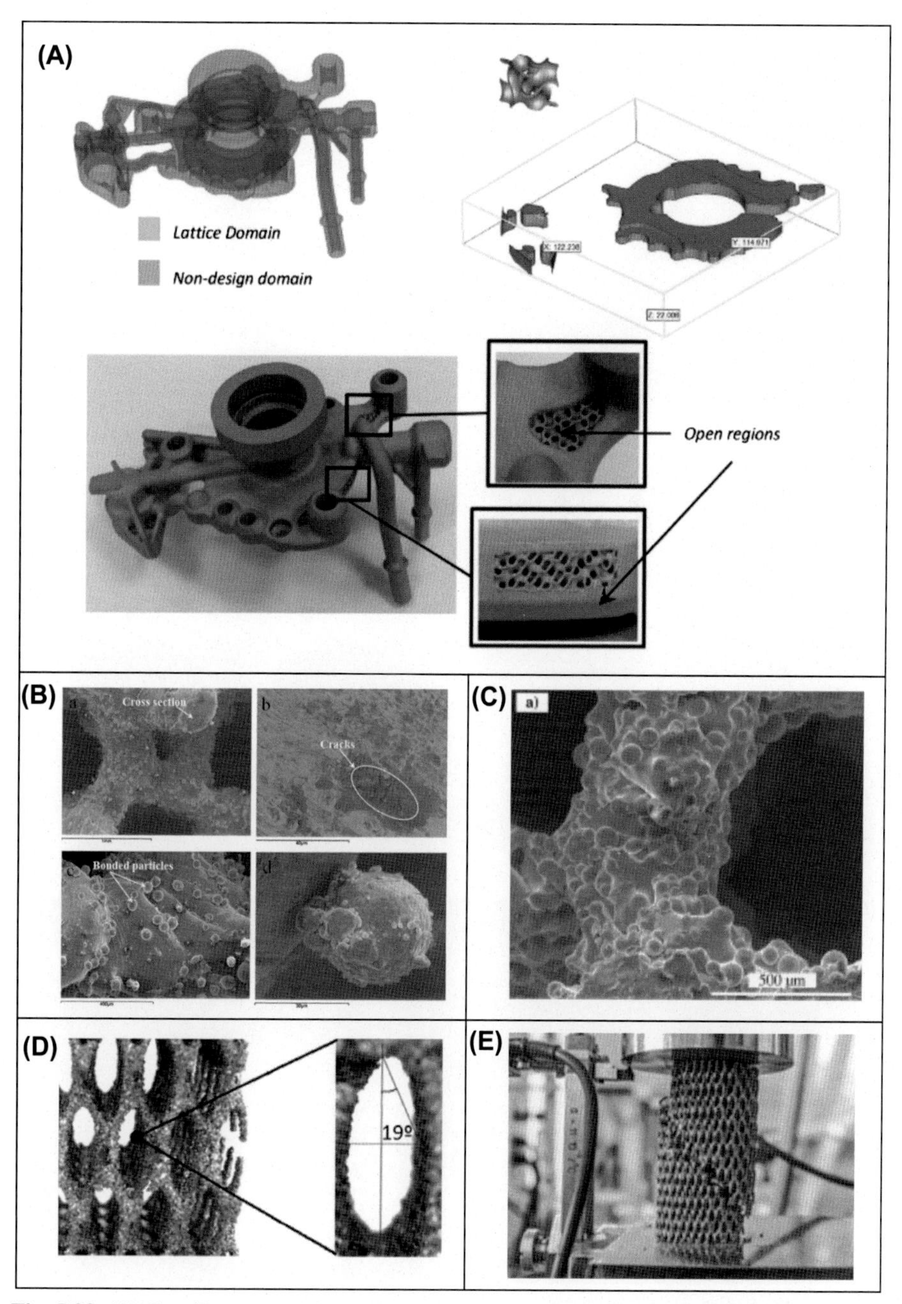

Fig. 5.23 (A) Gyroid supported internal lattice structures [10], (B) observation of as-manufactured SLM titanium specimens [16], (C) local geometric artefacts including attached particles result in anisotropic response [11], (D) effect of local geometry on AM manufacturability and associated mechanical properties [13], (E) mechanical response (compression and torsion) of cylindrical EBM specimens [12].

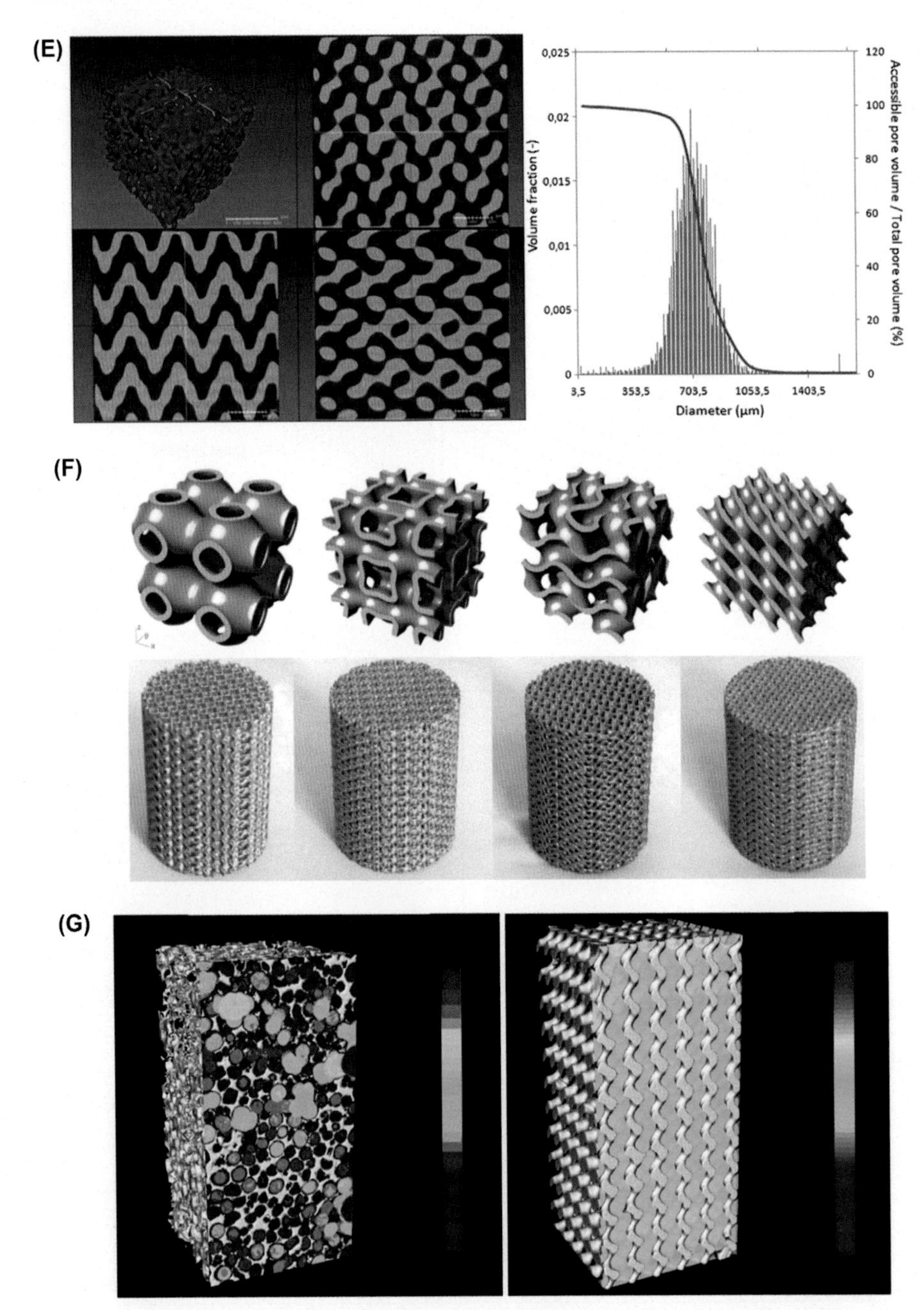

Fig. 5.24 (E) TPMS bio-scaffold manufactured by SLA and quantified in terms of the as-manufactured morphology [14], (F) morphology and mechanical response of TPMS specimens manufactured by SLM [15], (G) permeability simulation of traditional (left) versus SLA manufactured specimens (right) [17].

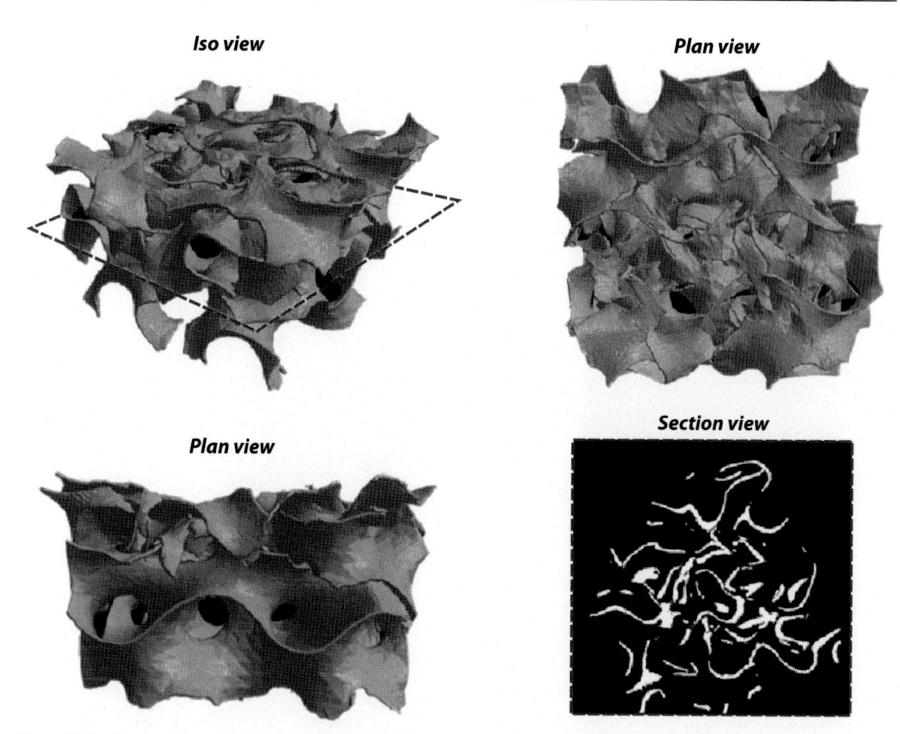

Fig. 5.25 Reconstructed micro-CT images of plastically deformed network Gyroid manufactured with SLM from titanium (Ti64) powder, including an internal cross-section view of the deformed structure. Reference data of this type is much needed to further understand fundamental deformation mechanisms and to verify numerical simulation results.

5.2.2 Technical application of TPMS

Minimal surfaces are known to be important biological constructs that are regularly observed in nature (Fig. 5.22). The manufacture of minimal surfaces for engineering applications has traditionally been highly challenging; however, AM technologies are eminently compatible with the associated geometric complexity. Consequently, the technical application of minimal surfaces fabricated with AM technologies has exponentially increased. The following research is offered as a non-exhaustive summary of the current state of the art of TPMS application enabled by AM; however, we can expect this unique application area to manifest in profound and surprising research outcomes and commercial offerings (Figs 5.23 and 5.24), for example:

- Aremu *et al.* demonstrated a voxel-based method for the generation of space filling geometry. This computationally efficient DFAM strategy enables the incorporation of TPMS within bulk AM structures as a mechanism for increasing mass-specific structural efficiency [10].

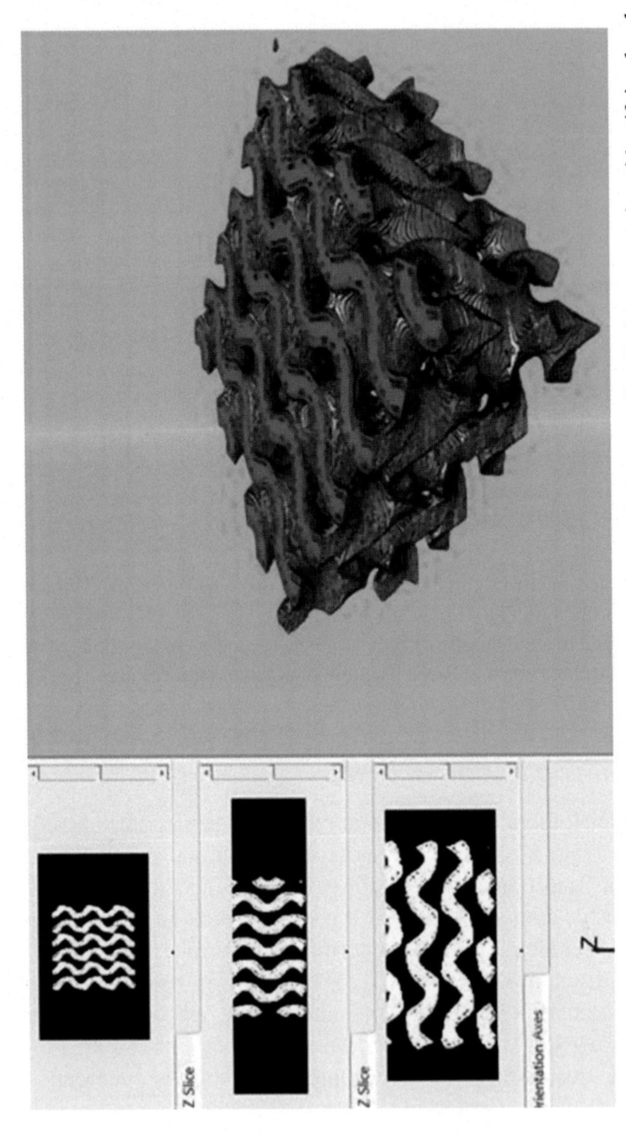

Fig. 5.26 Custom DFAM tool to predict the influence of tool path on manufactured part quality. Left: cross-sections identifying local porosity. Right: 3D visualization of as-manufactured AM part including predicted regions or porosity. These predictive DFAM tools enable pre-production planning of oncological and structural components prior to manufacture.

- Maskery et al. provide a detailed experimental evaluation of the mechanical response of *polymer* TPMS. Diamond, gyroid and primitive TPMS morphologies are investigated, including the effect of iso-value on density, the effect of number of cell repetitions on observed mechanical properties and the critical stress concentrating geometry associated with these TPMS [7].
- Ataee *et al.* manufactured Ti-6Al-4V Gyroid structures for bone implant applications using Electron Beam Melting (EBM, Chapter 11). They observed statistically anisotropic mechanical properties (Young's Modulus and yield strength). Specimens with an elastic modulus comparable to trabecular bone were generated. The observed dominant failure mode under compression was by orthogonal crush bands aligned at 45° to the loading axis [11].
- Yánez *et al.* analyzed the mechanical properties (compression and torsion) of cylindrical EBM specimens filled with both standard and distorted gyroid structures for $t > 0$. Experimental observations are supported by numerical analysis and suggest that these constructs can provide a technical basis for the manufacture of bone-substitute implants [12]. Furthermore, the manufacturability (and associated mechanical properties) for gyroid lattice structures ($t > 0$) can be increased by modifying the associated geometry [13].
- Blanquer *et al.* completed a comprehensive investigation into TPMS scaffold manufactured by steriolithography (SLA, Chapter 10) using a biocompatible polymer poly-trimethylene carbonate (PTMC). Eight distinct TPMS classifications were investigated, and individual specimens were designed to possess varying geometry while maintaining constant porosity and number of unit cells. These structures were quantified in terms of the associated Gaussian curvature as well as their as-manufactured morphology, including surface curvature, porosity, permeability and surface roughness. These fundamental findings provide a useful template for those considering research contributions in this space [14].
- Bobbert *et al.* characterized the morphology and mechanical response of variants of primitive, diamond, and gyroid TPMS specimens manufactured with Ti-6Al-4V by Selective Laser Melting (SLM, Chapter 11). Manufactured specimens are assessed in terms of their quasi-static mechanical response; this research is one of the few to report fatigue resistance data, with fatigue-life in excess of 10E6 cycles reported, as is required for orthopaedic applications [15].
- Yan et al. evaluated the manufacturability, microstructure and mechanical properties of TPMS Gyroid and Diamond lattices fabricated by SLM with Ti-6Al-4V. This fundamental analysis provides calibration data between CAD input and the observed micro-CT data of as-manufactured specimens. Artefacts and geometric variation inherent to the SLM manufacturing process was reported at microscopic (via SEM observations) and bulk (via micro-CT scan) levels. Specimens were manufactured with a range of porosity values, and stress-strain curves are explicitly reported. Young's Modulus comparable to that of human trabecular bone was observed [16].
- Melchels et al. compared traditional and AM manufactured TPMS bio-scaffolds fabricated using poly-DL-lactide (PDLLA). Stereolithography (SLA, Chapter 10) was found to generate structures with consistent mechanical properties that are tuneable by the application of graded structures. Numerical simulation results were validated with reference to experimental data. AM specimens provide superior permeability than do traditional methods of manufacture [17].

5.2.3 TPMS application fields and active research areas

TPMS provide unique mechanical and geometric properties that are observed in naturally occurring biological systems and can be engineered to the requirements of high-value technical applications. The potential commercial application of TPMS is very high and it is likely that the specific application areas are difficult to predict. However, a pre-requisite to the commercial application of these engineered TPMS is an enhanced fundamental understanding of the associated manufacturability, including the correlation between design inputs and the as-manufactured geometry and microstructure; experimental data to validate mechanical response; and robust numerical modelling to allow efficient DFAM of custom TPMS structures.

The existing literature in this space is rapidly expanding, particularly in the application of TPMS to biological systems; an outcome that is to be expected given the observed instances of TPMS in naturally occurring biological systems. Biological areas of interest include the application of TPMS to the generation of biocompatible engineered structures with mechanical properties compatible with naturally occurring human bone. The research summarized above identifies that both EBM and SLM systems are capable of generating TPMS structures that match the observed stiffness of human trabecular bone. In addition to the application of metallic AM systems, polymeric AM provides an opportunity for the design of bio-implants that can be functionally graded to engineer a biological response or to facilitate interaction with host biological systems. The reviewed research falls into several key areas, each associated with specific Technology Readiness Levels (TRL).

Fundamental studies of morphology and manufacturability (TRL 1–3)

These studies include analyses of as-manufactured AM geometry such that the effect of AM manufacture on the fabricated structure is quantified. This data is typically presented in terms of SEM or micro-CT data, for example [11,16]. The available data in this research space is scant, and much additional research data is required, particularly for emerging AM systems and for TPMS variants that are particularly suited to inherent AM manufacturability requirements and required mechanical response.

Mechanical and functional response of TPMS structures (TRL 4–6)

The mechanical response of as-manufactured TPMS can be challenging to predict numerically, and explicit mechanical testing is required. The available research data in this space is rapidly expanding. Of particular value in this space are research efforts that attempt to harmonize AM manufacturability requirements with observed mechanical response, for example [13]. When utilizing reported mechanical and functional response for design purposes, it is imperative that the quality of this reported data be rigorously appraised, as minor uncertainties in manufacturing, geometry and testing conditions can result in significant variation in mechanical response.

Application of TPMS to commercial structures (TRL 7–9)

As the understanding of TPMS applicability extends beyond the understanding of simplified test-coupon geometry, it is necessary to formalize the applicability of the data to that of full-scale manufactured structures. Such DFAM tools are emerging in the research literature, for example [10], however, this application-focussed research area is underrepresented in the available literature and provides a significant opportunity for commercially oriented research outcomes.

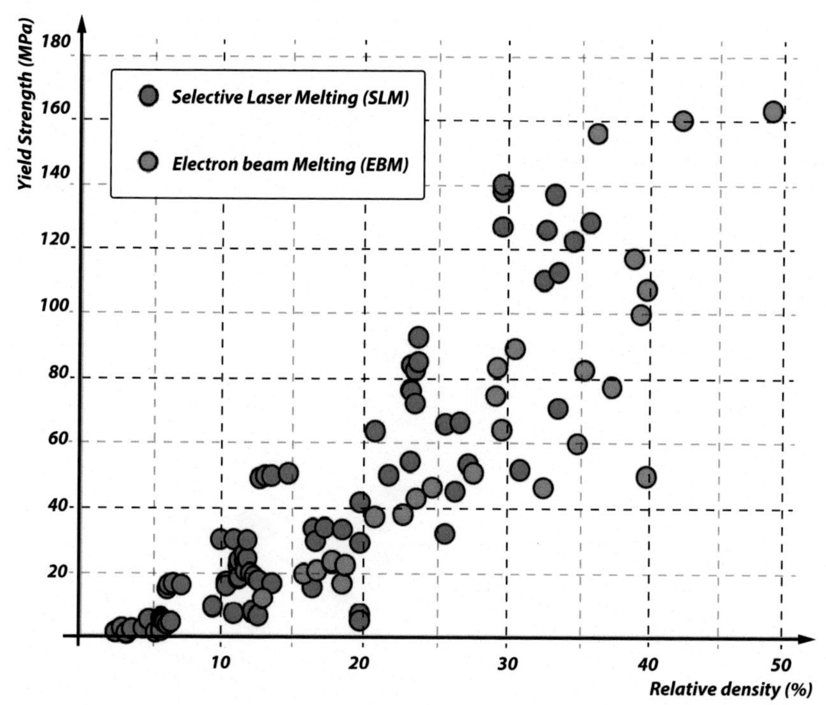

Fig. 5.27 Relative density versus (compressive) yield stress for lattice and TPMS structures fabricated by Metal AM systems as reported in the literature. Data from EBM and SLM systems identified.

5.2.4 Mechanical response data

It is apparent that exceptional technical opportunities exist in the application of TPMS enabled by AM. In response to this nascent opportunity, research publications are emerging that define the specific observed mechanical response for these structures. The majority of these data are associated with Metal AM (MAM) systems, notably Electron Beam Melting (EBM) and Selective Laser Melting (SLM) systems (Chapter 11). Laser and electron-beam system have inherent technical advantages and associated challenges. In particular, EBM geometry is often reported as providing higher build rates and reduced thermal distortion challenges than for equivalent SLM systems, although with a potentially larger minimum feature size. These reported advantages and challenges remain poorly characterized in the available literature; as are the required DFAM tools to mitigate potential design challenges for commercially focussed applications to enable prediction of the mechanical response of as-manufactured TPMS, for example Fig. 5.25.

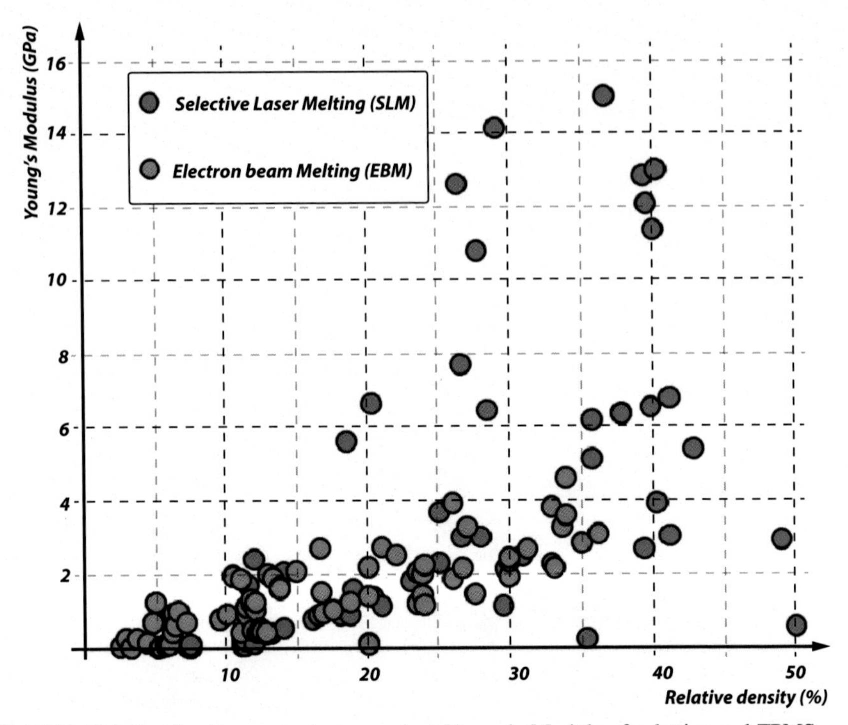

Fig. 5.28 Relative density versus (compressive) Young's Modulus for lattice and TPMS structures fabricated by Metal AM systems as reported in the literature. Data from EBM and SLM systems identified.

5.2.5 Challenges to TPMS application – tool path design

As identified in the DFAM rules of Chapter 4, the AM toolpath is an often-neglected potential source of defects in the as-manufactured geometry. TPMS provide an interesting DFAM challenge in terms of tools path design as the associated cross sections are often curvilinear, and therefore lend themselves to the skin-and-core strategy typical of many AM tool path generation algorithms. Conversely, these curvilinear geometries provide an opportunity to conceal internal defects – defects that if deployed in high-value structural or oncological components would increase the potential risk of failure. In response to this risk, custom DFAM tools are emerging to predict the occurrence of in-situ defects prior to manufacture (Fig. 5.26). These predictive DFAM tools provide a robust method to de-risk AM engineering projects prior to manufacture.

5.3 Summary of lattice and TPMS experimental response

To provide a coherent summary of the current understanding of the mechanical response of TPMS manufactured with MAM systems, experimental data of the mechanical

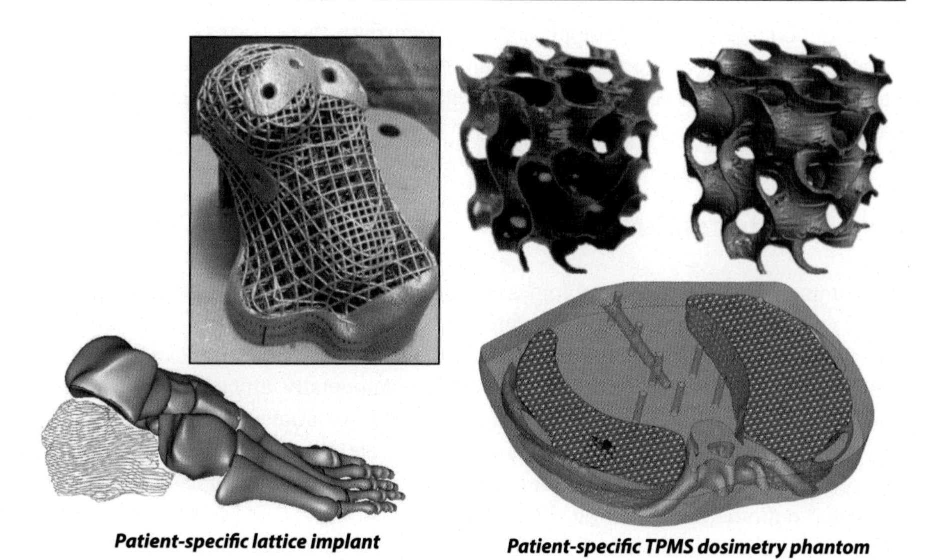

Fig. 5.29 Examples of AM lattice and TPMS structures for patient-specific clinical applications.

response of lattice and TPMS structures are presented for various published data (Figs 5.27–5.28). In this data the associated manufacturing system (either SLM or EBM) is identified; however, specific process parameters and methods to assess specimen density are omitted. As stated previously, it is imperative that the associated data reporting quality be critically appraised by the intended end-user, as much of the reported data on AM mechanical response omits data necessary for scientific repeatability.[9]

These data demonstrate the technical capability for EBM and SLM systems to fabricate robust engineering structures, as well as the versatility of these structures to be tuned by the associated relative density, ρ^*/ρ_s, to achieve the intended structural objective. Although not directly developed for this application, the Gibson-Ashby model of mechanical response appears to provide valuable insight into the mechanical response of TPMS; an observation that warrants further research investigation.

5.3.1 TPMS summary

Triply Periodic Minimal Surfaces (TPMS) are a unique topological classification observed in naturally occurring biological structures. TPMS are highly challenging to fabricate with traditional methods; however, they are readily manufacturable with AM technologies, and are consequently gaining in commercial application. TPMS fill space in three-dimensions according to crystallographic symmetries. Of these symmetries, cubic structures are presented here, although alternative symmetries exist and deserve research attention. Despite the exceptional engineering opportunities enabled by TPMS, there exists sparse research

[9] Robust scientific reporting requires that the experiment and associated conclusions be completely reproduceable by an independent assessor. Specific requirements for data quality therefore depend on the techniques used to generate the data, including manufacture, post-processing and testing.

data to support their commercial application. Consequently, there exists significant DFAM research opportunity within this space. Unfortunately, there is a risk that the available research is under-documented and consequently has poor reproducibility, a flaw that future researchers are urged to avoid [19].

5.4 Chapter summary

The compatibility of AM technologies with the geometric complexity of lattice and TPMS structures enables, for the first time, the commercial application of these highly efficient structures to high-value commercial applications. These applications include a range of scenarios that benefit from these fundamentally efficient topologies for structural and functional optimization. Pertinent examples include conformal patient-specific MAM lattice structures that replicate mechanical properties of the resected bone, and radiation dosimetry phantoms that emulate human tissues under imaging and radiotherapy radiation (Fig. 5.29).

Lattice structures (Section 5.1) and zero-mean curvature structures (Section 5.2) ground external loads with fundamentally high efficiency. Design tools exist to assist in the deployment and optimization of these structures such as the Maxwell stability equation, TPMS geometric relationships and the Gibson-Ashby model of mechanical response. Experimental data exists to corroborate and extend these models; however, there remains significant opportunity for novel research contributions and design-ready DFAM tools such that the commercial application of lattice and TPMS be enhanced.

Commercially relevant research outcomes have been identified for both TPMS and lattice structures. The experimental assessment of TPMS is somewhat less well developed than for similar lattice structures, and there exist significant research opportunities at various Technical Readiness Levels (TRLs). Fundamental studies of morphology and manufacturability are required (TRL 4–6), especially for variants of mathematically rigorous TPMS that benefit AM manufacturability constraints. For specific morphologies of interest, it is imperative that rigorously documented experimental data be generated for the associated mechanical and functional response (TRL 7–9).

As the understanding of TPMS morphologies with useful mechanical response extends beyond that of simplified test-coupon geometry, it is necessary to generate formal DFAM tools that can apply this experimental data to full-scale manufactured structures. This application-focussed research area requires the integration of numerical analysis methods with DFAM manufacturability and experimentally validated data of mechanical response. This multidisciplinary research area is underrepresented in the available literature and provides a significant opportunity for commercially oriented research outcomes for both lattice and TPMS systems.

References

Lattice structures in nature

First microscopic observation of cellular structure

[1] R. Hooke. Micrographia: or some physiological descriptions of minute bodies made by magnifying glasses. With observations and inquiries thereupon. January 1665. The Royal Society. Great Britain.

Observations on trabecular bone density

[2] Qin YX, Lin W, Mittra E, Xia Y, Cheng J, Judex S, Rubin C, Müller R. Prediction of trabecular bone qualitative properties using scanning quantitative ultrasound. Acta Astronautica 2013;92(1):79–88.

Lattice structure response

Maxwell's stability criterion

[3] Maxwell JC. L. on the calculation of the equilibrium and stiffness of frames. The London, Edinburgh, and Dublin Philosophical Magazine and Journal of Science 1864;27(182): 294–9.

Fundamental analysis of the mechanical response of cellular structures

[4] Gibson IJ, Ashby MF. The mechanics of three-dimensional cellular materials. Proceedings of the Royal Society of London. A. Mathematical and Physical Sciences 1982;382(1782):43–59.
[4a] Ashby MF. The properties of foams and lattices. Philosophical Transactions of the Royal Society A: Mathematical. Physical and Engineering Sciences 2005;364(1838):15–30.

Experimental observation and numerical prediction of the mechanical response of SLM Inconel

[5] Leary M, Mazur M, Williams H, Yang E, Alghamdi A, Lozanovski B, Zhang X, Shidid D, Farahbod-Sternahl L, Witt G, Kelbassa I. Inconel 625 lattice structures manufactured by selective laser melting (SLM): mechanical properties, deformation and failure modes. Materials & Design 2018;157:179–99.
[5a] Lozanovski B, Leary M, Tran P, Shidid D, Qian M, Choong P, Brandt M. Computational modelling of strut defects in SLM manufactured lattice structures. Materials & Design 2019;171:107–671.

Introduction to inertial effects with respect to impact

[6] Calladine CR, English RW. Strain-rate and inertia effects in the collapse of two types of energy-absorbing structure. International Journal of Mechanical Sciences 1984;26(11–12):689–701.

Generalisable insights into the mechanical response of polymer AM TPMS

[7] Maskery I, Sturm L, Aremu AO, Panesar A, Williams CB, Tuck CJ, Wildman RD, Ashcroft IA, Hague RJ. Insights into the mechanical properties of several triply periodic minimal surface lattice structures made by polymer additive manufacturing. Polymer 2018;152:62–71.

Biological minimal surfaces in nature

[8] Galusha JW, Richey LR, Gardner JS, Cha JN, Bartl MH. Discovery of a diamond-based photonic crystal structure in beetle scales. Physical Review E 2008;77(5):050904.
[9] Michielsen K, Stavenga DG. Gyroid cuticular structures in butterfly wing scales: biological photonic crystals. Journal of The Royal Society Interface 2007;5(18):85–94.

TPMS studies in AM

Computationally efficient TPMS within voxel-based CAD representations

[10] Aremu AO, Brennan-Craddock JPJ, Panesar A, Ashcroft IA, Hague RJ, Wildman RD, Tuck C. A voxel-based method of constructing and skinning conformal and functionally graded lattice structures suitable for additive manufacturing. Additive Manufacturing 2017;13:1–13.

Observed anisotropy in EBM manufactured matrix gyroid structures ($t = 0$)

[11] Ataee A, Li Y, Fraser D, Song G, Wen C. Anisotropic Ti-6Al-4V gyroid scaffolds manufactured by electron beam melting (EBM) for bone implant applications. Materials & Design 2018;137:345–54.

Mechanical response of EBM manufactured gyroid lattice structures ($t > 0$)

[12] Yánez A, Cuadrado A, Martel O, Afonso H, Monopoli D. Gyroid porous titanium structures: a versatile solution to be used as scaffolds in bone defect reconstruction. Materials & Design 2018;140:21–9.

EBM Manufacturability of lattice gyroid (with $t > 0$)

[13] Yánez A, Herrera A, Martel O, Monopoli D, Afonso H. Compressive behaviour of gyroid lattice structures for human cancellous bone implant applications. Materials Science and Engineering: C 2016;68:445–8.

Fundamental assessment of geometry and mechanical response of as-manufactured TPMS

[14] Blanquer SB, Werner M, Hannula M, Sharifi S, Lajoinie GP, Eglin D, Hyttinen J, Poot AA, Grijpma DW. Surface curvature in triply-periodic minimal surface architectures as a distinct design parameter in preparing advanced tissue engineering scaffolds. Biofabrication 2017;9(2):025001.

Mechanical testing of SLM fabricated titanium TPMS for orthopaedic implant applications

[15] Bobbert FSL, Lietaert K, Eftekhari AA, Pouran B, Ahmadi SM, Weinans H, Zadpoor AA. Additively manufactured metallic porous biomaterials based on minimal surfaces: a unique combination of topological, mechanical, and mass transport properties. Acta Biomaterialia 2017;53:572–84.

As-manufactured geometry and mechanical response of SLM manufactured titanium

[16] Yan C, Hao L, Hussein A, Raymont D. Evaluations of cellular lattice structures manufactured using selective laser melting. International Journal of Machine Tools and Manufacture 2012;62:32–8.

SLA manufactured TPMS bio-scaffolds

[17] Melchels FP, Bertoldi K, Gabbrielli R, Velders AH, Feijen J, Grijpma DW. Mathematically defined tissue engineering scaffold architectures prepared by stereolithography. Biomaterials 2010;31(27):6909–16.

Tensegrity structures provide insight into the complexity of static determinacy

[18] Calladine CR. Buckminster Fuller's "tensegrity" structures and Clerk Maxwell's rules for the construction of stiff frames. International Journal of Solids and Structures 1978;14(2):161–72.
[19] Maconachie T, Leary M, Lozanovski B, Zhang X, Qian M, Faruque O, Brandt M. SLM lattice structures: Properties, performance, applications and challenges. Materials & Design 2019. 108137.

Topology optimization for AM

6

Topology optimization (TO) refers to the algorithmic identification of geometric connectivity that optimizes structural response. Manual topology selection often results in sub-optimal design outcomes due to preconceptions of the designer, and brute force solution methods are computationally limited due to the high dimensionality of possible solutions. This chapter provides a synopsis of algorithmic TO methods, including Bidirectional Evolutionary Structural Optimization (BESO), Solid-Isotropic Material with Penalization (SIMP), ground-structures, and Level-Set Methods (LSM). By summarizing these methods, practical DFAM tools are developed and their associated challenges identified; in particular, the challenges inherent in the specific TO method, and those that occur when TO methods are integrated with AM manufacturability constraints.

Two fundamental philosophies exist for *TO integrated DFAM*: namely, methods that *modify* TO outcomes to satisfy AM manufacturability requirements, and methods that *retain* the TO structure, and add features as required. Both philosophies are presented in detail and summarized in the context of their applicability to commercial AM design, and with reference to emerging AM technologies and methods. A brief case study is presented with a focus on commercial best practice for AM topology optimization for a high-value non-stationary aerospace component using topology and parametric optimization (Section 6.7), and the concepts developed in this chapter are also applied in case studies in other chapters. This chapter concludes with a review of the commercial and research opportunities for enhanced TO integrated DFAM tools.

6.1 The motivation towards topology optimization for AM design

AM design typically requires that external loads be grounded while retaining the intended design functions. This requirement for structural integrity is achieved by the strategic selection of materials, methods of manufacture, and associated component geometry. As discussed in Chapter 3, component geometry may be defined by various methodologies according to the intended level of geometric abstraction. For example, highly abstracted geometry may be defined in terms of the topological connectivity of various locations within a design volume of interest. At lower levels of abstraction, the shape of specific geometric elements may be formally defined according to measurable geometric parameters (Fig. 6.1).

The selection of component geometry directly influences the reliability of the designed component to successfully ground the anticipated service loads with an acceptable factor of safety. Successful geometric design results in efficient material utilization with lightweight structures that achieve their intended function with

Design for Additive Manufacturing. https://doi.org/10.1016/B978-0-12-816721-2.00006-3

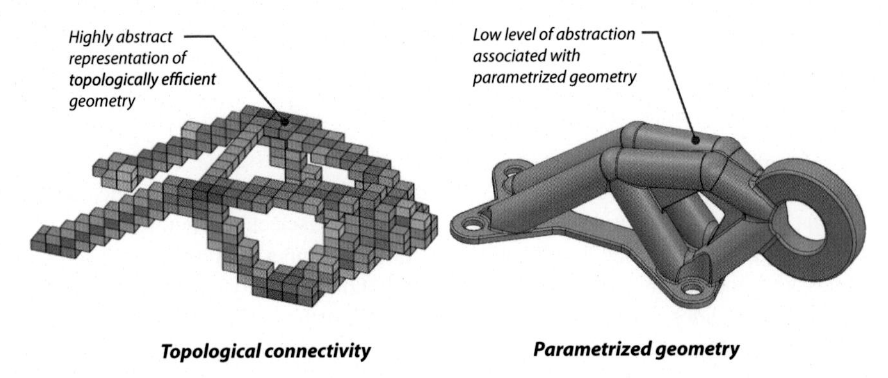

Fig. 6.1 Component geometry may be defined at various levels of abstraction as required, including abstract definitions of topological connectivity or parametric definitions of measurable shapes. This chapter is primarily involved in the identification of efficient connectivity, which may then be refined into a specific geometric form.

minimal structurally redundant material. Conversely, sub-optimal design results in material distributions that include technically redundant materials and geometric features that reduce the efficiency of the geometry to achieve its intended function. Sub-optimal design outcomes can occur for a multitude of reasons, including limitations of the selected manufacturing process in generating high-efficiency geometry; constraints associated with assembly and interaction with mating components[1]; an incomplete understanding of the associated loading scenario; or, insufficient design time or understanding to implement optimal design outcomes. TO methods address the latter challenge by providing a rapid understanding of structurally efficient material distributions such that the available design time can be more efficiently utilized.

The criticality of efficient material utilization ultimately depends on the economic and technical attributes of the intended design. For example, mass-critical applications such as non-stationary aerospace and automotive applications benefit significantly from mass reduction, resulting in *a high commercial value for lightweight structures*. For stationary applications (such as jigs and fixtures, moulding tools, and commodity engineering structures) the technical and economic benefit of mass-reduction is relatively low. However, although these applications are not mass-critical, the economic value of rapid product deployment and reduced manufacturing cost is typically significant. In contrast with other manufacturing methods, the cost of AM fabrication is strongly proportional to component volume, due to the associated implications on

[1] Stress concentrations associated with component interaction and clearance for assembly tools can introduce failure modes that significantly compromise the integrity of structural systems. AM provides exceptional opportunities for enhanced structural efficiency by the consolidation of components to remove the failure modes associated with component interaction and negating the need for assembly tool access (Chapter 3).

material consumption and manufacturing time.[2] This observation directly translates to *a high commercial value for low-volume structures.*

Interestingly therefore, scenarios that benefit from AM will almost always benefit from the optimization of structural material utilization, either by being intended for mass-critical applications that benefit directly from *lightweight structures*; or by enabling cost reduction and reduced manufacturing time by the deployment of *low-volume* structures. The optimization of structural efficiency requires a design effort (an act that itself incurs a cost), with a potentially diminishing increase in structural efficiency in return for this effort. It is therefore imperative that the applied design effort be matched to the economic value of the structural optimization benefit. This chapter presents a broad overview of commercially relevant structural optimization techniques. Of these techniques, topology optimization is of particular importance to AM, and is therefore the primary chapter focus, especially the commercial and research opportunities for associated DFAM tools. Aspects of parametric optimization methods are also presented where they complement the commercial benefit of TO methods in AM design.

6.2 Efficient optimization of load bearing structures for AM scenarios

The optimization of engineering structures is of significant technical and economic value for commercial AM applications. As discussed, this value exists for structures that intentionally minimize mass, such as the mass-critical aerospace structure assessed in Section 6.7, as well as for structures that intend to be economically competitive with traditional manufacturing methods.

The optimization of engineering structures requires design effort. This effort incurs direct costs associated with labour, as well as indirect costs associated with software acquisition; mechanical and metallurgical testing; and other technical tasks. To achieve efficient design optimization for AM, it is imperative that the applied design effort be commensurate with the associated economic opportunity. Fig. 6.2 provides an idealized representation of the diminishing return on effort that is typical for potential structural optimization strategies, including:

- closed form expressions of structural response
- reference to precedent of optimized structures, including naturally occurring phenomena and engineered systems and structures
- topology optimization methods.

For each of these optimization efficiency strategies, parametric optimization methods may be applied to further increase the associated performance. The application of parametric optimization methods is a commercial best practice DFAM method and is presented in more detail in Section 6.5.

[2] This phenomena of reduced component cost by the reduction of physical volume is discussed from an economic perspective in Chapter 2, and a design perspective in Chapters 3 and 4.

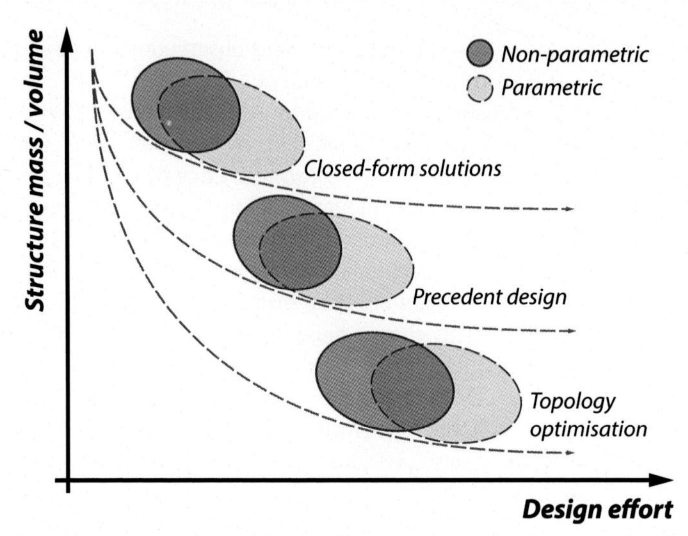

Fig. 6.2 Schematic representation of design effort versus degree of structural optimization for potential optimization strategies. Note that this schematic is applicable to sophisticated design scenarios with no direct available precedent. For scenarios with available closed-form solutions or an existing design precedent, topology optimization methods may provide no distinct technical benefit.

These design philosophies are presented in more detail below, further references are provided in Section 6.8 for the benefit of interested readers.

6.2.1 Closed-form expressions of structural response

An extensive range of closed-form algebraic solutions are available for industrially relevant structural design scenarios. These closed-form solutions are often developed for relatively simplified geometries, but can be combined with empirical solutions for relevant stress concentration factors in order to rapidly define optimal geometries for industrially relevant engineering scenarios.[3] The application of closed-form expressions provides an efficient mechanism for the design of highly efficient structures, and for geometries that are compatible with the available solutions, more complex solution strategies may provide little further technical benefit (Fig. 6.2). However, for many loading conditions of relevance, there may be no directly applicable closed-form reference, resulting in sub-optimal design outcomes (Fig. 6.3).

6.2.2 Precedent of existing optimized structures

A vast library of design precedents exist that can provide deep insight into structurally efficient design. This library includes extant engineering structures and natural forms

[3] Seminal references for closed-form solutions and stress-concentration factors are provide by Roark [1] and Peterson [2], respectively.

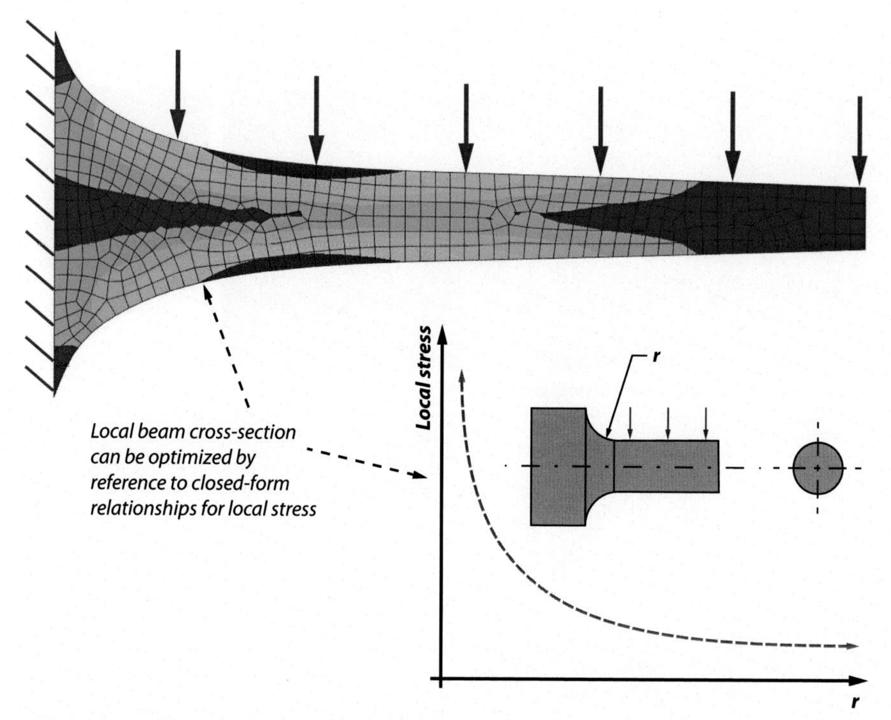

Fig. 6.3 Application of closed-form solution to classical beam bending equation. Local geometry of beam root is optimized by reference to stress-concentration curves for fillet radii (inset). For this relatively simple structure the closed-form solution provides an efficient method to optimize structural outcomes.

that have been subject to extensive design effort and evolutionary optimization processes. For example (Fig. 6.4), sandwich structures of cortical (hard) bone separated by a trabecular (soft) bone core to enable a mass efficient structure with high strength and stiffness; the combination of a large diameter structural annulus with internal reinforcing rings to resist gross and local buckling in the bamboo plant; the efficient distribution of material in a structural I-beam to maximizes stiffness and strength by the use of a web to offset large cross-section flanges in compression and tension; the design of an archway structure to allow a self-supporting aperture by redirecting self-weight and external loads; rib reinforced structures observed in naturally stiff plants and engineered bulkheads; and the use of a triangulated network of axially loaded elements in an engineered truss structure to efficiently ground external loads.[4]

Reference to these existing optimized structures provides a greater range of design opportunities than does closed-form solutions and can enable the design team to rapidly identify highly efficiency structures for commercially relevant applications.

[4] Although these examples are structural in nature, there exist ample resources that summarize efficient embodiments for other scenarios such as machine design and kinematic systems, for example [3–5].

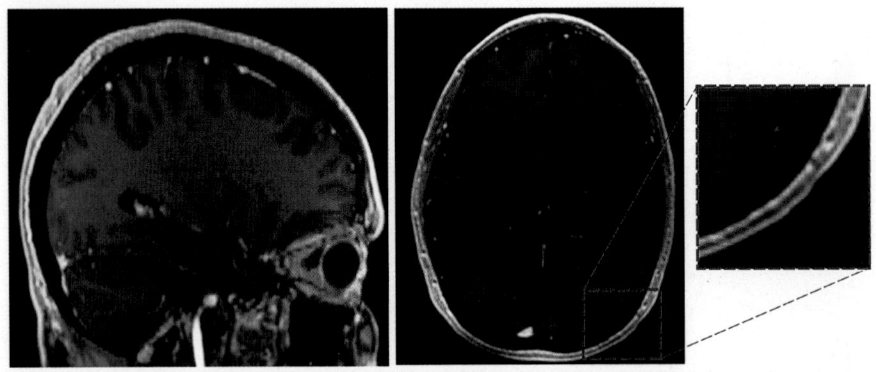

Skull with trabecular (soft) and cortical (hard) bone

I-beam girders utilized in space frame truss

Cellular structure of honeycomb

Archway avoids tension on ceramic materials

Space-frame archway suspension bridge

Fig. 6.4 Extant engineering structures and natural forms provide inspiration for efficient design, including: naturally occurring bone and honeycomb structures; I-beam girders that maximize mass-specific stiffness; archway structures that ground self-weight and external loads by compression loading; and structures that provide high efficiency with graded space frames at multiple scales.

For scenarios that require some refinement of the existing structure, parametric optimization methods can be applied, as described in Section 6.5.

6.3 Topology optimization methods

A distinct challenge exists for the efficient commercialization of engineering structures that are not sufficiently compatible with either closed-form solutions, or with reference to existing structural embodiments. For example, for non-trivial and revolutionary design scenarios[5] there is a relatively high likelihood that typical design methods may not be suitable. For these scenarios the designer may decide to utilize precedent or closed-form solutions as best as possible or may design the structural system based on prior experience or intuition. This manual selection of topology is subjective and is often sub-optimal due to preconceptions of the designer and the often counter-intuitive nature of optimal solutions. To provide effective design outcomes it is desirable that structural optimization methods be *systematic*, i.e. independent of the specific designer's intuition, and allow identification of effective geometric connectivity for complex loading scenarios. Such strategies are referred to as *topology optimization* methods.

Topology optimization (TO) methods have received significant attention by both commercial and research communities. Numerous practically useful TO methods have been developed, a selection of which are reviewed for their applicability as TO integrated DFAM tools in Section 6.4.1. These TO methods provide both challenges and opportunities for commercial design applications; the following considerations are relevant to the development of TO integrated DFAM tools:

- TO methods are particularly valuable for AM scenarios for two distinct reasons. Firstly, commercial AM outcomes inherently benefit from structural optimization due to the benefits of either mass reduction or volume reduction (Chapter 3). Secondly, topological optimization methods often result in geometry that is not compatible with traditional manufacturing but is potentially compatible with AM technologies.
- Two distinct DFAM philosophies exist for the topology optimization of AM structures: methods that retain optimized TO geometry and methods that modify TO geometry to satisfy AM manufacturability requirements. Both philosophies are presented in this chapter in detail and summarized in the context of their applicability to practical AM design, and with reference to emerging AM technologies and methods. In addition, case studies are presented briefly with a focus on commercial best practice and TO aspects of relevance to fundamental DFAM principles.
- The curse of dimensionality (Chapter 3) is particularly relevant for engineering design efforts that apply TO strategies. This outcome occurs because TO strategies utilize computationally expensive algorithms in an iterative manner, thereby compounding the associated design costs. A DFAM strategy to systematically accommodate this challenge is presented whereby previously generated data is used to predict TO solution time for a given model size, thereby allowing the design team to match the available computational resource with the intended model size (Section 6.4).

[5] A design project is defined as revolutionary if it attempts to resolve a particular technical challenge that is outside the experience of the design team. Revolutionary design projects present particular risks to successful commercialization (Chapter 3).

- TO strategies are highly effective in identifying regions of high performing local optima but are subject to inherent challenges in converging on this optimal point. Parametric optimization is not practically useful in identifying functionally efficient topologies, but it can provide robust solutions that can optimize identified local optima. Commercial best practice is to apply topological and parametric optimization in a complementary fashion according to their unique merits.
- For each of the TO methods investigated, relevant examples are presented with a focus on the unique attributes of each method and the effect of specific input variables on TO outcomes. DFAM relevant outcomes are presented with respect to fundamentals as well as: methods to accommodate the curse of dimensionality, the selection of optimization methods for minimal-mass or manufacturability, and the utilization of parametric optimization. These aspects are applied in a case study for the redesign of a high-value aerospace component in Section 6.7.

6.3.1 Topology optimization definition and methods

Significant challenges exist for designers who intend to optimize for design scenarios that are not directly represented by existing closed-form solutions, and who do not have an existing design example that can be used as a reference for structurally efficient design. For such scenarios, it is highly desirable to have access to systematic methods that can, with reasonable computational efficiency, identify structurally efficient distributions of material. These strategies, formally known as topology optimization (TO) methods, are a significant enabling technology for the commercial application of AM technologies.

This chapter presents TO contributions that have been utilized in commercially relevant applications, as well as those that show potential for DFAM research. These existing TO methods are then analyzed according to their inherent opportunities and challenges for commercial AM applications. Based on this understanding of TO methods, existing DFAM strategies are identified, and specific opportunities for commercially relevant research contributions are identified.

As previously discussed, topology refers to spatial connectivity within a particular physical domain of interest. Topology is therefore distinct from local shape, which is defined in terms of parametrically measurable dimensions associated with a specific local geometry. Commercial best practice is to apply topology and parametric optimization in concert; with TO used to identify material distributions with high structural efficiency, and parametric optimization used to refine these regions into an optimized and parametrically documented geometry (Section 6.7).

6.3.2 Existing TO methods

In practical terms, TO refers to the identification of efficient distributions of material for a given engineering function. This function is typically structural – although it may refer to heat transfer, fluid flow, or any measurable physical outcome – and is referred to as an *objective function*, O. This objective must be satisfied subject to specific design constraints and within a specified physical domain, referred to as the *design space*, Ω. This design space defines the allowable external volume of the solid material, and includes any internal voids required for interaction with other systems, as

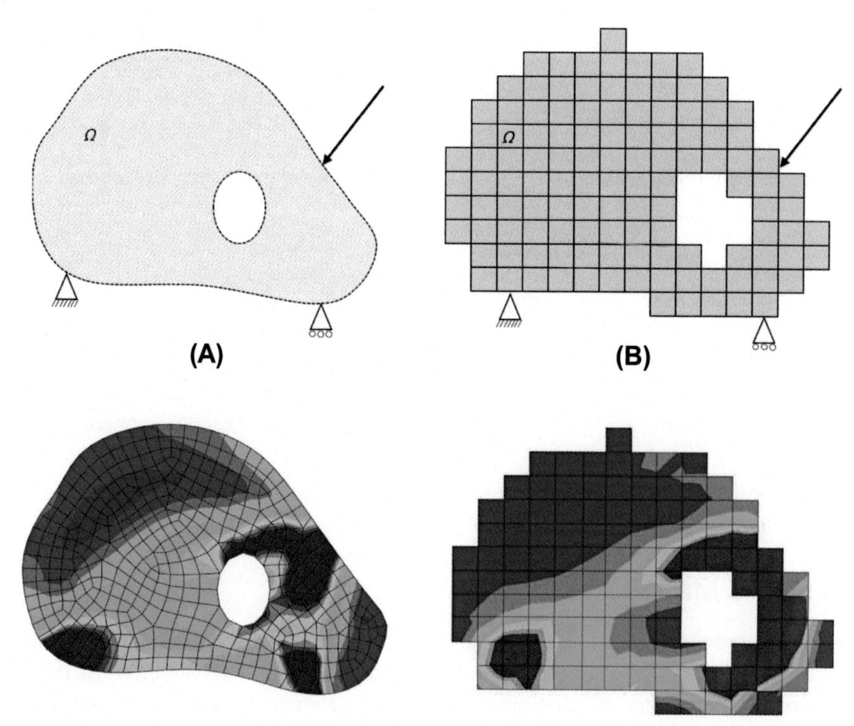

Fig. 6.5 Generalized representation of TO problem (A), whereby a distribution of material is sought to optimize some objective function, O, (typically minimizing compliance) while avoiding identified constraints (often deflection) within an allowable design space, Ω. The problem is typically discretized with a voxel field (B), however this discretization fundamentally alters the associated structural response.

well as fasteners and assembly hardware. The material distribution must avoid defined failure modes or undesirable states, specified as constraints (Fig. 6.5).

The challenge of topology optimization has been an active research problem for over a century, and numerous distinct solution strategies have been proposed. This review focusses on proposed TO methodologies that have found strong industrial application, as well as those showing promise for commercial and research DFAM activities, specifically Michell truss, ground structures, discrete (voxel) methods, and level-set methods. This initial review of TO challenges and opportunities is made independently of specific DFAM requirements.

Based on this review, specific DFAM requirements are then introduced, followed by a discussion of commercial best practice and active fields of research that address these requirements. The outcomes of this review are then applied in a case study that demonstrates the practical application of TO methods for the optimization of a mass-critical structural component subject to spatial constraints and multiple loading scenarios (Section 6.7).

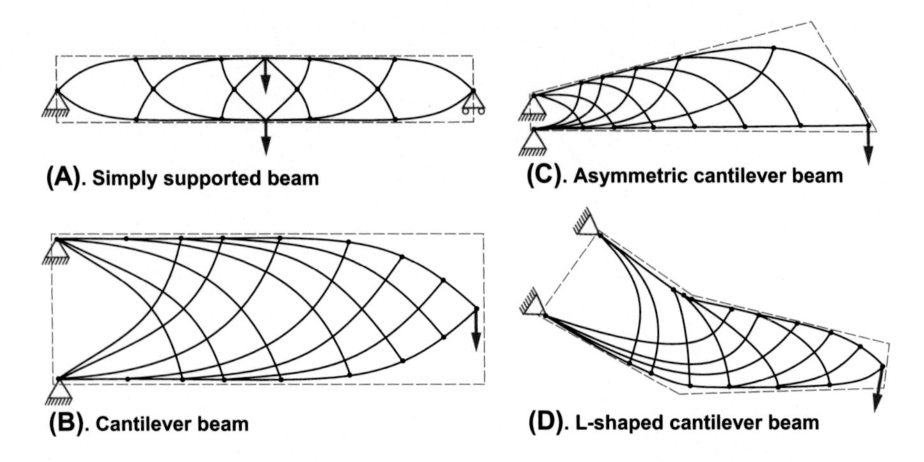

Fig. 6.6 Solution forms for the Michell truss TO strategy, including simply supported and cantilever beam as well as solutions with irregular design space (*red dashed line*).

6.3.3 The Michell truss

The contributions of *Michell* in the early 1900s represent a seminal founding work in the field of TO [6]. Michell proposed that a deflection-limited planar structure of minimal mass could be generated by aligning truss structural elements along vectors of principal strain. Michell proposed solutions to planar structures: more recently these outcomes have been extended to accommodate asymmetry and L-shape geometry (Fig. 6.6). The Michell truss method has been extended to 3D structures and has been applied as a DFAM tool to accommodate AM manufacturability requirements (Section 6.4).

6.3.4 Ground structures

Ground-structure methods rely on a predetermined truss structure assembly in two-or three dimensions [7,8]. This predetermined truss is often based on a self-tessellated unit-cell of connection elements, as well as permutations of node interactions. External

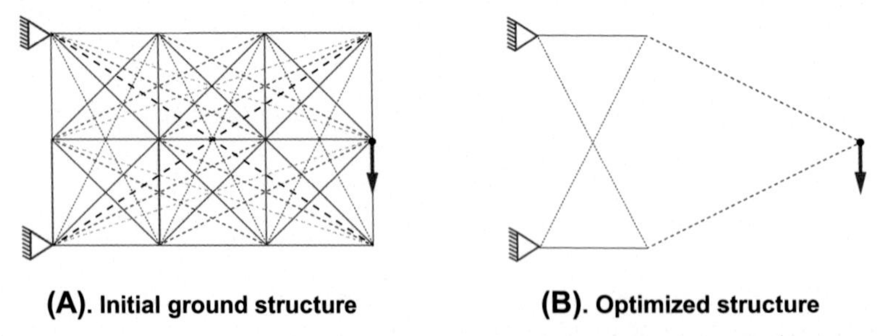

Fig. 6.7 Representative solution for two-dimensional ground structure strategy indicating (A) ground structure with multiple connective elements (identified by unique line markings) and, (B) optimized structure based on a subset of these connective elements.

loading is then applied, and the local strain of connection elements is assessed by numerical methods. Unit-connectors that inefficiently contribute to the associated objective function are iteratively deleted from the structure until only high efficiency elements remain (Fig. 6.7). Ground structures are technically robust and have provided a reference by which to compare the convergence of other TO methods; however, the number of truss permutations associated with non-trivial design domains can result in relatively low computational efficiency.

6.3.5 Level-set methods

Level-set methods represent the structural boundary between solid and void by some explicit mathematical expression. Various methods exist for defining this expression. One such method is to represent the sensitivity of material removal from the design domain in a closed-form function [9]. By intersecting a plane with this *sensitivity* function, a level-set representation of the optimal geometry is obtained (Fig. 6.8). Level-set methods enable significant advantages by allowing a mathematical definition of structure boundary and have been demonstrated for 3D geometry and for numerous

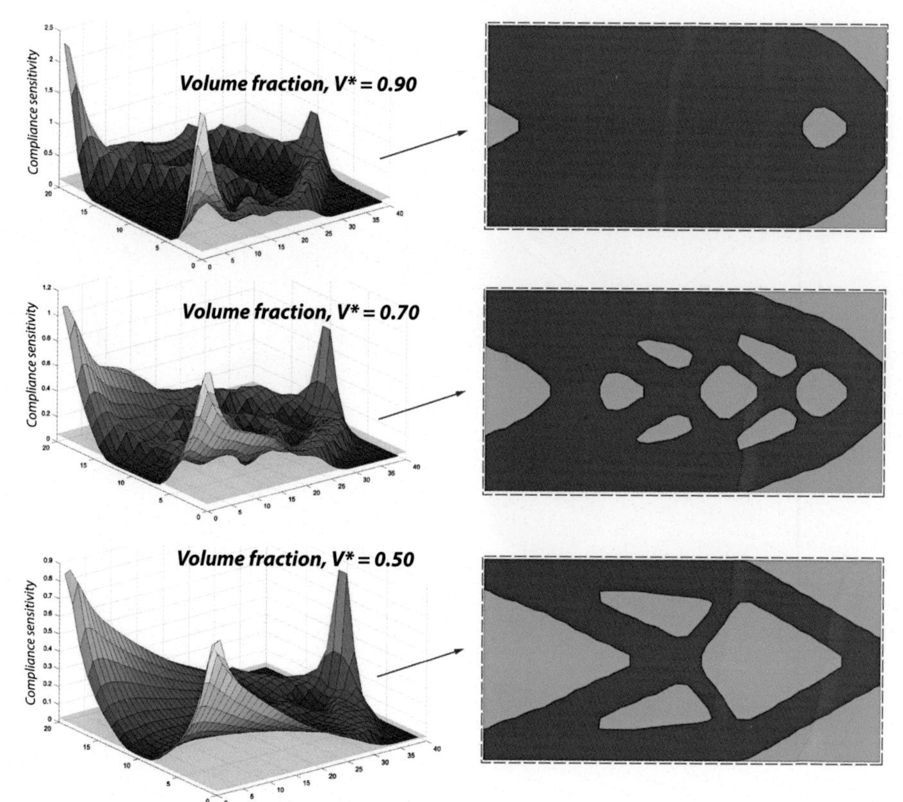

Fig. 6.8 Level-set TO strategy applied to 2D cantilever beam for various volume fractions, V^*. Images generated with the method of [9].

physical phenomena. Despite the opportunities inherent with the level-set TO strategy, it has received relatively sparse attention from the research community and extensions of this method to AM applications represents a strategic opportunity for DFAM researchers.

6.3.6 Discrete (voxel) methods

Commercially applied TO strategies are typically based on a discrete representation of a continuum of interest. Although any geometric discretization may be acceptable, discretization to square or cubic geometries is common - hence the reference to these strategies as voxel methods. The continuum design domain is typically discretized by n_x, n_y, n_z voxels in 3D space, resulting in a total of N unique voxels (Fig. 6.9). Numerical methods are then applied to this voxel array to assess the state of stress and strain at individual voxels as well as to assess the associated performance measures, typically structure mass and deflection. Of the available voxel TO methods, BESO and SIMP dominate the literature.

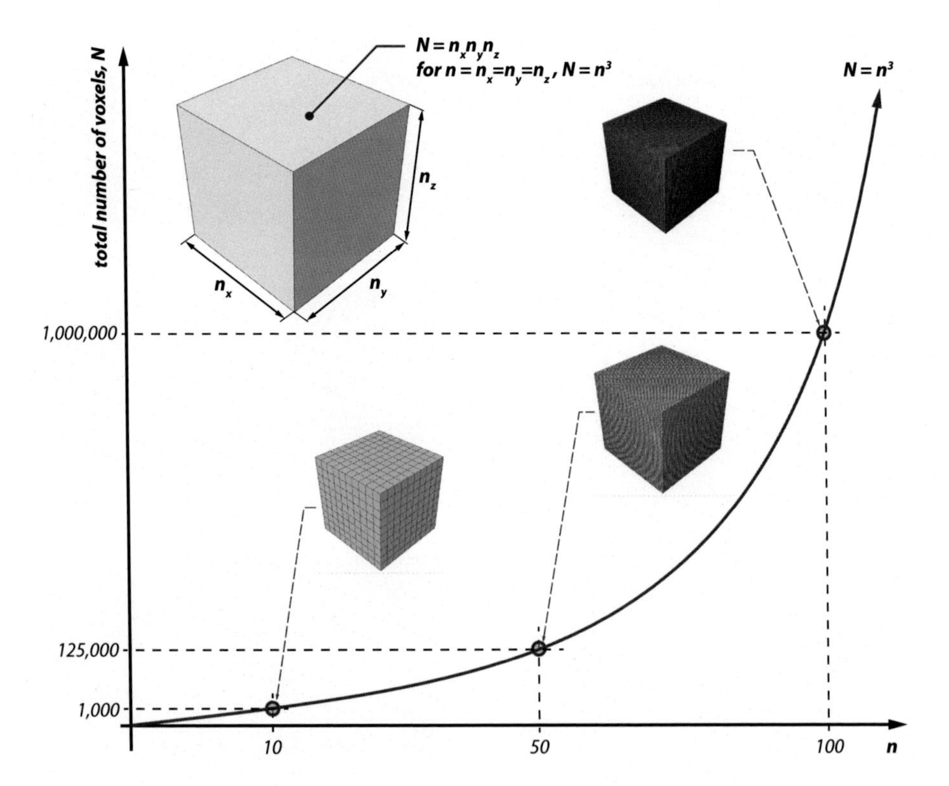

Fig. 6.9 Voxel discretization of continuum design domain. This discretization is subject to the curse of dimensionality described in Chapter 3, implying that the total number of voxels, N, increases exponentially with n.

6.3.7 Bidirectional evolutionary structural optimization method

The Bidirectional Evolutionary Structural Optimization (BESO) method of topology optimization progresses by an initial application of a numerical analysis to a voxel discretization of the available design space. The results of this analysis are then evaluated by a representative characteristic, for example stress or energy density. Based on relative values of this characteristic, discrete voxels are either *added* or *removed* such that the strain energy is minimized for a specific volume fraction constraint, V^*. The structure iteratively evolves from the initial form to a more efficient topology until some convergence criteria is achieved [10,11]. This method accommodates control factors including the allowable rate of evolution, *ER*, and a filter radius, r_{min}, to control undesirable solutions (Fig. 6.10). The BESO method has been implemented in both *hard* and *soft* variants, whereby the soft variant avoids discontinuities associated with zero density voxels. The method is applicable to a range of objective functions of relevance to commercial engineering challenges including thermal and vibrational problems [12]. BESO is well documented and numerous publications provide guidance on maximization of TO outcomes for a given computational resource. Robust, open-source code is available [13], making the BESO method a capable platform for the development of commercial code and for research contributions on the application of TO as a DFAM tool.

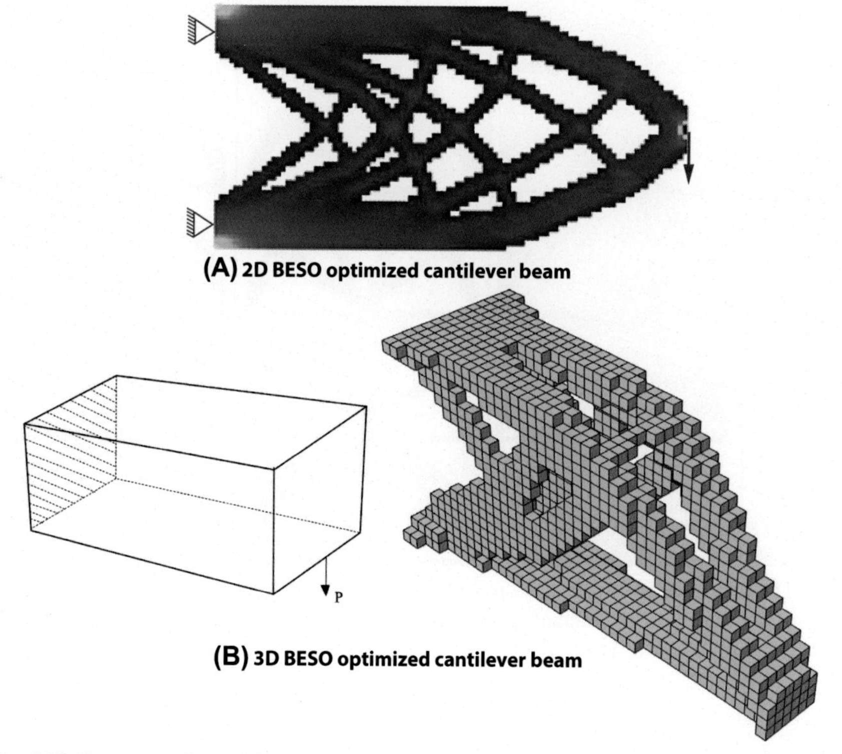

Fig. 6.10 Representative BESO solutions implemented with (A) open-source code for cantilever beam in 2D [10] and (B) an example of a 3D implementation [13].

6.3.8 Solid isotropic material with penalization method

The Solid Isotropic Material with Penalization (SIMP) method is initialized with a solid voxel array according to the allowable design volume, Ω. Numerical analysis is then applied to identify the local mechanical response and global objectives of interest, often compliance minimization. The local mechanical response (often in terms of strain) is then used to determine the local material density. This strategy results in a continuous distribution of voxel density that optimizes the solution to the associated TO problem [14,15]. This distribution is mathematically valid; however, it is not manufacturable with homogenous materials. To overcome this limitation, a penalization factor is applied to discourage intermediate densities, such that voxels tend toward either solid or void states (Fig. 6.11). Penalization is successful in avoiding intermediate densities and thereby enabling manufacturable TO outcomes.[6] However, technical challenges remain in generating commercially useful topologies, including checkerboarding, solution dependence and geometric discontinuities.

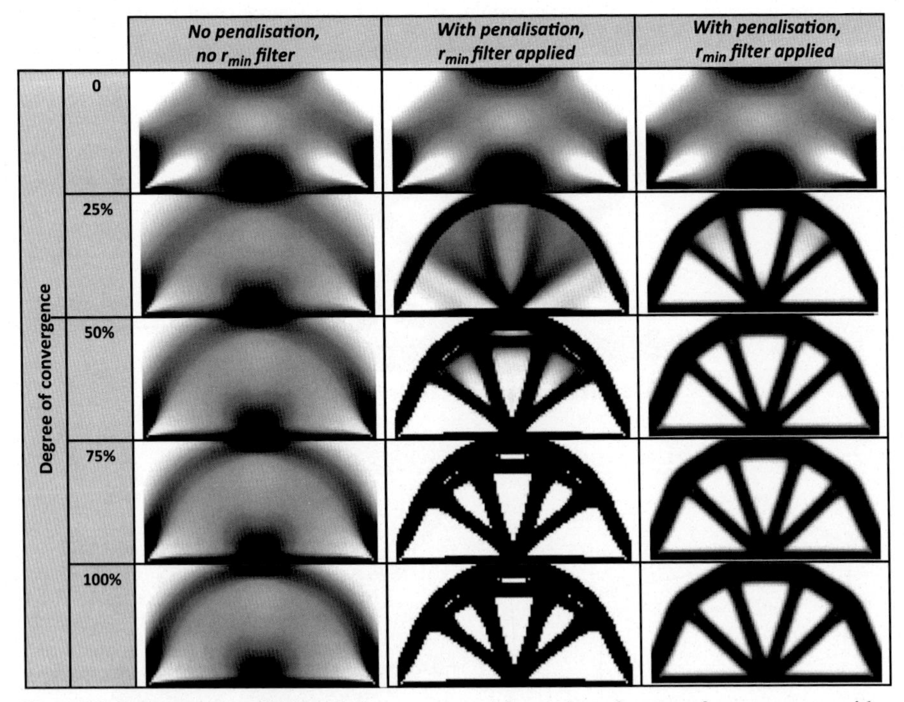

Fig. 6.11 Representative SIMP solutions presented for various degrees of convergence with varying control parameters, including the effects of penalization and filter radius, $r_{\min}$. Images generated with the open-source algorithms of [14].

[6] It is interesting to note that a selection of AM methods are compatible with the direct fabrication of structures with discontinuous density (for example Material Jetting, MJT, Chapter 8).

The *checkerboarding* phenomenon occurs when adjacent voxels alternate between solid and void state in order to emulate a solution which, without penalization, would be of intermediate density. Checkerboarding is overcome by the application of a filter applied to the state space, $r_{\min}$, which essentially smooths the local function solution prior to penalization (Fig. 6.11).

The second challenge is associated with the dependence of the TO outcome on the associated control parameters. For example, Fig. 6.11 indicates several design solutions with common boundary conditions and loading solved with varying SIMP control parameters, resulting in varying topological solutions. This outcome demonstrates that, as for BESO, SIMP is not guaranteed to identify a global optima and the specific solution generated is a function of the associated input parameters. However, much research and practical guidance is available that provides strategies to assist in the systematic identification of high performing optima. One practical method is to iteratively solve the TO problem while parametrically varying the input parameters of interest. This approach, sometimes known as *extension*, is useful and readily implemented; however, it is an exhaustive search and thereby compounds the required execution time.

By their fundamental nature, discrete TO methods result in geometric discontinuities and roughness at the void/solid boundary. Various methods have been proposed to algorithmically accommodate these challenges, including local mesh refinement and local geometry smoothing. These methods potentially enhance geometric outcomes but can be computationally expensive, can fail to provide technically robust solutions and may be incompatible with commercial documentation requirements. For these reasons typical commercial best practice requires that parametric optimization strategies are applied to the TO outcomes, as discussed in Section 6.5.

Fig. 6.12 represents feasible solutions of Michell truss, ground structure, level-set and voxel methods (BESO and SIMP) for a point loaded cantilever beam. From this data it is apparent that in practical terms the generated solutions enabled by these methods are equivalent. This outcome may seem surprising given the extensive debate within the research community on the relative merit of the available TO methods. For the practicing designer with a mandate to effectively deploy TO for the identification of efficient geometries, it is sufficient that the design team select a TO method that suits their specific preferences. In addition, they must be aware of the limitations inherent in the TO methods in general, as well as specific challenges associated with the selected TO method. The following section summarizes current research and commercial best practice applications of TO in the specific domain of AM.

6.4 Opportunities for TO applied to AM

Although topology optimization has been an active research area for over a century, its practical application had been reserved to a relatively small group of engineering optimization specialists. More recently, and in particular since the development of the robust and generally applicable TO methods of Section 6.3, topology optimization has found increasing application by generalist engineering practitioners and has become well integrated within non-specialist commercial engineering design tools.

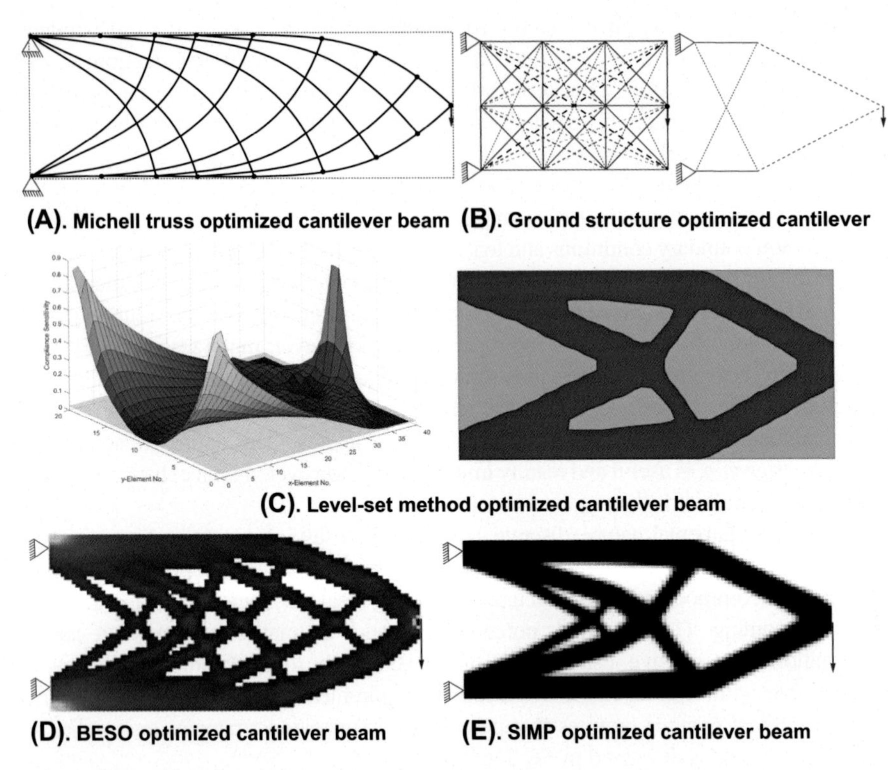

Fig. 6.12 A comparison of the solutions generated by the TO reviewed here suggests that, (despite ongoing debate within the research community) for the practicing engineer, there is little practical difference in structural insights enabled by Michell truss, ground structures, level-set, BESO or SIMP.

A particular challenge associated with the commercial utilization of topology optimization is the *potential incompatibility between TO outcomes and associated manufacturing constraints*, especially for traditional manufacturing methods. Although, as discussed in detail in Chapter 4, the description of AM as *constraint-free* manufacture is incorrect and misleading, AM does enable the direct manufacture of high complexity geometry and is therefore often *more compatible with the outcomes of TO methods than traditional manufacture*. This synergy between TO and AM has spurred a range of commercial applications of TO for AM as well as associated DFAM research contributions and TO integrated DFAM tools, including:

- accommodation of specific AM manufacturability constraints
- allowing systematic compromise between TO outcomes and AM manufacturability
- methods to predict and mitigate the computational costs of TO simulation.

A summary of current DFAM approaches and potential research contributions associated with these technical challenges is presented below with reference to commercial best practice design methods and the case study of Section 6.7.

6.4.1 TO integrated DFAM tools

As discussed in Chapter 4, there exist significant challenges in developing all-encompassing definitions for AM manufacturability. These challenges exist because AM failure modes are often subject to complex underlying phenomena that are challenging to predict, stochastic in response, and are validated by limited experimental data. Nonetheless, AM manufacturability constraints have been defined for a range of potential failure modes. The development of TO integrated DFAM tools has progressed significantly, resulting in a useful body of research, and commercially useful methods for integrating AM manufacturability requirements within TO outcomes. However, there remains much opportunity for the development of advanced TO integrated DFAM tools, as is demonstrated by the following summary of current approaches.[7]

Of the many applicable DFAM failure modes (Chapter 4), the potential limit on inclination angle is the most commonly understood measure of AM manufacturability, and is the most commonly implemented within TO integrated DFAM tools. Several such tools have been presented within the literature that intend to modify a planar TO outcome such that allowable AM inclination angle constraints are satisfied (Fig. 6.13). These DFAM tools provide useful design insight but they remain limited in the potential DFAM failure modes that are explicitly accommodated, and they are not necessarily transferrable to 3D structures.

Fewer TO integrated DFAM tools that can accommodate 3D space fields exist than for the equivalent 2D scenario; however, innovative methods have been proposed in the research literature (Fig. 6.14). These methods are primarily based on the avoidance of specific inclination angles: note that some of the proposed 3D methods utilize a diagonal voxel field (equivalent to a 45 degree angle) to represent the allowable inclination angle and cannot be modified to accommodate alternate inclinations. However, more advanced methods are emerging that can accommodate integrated requirements for support structure optimization, including combined requirements for support-free inclinations, as well as methods to minimize the overall volume of support material required.

Furthermore, methods that accommodate AM manufacturability constraints by the application of self-supporting internal trusses have been proposed for 2D and 3D scenarios (Fig. 6.15).

Despite their inherent commercial value, relatively few DFAM tools exist that accommodate AM manufacturability constraints. The existing tools reported here utilize only inclination angle and, to a lesser extent, minimum feature size as the measures

[7] In this exciting and highly useful field of DFAM it is difficult to write a useful state of affairs as it will likely change quickly. Best efforts have been made not to omit any relevant research directions.

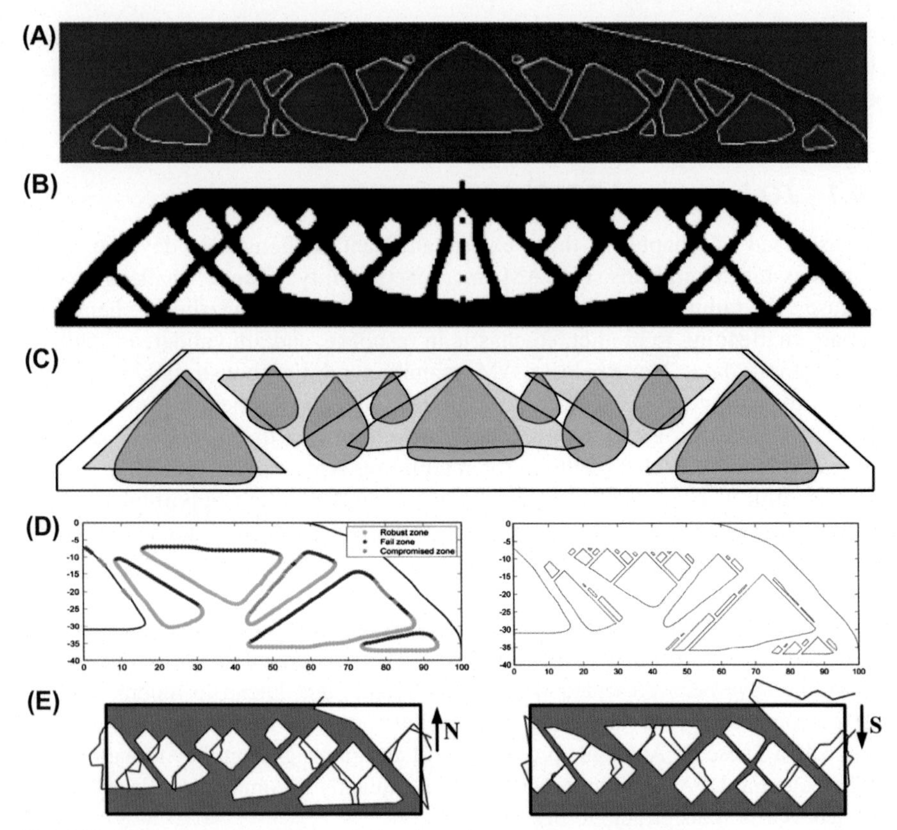

Fig. 6.13 TO integrated DFAM tools that intend to modify a planar TO outcome such that allowable AM inclination angle constraints are satisfied. (A) [16], (B) [17], (C) [18], (D) [19], (E) [20].

that characterize manufacturability. It is apparent then that significant opportunities exists in the design of TO strategies that accommodate DFAM constraints. In particular, existing methods do not accommodate temperature field as a TO constraint; however, the avoidance of excessive local temperatures is critically important for high-value product manufactured with thermal AM systems.

6.4.2 Compromise between TO outcomes and AM manufacturability considerations

As for all manufacturing technologies, AM presents a series of unique technical manufacturability challenges; for example, minimum feature size, self-supportable inclination angle, thermal conductivity, and material entrapment (Chapter 4). However, TO outcomes are typically more compatible with the manufacturability limitations of

Fig. 6.14 TO integrated DFAM tools that accommodate AM inclination angle constraints in a 3D space field to optimizes the standard TO outcome (left) for advanced AM manufacturability (right), including (A) restriction of allowable overhang [21], and (B) outcomes that minimize support volume [9].

AM than with traditional manufacture. In response to the synergy between AM and TO, numerous *TO integrated DFAM tools* have been proposed, and this field remains a highly active area of research and commercial innovation. These strategies can be classified as either strategies that *modify* the TO optimization outcomes such that AM manufacturability constraints are satisfied; or, strategies that *retain* the optimized TO geometry but provide additional material, modified processing or active support structures in order to satisfy AM manufacturability.

Strategies that modify TO outcomes to satisfy manufacturability constraints are advantageous for applications that intend to be implemented algorithmically, as they generate output topology that can be directly manufactured without requiring additional support or geometric post processing.[8] These strategies compromise the structural optimality of the manufactured topology in favour of direct manufacturability. For scenarios where manufacturability is less important than overall structural efficiency of the manufactured product, and for components that are not strictly net shape (for example structures that include some post-AM machining), methods that include

[8] It is likely that the current state-of-the-art of TO integrated DFAM tools (Section 6.4.1) are not directly compatible with net-shape outcomes, and that some degree of post-processing is required.

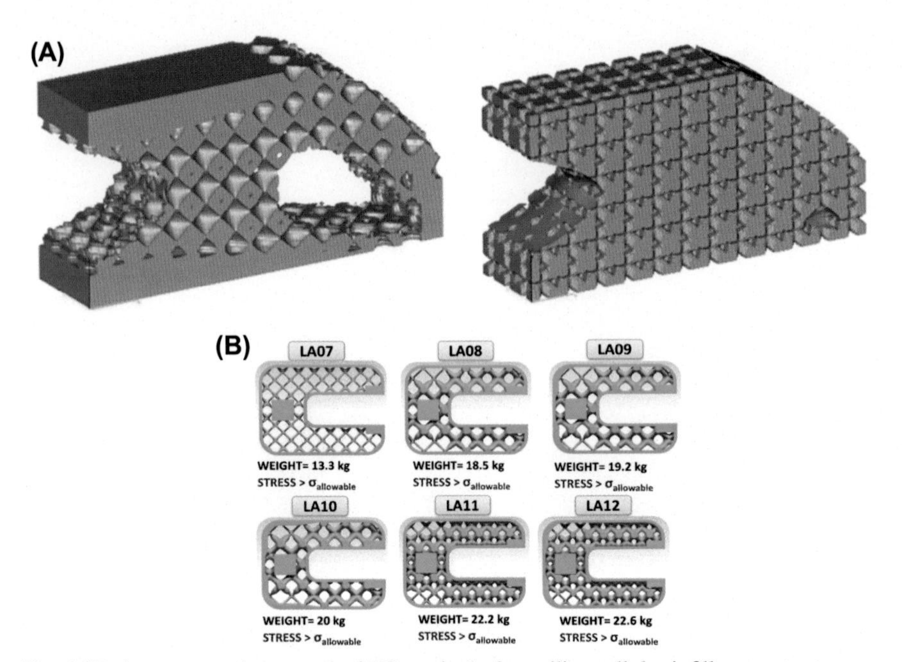

Fig. 6.15 A representative survey of TO methods that utilize cellular infill geometry to accommodate AM manufacturability limits (A) [22], (B) [23].

additional supporting geometry to retain the optimal topology may be preferable. Fig. 6.16 indicates an example of a cantilever beam processed by strategies from both categories.

The former strategy tends to be of interest to theorists who intend that manufacturability constraints be accommodated directly within the TO algorithm. The latter strategy tends to be the focus of commercially motivated designers, where it is understood that post-processing is required, and that additional material removal is typically less important than structural efficiency of the delivered component. The majority of TO integrated DFAM tools presented in the literature favour modified TO outcomes for optimal manufacturability rather than structural efficiency of the manufactured component. It is apparent that there remain significant commercial and research opportunities for the development of TO integrated DFAM tools that retain structural efficiency.

6.4.3 Challenges associated with computational costs of TO simulation

A fundamental technical challenge associated with the practical utilization of TO is the compounding of computational complexity as a function of associated problem size. This challenge is often referred to as the *curse of dimensionality* (Chapter 3), whereby a linearly increasing problem size can rapidly become prohibitively expensive to solve.

In fact, the *curse of dimensionality* is exacerbated for the TO methods presented in Section 6.3, as these TO strategies must numerical solve a finite-element problem in an

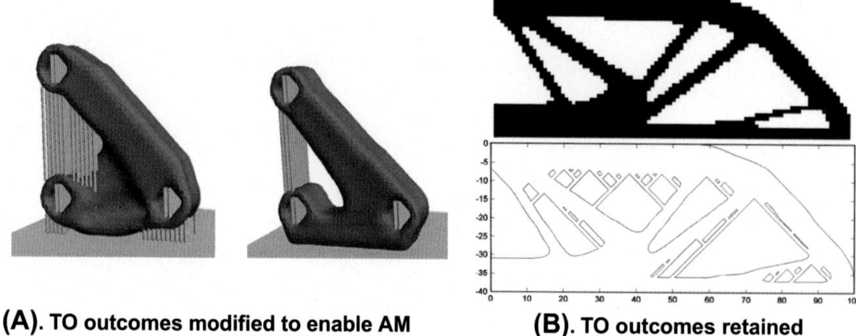

Fig. 6.16 Example of TO integrated DFAM tools that: (A) *modify* the TO optimization outcomes to satisfy AM manufacturability constraints [9], and (B) *retain* the optimized TO geometry [19].

iterative manner. Specifically, TO strategies apply numerical analysis techniques to solve for the local structural field variable (for example peak displacement). Numerical methods such as the Finite Element Method (FEM) can solve this problem for generalisable geometry, boundary conditions and loading; however, FEM is dimensionally inefficient, even for single iteration solutions, and with a linearly increasing number of voxels,[9] N, computational cost exponentially increases to values that exceed the data processing limit of the available computational hardware. For TO methods, this phenomenon is exacerbated as the numerical solution is iterated to allow convergence on a specific topological solution (Fig. 6.17).

Pragmatic design strategies that utilize TO methods must accommodate their inherent computational inefficiency[10]. For example, charts such as Fig. 6.17 of computational cost versus number of voxels can be utilized to define the feasible limit on the allowable number of voxels within the solution space based on the practically allowable computational cost[11] – this approach can reduce the risk of failing to generate solutions within the available design time and is demonstrated in the case study of Section 6.7.

6.5 Parametric optimization

The optimization methods identified above provide insight into efficient methods to generate solutions for explicit design scenarios, either by reference to closed-form solutions, design precedent, or topology optimization. These optimization methods are

[9] Assuming each voxel is represented by a unique numerical finite element.

[10] Much literature exists on the relative computational costs of proposed TO methods. These specifics will not be addressed here.

[11] As the specific number of iterations (and specific solution time of each iteration) is a function of the specific loading scenario, it is not possible to explicitly define the exact solution time associated with future optimization projects, however previous data can provide useful insights.

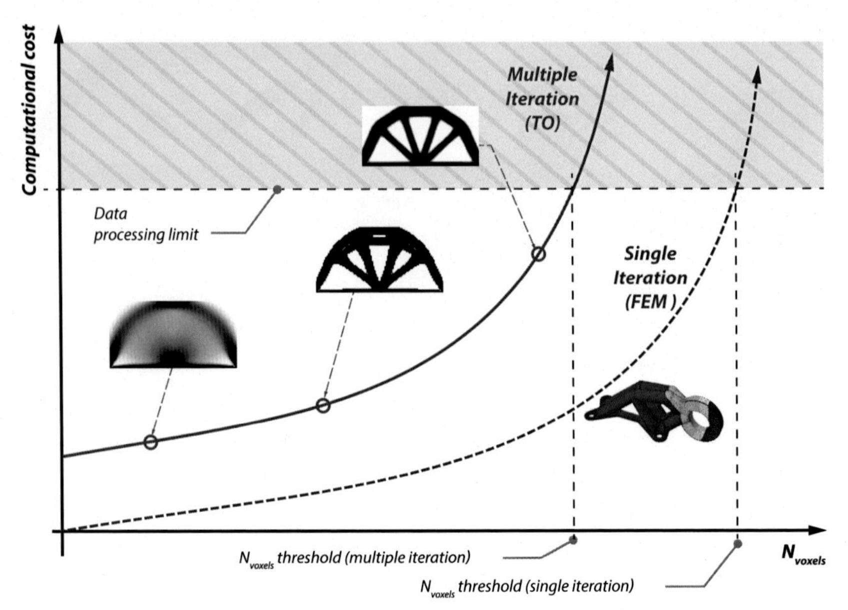

Fig. 6.17 Simulation data indicating solution time versus number of voxels, N_{voxels}. Computational cost increases non-linearly for a single iteration numerical analysis. For TO methods, computational costs are exacerbated as multiple iterations of the underlying numerical analysis are required to allow convergence on a specific TO solution. With increasing N_{voxels}, computational costs rapidly exceed the technical data processing limit of the available computational hardware.

often highly efficient in identifying globally optimal structural topology, but due to imperfect assumptions or limitations of the solution method, may result in sub-optimal refinement of the local geometry. For these scenarios, parametric optimization provides a complementary optimization method.

Parametric optimization refers to techniques that optimize some *objective function* (typically a single value representation of desired performance) in terms of the parametrically defined control factors that represent local geometry. Parametric methods allow optimization of these local geometric variables by various methods including *brute force* and *sequential optimization* methods.

6.5.1 Brute force methods

Brute force methods, also known as *exhaustive search* refer to optimization methods that assess solutions for various permutations of the design space, in this case the parametrically defined *control factors*. Once the specified permutations of control factors are assessed, they are compared in terms of the relevant objective function to allow identification of high performing solutions. To increase their effectiveness, systematic brute force methods have been proposed, for example, full-factorial design of experiments (DOE), whereby all feasible permutations of a particular design space are assessed. To reduce the dimensionality of full-factorial DOE, partial-factorial DOE methods exist that assess some subset of the full-factorial data without losing insight into efficient parametric solutions.

6.5.2 Sequential optimization methods

Sequential optimization methods differ from brute force methods in that the results of previous evaluations are used to inform the selection of future evaluations, potentially reducing the number of solution iterations required to converge to the optimal solution. Extensive research exists on these iterative optimization methods and numerous algorithms have been proposed. Of the available algorithms, *gradient* and *Nelder-Meade simplex* methods will be discussed in more detail as they provide solutions to a general category of problems, are well documented in the literature, and are readily applied to commercially relevant design problems.

Gradient methods evaluate some objective function and associated rates of change (gradient) with respect to the parametric variables of interest. Based on the local gradient, parametric variables are then selected for the next solution to be evaluated. The step size is modulated by an experientially selected *learning rate.* This rate may be tuned according to the observed rate of change of the objective function: too low and convergence may not occur, too high and local optima may be overshot. Gradient methods are simple to implement and are robust in enhancing performance in the region of a local optima. Acquisition of the local gradient can be computationally expensive as the effect of each control factor must be evaluated where each evaluation requires an additional numerical analysis.[12]

Nelder-Meade simplex methods are an industrially useful class of optimization methods that are gradient free. The method involves the initialization (usually arbitrarily) of a simplex[13] of potential solutions. The simplex solution with lowest performance is deleted, and a replacement solution is then generated, typically by a sequence of reflection, extension and subtraction. *Reflection* involves moving from the deleted point to the centroid of the remaining solutions and then continuing this trajectory for the same distance again. If this solution is superior to any other existing solution, *extension* of this trajectory is applied. Conversely, if this solution is the least optimal of the existing solutions, the trajectory is reduced to a value half-way to the identified centroid, known as *subtraction*.

These methods have proven to be remarkably effective in the optimization of complex industrial problems, and as the optimization method is gradient-free it has the advantage of reducing the number of numerical evaluations required per iteration (Fig. 6.18).

[12] The Curse of Dimensionality (Chapter 3) results in an exponential increase of the number of required simulations with the number of control factors. Control factors with an independent effect on the objective function can be concurrently evaluated, potentially reducing the required computation resource.

[13] A simplex is a polygon with one more node than the dimension of the problem. For example, in the two dimensional problem of Fig. 6.19, the simplex consists of 3 points.

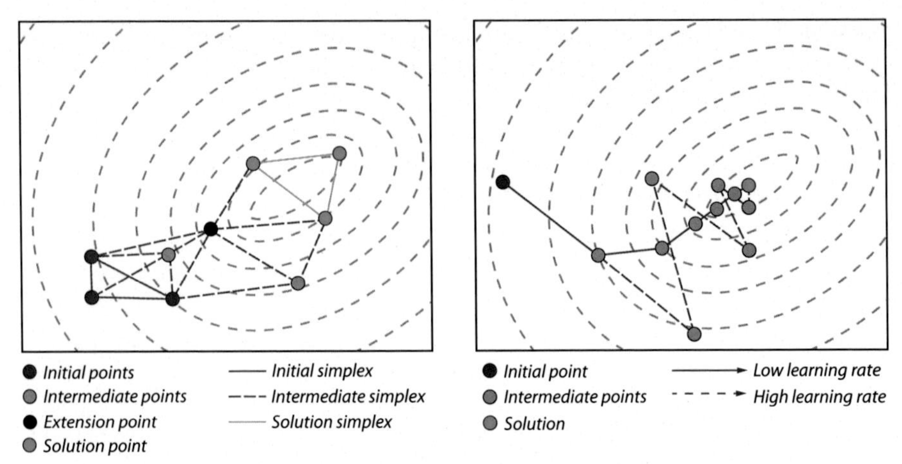

Fig. 6.18 Schematic representation of Nelder-Meade simplex methods (left) and steepest descent gradient methods (right) applied to an objective function in two dimensions. Both methods provide a mechanism to converge on local optima, but once converged are not capable of identifying neighbouring optima.

6.5.3 Practical application of brute force and iterative methods

It is apparent that parametric optimization methods are useful at refining local optima and are therefore a complementary capability to TO methods (which are successful at identifying local optima but are not robust in refinement). To implement a strategy of parametric optimization, the design team must select between brute force and iterative methods. The design merit of iterative methods is described briefly above and in detail within numerous optimization texts [24,25]. Although brute force methods often receive dismissive reviews within mathematically focussed optimization literature, they do provide important advantages for the pragmatic outcomes required in commercial engineering practice. These advantages include solution parallelization, robustness to flawed analysis, and robustness to discontinuous solution space.

For brute force methods, the intended simulation permutations are independent and are known *a priori*. Consequently, multiple computational simulations can be evaluated concurrently, either in parallel on a specific workstation, or by delegating simulations across multiple computational resources. This opportunity for solution parallelization is not available for sequential optimization methods as the associated solutions are not independent. Consequently, brute force methods present an important computational opportunity, especially for engineering enterprises that have access to the computational resources necessary for simulation parallelization.

A significant challenge to the automated optimization of engineering systems is associated with the robustness of parametric models. Parametric models typically include numerous control factors that interact in complex ways that can be difficult to predict, including states that are not numerically robust. Furthermore, numerical

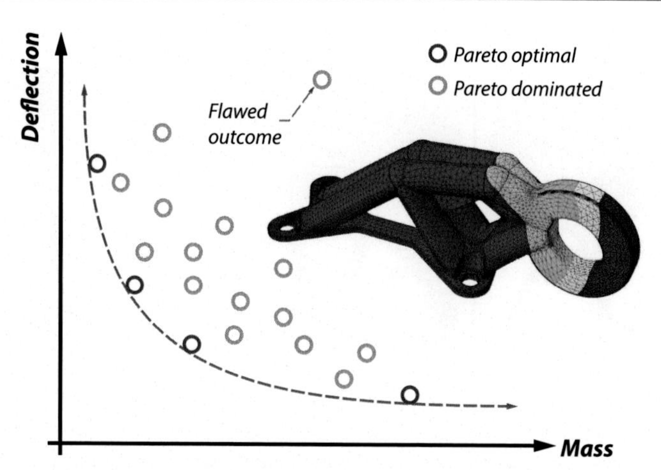

Fig. 6.19 Parametric models ideally should generate robust geometry and finite-element mesh for any feasible parametric input. These flaws are technically the fault of the design team; however, for non-trivial design scenarios, it can be challenging to ensure that parametric models are robust for all feasible input states.

models require that these geometries are compatible with finite element meshing tools, further increasing the potential for flawed simulation outcomes.

Brute force methods are robust to the flawed regeneration of the parametric model: such a flaw will simply result in a failure to generate a solution, or the generation of an incorrect solution.[14] Neither of these results will contaminate the results for other permutations sheduled to be assessed (Fig. 6.19). Conversely, sequential optimization methods are not robust to flawed parametric regeneration and will either terminate the optimization process, or will attempt to iterate with flawed data, resulting in incorrect convergence.

Similarly, brute force methods are robust to discontinuities within the allowable solution space. Furthermore, where control-factors are ill-defined, the brute force DOE can be modified to omit these solutions, thereby eliminating the cost associated with a failed or invalid solution execution. From a practical point of view for commercial application of optimization methods, this is possibly the most significant imperative toward the use of brute force methods for models that have uncertainties in robust regeneration.

6.6 Topology optimization and generative design (BC2AM)

A core commercial advantage of additive manufacturing methods is the opportunity to rapidly manufacture commercial grade components based on a digital data representation (Chapter 3). To enable commercial outcomes requires that the digital data for a

[14] In practice, this flawed result will be identified and corrected by the manual examination that is required prior to manufacture.

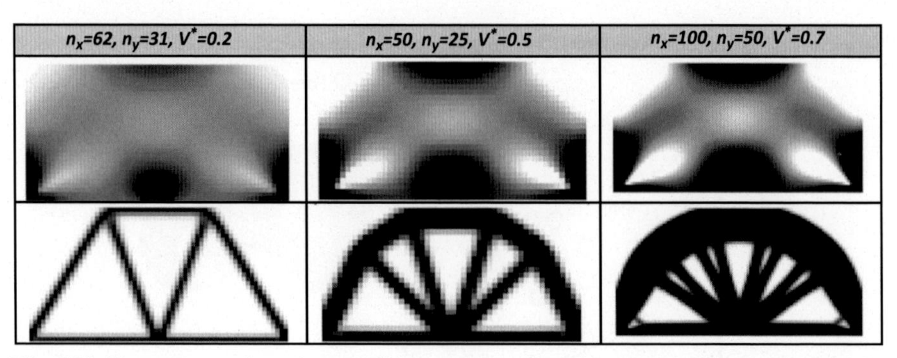

Fig. 6.20 For a common boundary condition, a range of candidate topologies are generated for varying number of voxels, n, and volume fraction, V^*. Results are reported for intermediate (upper) and final convergence (lower). These topologies can provide the basis for generatively designed structures. TO outcomes generated with the open-source algorithms of [14].

specific design outcome be generated with minimal manual effort, preferably by *generative methods*[15], where the final design is prescribed algorithmically with the design engineer providing oversight of these algorithms rather than providing specific design input. An overview of these generative methods is provided in Chapter 7, where it is applied to scenarios where the structure deployment includes parametric variants of a specific design requirement.

Topological optimization provides a profound opportunity for the generative design of engineering systems that are subject to sophisticated design requirements. For example, the outcomes of Fig. 6.20 indicate a range of potential solutions for a compliance limited structural design subject to specific loading conditions. These topological variants are generated by manipulating the control variables associated with the number of voxels within the design domain, n_x and n_y, and the intended volume fraction, V^*.

Despite the significant commercial opportunities for TO in enabling generative design, there exist significant barriers to its implementation as a commercial DFAM tool. Commercial best practice in this space involves the use of TO for the generation of optimal topologies, which are then manually parametrized prior to production. This outcome allows the benefits of TO and parametric optimization to be integrated within the design process, as well as allowing the designer to add value from their experience and intuition by manually determining elements of the design; however, this manual intervention is not compatible with the autonomous implementation required for generative design. Specific research opportunities exist for methods that can

[15] The term generative design has become increasingly applied, and has various meanings within various design communities. In this text a broad definition is used to encompass many of these variants in meaning; specifically that generative design enables the designer to solve a problem in terms of associated functional requirements rather than by controlling the specific design details that achieve these functional requirements (Chapter 7).

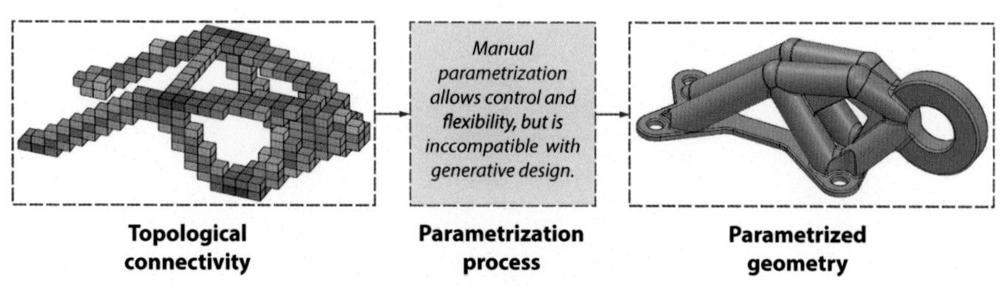

Fig. 6.21 Conceptual representation of parametrization techniques that are typically implemented manually, thereby providing an opportunity for the development of algorithmic DFAM tools.

autonomously extract a parametric representation directly from the TO outcome, as well as methods that can *smooth* the TO outcome such that is can be directly fabricated without the geometric discontinuities inherent to the voxel representation. The motivation of these methods is to enable robust AM outcomes to be achieved algorithmically from a specification of the functional boundary conditions (BC), therefore the acronym *BC2AM* may be useful in describing this emerging field of DFAM research.

6.6.1 Automated extraction of parametric data from TO outcomes

Commercial imperatives require that production geometry be parametrically defined. As illustrated by the examples presented in this chapter, TO outcomes include geometric features that are not directly feasible for manufacture and require subsequent refinement. Consequently, a strong commercial motivation exists towards the automated extraction of robust parametric data directly from the TO results. Despite this opportunity, there exists few robust research outcomes in this space.

Fig. 6.21 conceptually demonstrates a commercial challenge to BC2AM implementation. The TO outcomes provide robust insight into efficient geometry, but to be assessed with gradient optimization methods and for production documentation, a parametric representation is required. Commercial best practice is to manually parametrize these structural elements by inspection; however, algorithmic tools that can assist in this parametrization process are much required for commercial practice.

An alternate method observed in the literature involves the extraction of representative splines from a smoothed representation of the discretized solid boundary (Fig. 6.22). This method is not directly compatible with 3D structures but does indicate the potential for such DFAM implementations to enable automation of practical AM design workflows, especially as is associated with generative methods. These methods appear to have received little research attention in the DFAM relevant literature: an omission that is potentially due to the multidisciplinary nature of the underlying research required to execute this significant DFAM opportunity.

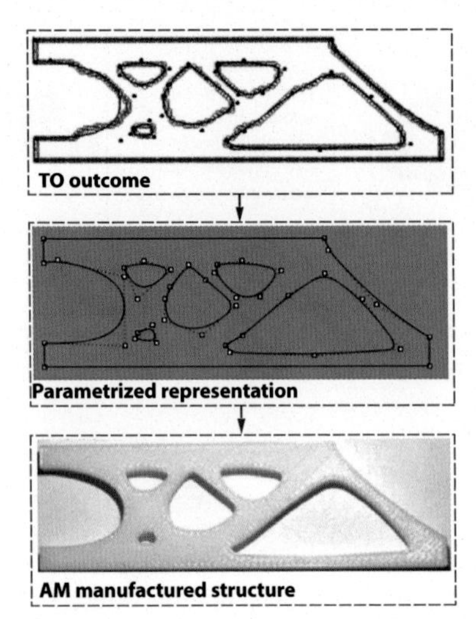

Fig. 6.22 A method for extracting parametric data directly from the TO outcome. Figure based on [26].

6.6.2 Smoothed TO outcomes

Topology optimization typically results in discontinuous data that is not directly compatible with AM manufacturability requirements. In order to achieve AM compatible outcomes algorithmically, it is required to modify this non-smooth data. Various methods have been proposed to achieve this BC2AM outcome. These methods include smoothing of the discretized TO outcome, dynamic re-meshing of the TO design domain such that resolution is enhanced locally as required, and methods that overcome singularities such that the discretized representation is geometrically conformal (Fig. 6.23).

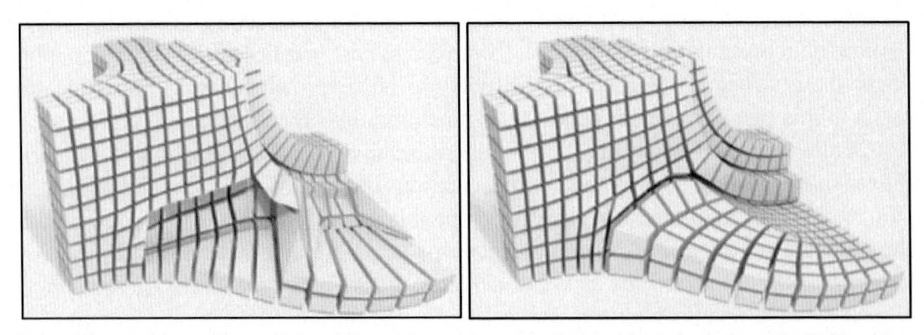

Fig. 6.23 Methods of increasing the conformance of traditional hexahedral mesh (left) with a less distorted meshed solution (right) [27].

These methods provide useful DFAM outcomes, but, as for the automated extraction of parametric data, Section 6.5, little published data is available for formal application of these methods to achieve BC2AM outcomes. Again, this limitation is potentially due to the multidisciplinary research inputs required and provides a significant commercial and research opportunity for those capable of integrating mesh smoothing methods from the computer science domain to the practical requirements of commercial DFAM tools.

6.7 Case study: optimization of high-value non-stationary aerospace component

Aerospace structures are non-stationary, meaning that propulsion energy must be consumed during flight in proportion to the associated mass. For such structures, mass-reduction enables commercial value in terms of increased flight capabilities, increased load carrying capability or as reduced fuel consumption. AM provides a technically and economically strategic opportunity for the commercialization of such high-value applications. As an aside, aerospace structures are often safety-critical, and are therefore subject to rigorous certification requirements, as discussed in Chapter 7.

As an example of how topology optimization can be strategically applied to increase the value of non-stationary applications, a high-value aerospace application is presented below, and is structurally optimized subject to AM manufacturability requirements using the topology and parametric optimization methods presented in this chapter (Fig. 6.24).

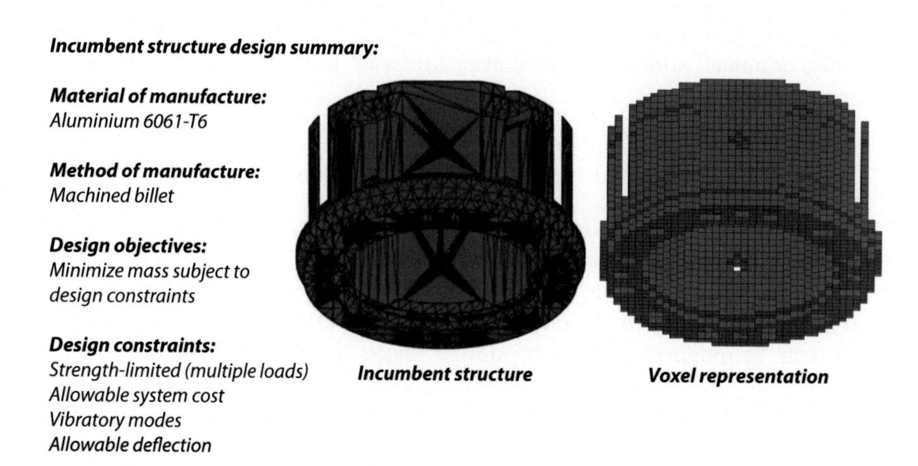

Fig. 6.24 Redesign of incumbent design according to defined statement of requirements.

6.7.1 Strategy for the efficient application of TO to the AM scenario

Based on the opportunities and limitations inherent in topology optimization methods (Section 6.3), a systematic strategy needed to be applied for the pragmatic optimization of AM for the mass-critical aerospace application specified in Fig. 6.24. This strategy includes the following phases:

- ***Define initial conditions:*** such that there is no ambiguity associated with the intended technical function and constraints of the proposed system.
- ***Identify spatial requirements*** and available design volume (including access for assembly and fasteners): such that the design domain can be systematically scheduled for topology optimization compatible with the available computational budget.
- ***Apply TO methods:*** to acquire insight into efficient material distributions.
- ***Accommodate AM manufacturability:*** this may include reference to integrated DFAM tools, but, given the current state-of-the art, commercial best practice typically requires manual input to accommodate the diverse potential failure modes associated with AM technologies.
- ***Generate parametric representation of the preferred topology:*** emerging BC2AM tools can be useful in enabling generative design at this stage. Once parametric models are specified, parametric optimization methods can be applied to optimize local geometry and document the intended design.

6.7.1.1 Define initial conditions

According to the overarching design philosophies presented in Chapter 3, it is imperative that design understanding be maximized early in the design phase, where there remains time and money to act on this understanding. This understanding is made with respect to a formal design specification (Fig. 6.24). However, when considering AM as an alternative to traditional manufacturing methods, it is important that latent failure modes are not overlooked. For example, optimized AM structures can achieve significant mass reduction, but may result in stiffness and vibration constraints becoming dominant which may have been irrelevant to the higher mass incumbent design.

6.7.1.2 Identify spatial limits and the available design volume

Existing CAD data, if it exists as for this re-design problem, provides a high efficiency shortcut to define the allowable design geometry (Fig. 6.24). However, it is important to note that this CAD data may include clearance geometry that exist to enable traditional manufacturing and may be functionally redundant for AM. This clearance geometry can be reintroduced to the design space of the AM structure, thereby increasing structural efficiency. By reference to a previously generated relationship of computational cost versus problem size (for example, Fig. 6.17) it is possible to quantify the allowable voxel resolution for the available design time. This technique is imperative to manage the risk of timely convergence of TO methods.

6.7.1.3 Apply TO methods

The intended TO strategy is applied to the design space of interest. At this stage it is important to confirm that the numerical model utilized by the TO model does correctly represent the intended loading scenario and that it is numerically converged. This confirmation is defined as verification and validation respectively and is presented in the context of medical implant design in Chapter 7. TO methods require numerous iterations of an expensive numerical analysis and are therefore computationally expensive. To manage this computational cost, it may be advisable that the numerical model be simplified and that non-linear effects such as contact and local plasticity are avoided.

6.7.1.4 Accommodate AM manufacturability

Despite the manufacturability advantages of AM, TO may result in geometry that introduces manufacturability challenges. As discussed in Section 6.4, two strategies exist in response to this challenge. The first is to apply TO methods that algorithmically accommodate AM manufacturability constraints. This strategy is useful as it can potentially be automatically implemented, thereby minimizing design effort. Conversely, such AM-manufacturability optimized structures may compromise the mass of the finished product. The second strategy enables the design of minimal mass structures, and then applies supporting structures as required to accommodate manufacturability. This strategy can enable greater mass optimization but can require the use of additional supporting structures.

For the aerospace component of this case study, a strategy of mass optimization was applied; however, care was taken to ensure powder removability and to avoid entrapped support structures. Specific design features of the TO outcome include (Fig. 6.25):

- more direct load transfer paths
- redundant material identified in central ring feature
- necessity for support material around grounding features identified
- over-design of upper ring feature identified.

These insights were then applied to manually define a parametric representation of the salient structural features.

6.7.1.5 Generate a parametric representation of the preferred topology

Although TO strategies provide highly effective design guidance on the identification of structural optima, they are relatively ineffective in terms of local shape optimization. Parametric optimization methods enable a complementary capability and can efficiently converge to the optimal local shape. The selection of specific parametric embodiment for optimization remains an open research challenge (Fig. 6.21). Commercial best practice involves the geometric interpretation of the TO outcomes by experienced AM design engineers in order to define a parametric representation of the component that embodies the structural efficiencies identified by the TO strategy,

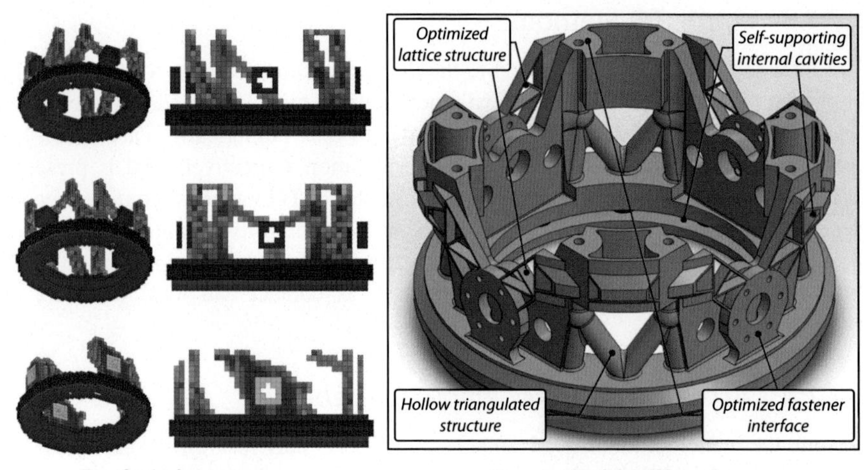

Fig. 6.25 Systematic strategy for the efficient application of TO and parametric optimization methods in the design of a high-value lightweight aerospace structure manufactured at the RMIT University Advanced Manufacturing Precinct.

while maintaining a parametrically efficient representation. These parameters are either manually optimized by the design team, or by methods suitable to local optimization. For this scenario, the following specific parametric representations were selected (Fig. 6.25).

- Triangulated structure to ground load-paths. These structural elements are hollow to increase structural efficiency in bending, and to allow powder removal, and are inclined to be structurally self-supporting to avoid entrapped internal support material.
- Self-supporting internal cavities were defined to increase the efficiency of the central ring feature which ensuring AM manufacturability. Powder removal was accommodated by physical conduits within low stress regions.
- Surrounding material included to allow accommodate external fasteners, but adjacent material removed in self-supportable manner.
- Lattice structure applied to enable connection of the locating ring.

6.7.2 Design outcomes

Based on the opportunities and limitations identified as inherent to topology optimization methods, a systematic strategy is proposed for the efficient application of TO and parametric optimization methods in the design of high-value lightweight applications. This strategy presents a pragmatic approach that utilizes available algorithmic DFAM techniques and manual design methods to generate commercial design outcomes within a reasonable time budget. For example, this allowed the rapid deployment of a functional prototype that was robustly fabricated using Selective Laser Melting (SLM, Chapter 11) with a mass reduction of over 50% when compared with the incumbent structure. This strategy is not a one-size-fits-all solution but is presented as a useful case study of commercial best practice that can be modified to suit the needs of specific design projects, and should be evolved as the associated DFAM technologies develop. This case study also provides a practical basis for DFAM researchers to be aware of commonly applied design workflows, and to encourage the development of practically useful DFAM tools that enhance these workflows.

6.8 Chapter summary

The optimization of engineering structures requires significant design effort. To achieve efficient design optimization for AM, it is imperative that this optimization effort be efficiently applied and commensurate with the associated economic benefit. Chapter 3 introduces the techno-economic motivation for the functional optimization of AM structures, and finds that, due to the unique attributes of the AM process, all AM design scenarios benefit significantly from functional optimization, even those that do not have a specific objective of mass reduction.

In response to this identified opportunity, a range of overarching strategies for design optimization are presented, including closed-form solutions and precedent design. These strategies embody various advantages and disadvantages and should be considered on their merit for the specific design scenario of interest. For example, closed-form solutions can provide an efficient pathway to the design of optimized structures. Similarly, extant engineering structures, including naturally occurring phenomena and engineered systems, are the result of extensive optimization processes that can be re-purposed for the rapid design and deployment of optimized commercial AM structures. The commercial AM design opportunities enabled by closed-form and extant structures have received little research DFAM attention and remain a highly relevant commercial and research opportunity.

Closed-form solutions and extant structures do not provide useful design insight for many complex loading scenarios. Historically, these scenarios were largely intractable and relied on the skill, experience and intuition of the designer to offer up candidate solutions. More recently, a number of readily implementable topology optimization (TO) algorithms have been developed that are highly generalisable and readily applied, enabling profound insight into the optimal structural deployment for commercially relevant design scenarios.

The systematic optimization outcomes enabled by TO can be applied to any manufacturing technology. However, the curvilinear geometry typically associated with TO can be technically challenging for traditional manufacturing methods, but is often compatible with the geometric flexibility inherent to AM. Furthermore, as identified in Chapter 4, all AM structures benefit from functional optimization due to reduced mass (lightweight structures) and physical volume (low-cost structures). Consequently, AM and TO are highly synergistic, and much commercial benefit can be achieved by their mutual application.

The technical implementation of TO methods is well documented in the public literature and numerous open-source codes are available, as well as implementations within standard commercial CAE software platforms. Due to the inherent challenges of dimensionality, there is arguably little technical advantage of commercial TO software over open-source implementations; thereby providing a commercial opportunity for AM design bureaus that have insufficient resources for the maintenance of commercial software licenses.

While TO methods are well documented in the research literature, attention is often focussed on their fundamental implementation and theoretical limits, rather their practical application. This chapter presented a brief summary of the challenges to the application of TO within commercial workflows; which can be categorized as either: challenges inherent to the TO method in the identification of an optimal solution, or, challenges in applying this identified solution to AM. It is imperative that commercial design engineers are aware of these challenges such that they can avoid risks to the timely generation of robust design outcomes. For research engineers, these challenges directly correlate with opportunities for the development of commercially valuable DFAM tools, including:

- Voxel based TO methods are not suited to the convergence on local optima. These difficulties can be mitigated by modified TO strategies, including mesh refinement and iterative methods. However, it is imperative that the design team be aware that further refinement of the TO outcome will be required for commercially optimized design.
- The selected discretization and solution method can result in geometric artefacts (such as checkerboarding) or manufacturability challenges (such as partial density regions) that may compromise structural response. These challenges are typically mitigated by the application of filtering methods and penalization factors that attempt to elicit technically useful

solutions. These techniques should be applied with an understanding of how they influence the associated design outcomes.

- The computational costs associated with TO methods are potentially high. These costs are incurred as the numerical analysis required to evaluate the system response must be iterated for TO, thereby compounding the overall computational costs. This challenge has the potential to dramatically restrict the range of feasible solutions that can be assessed within practical time limits. For a given TO method, overall computational cost is principally a function of the number of elements in the discretized design space. A method is presented here, where previously generated data on computational cost is used to predict the allowable TO discretization for the available design time. This method can be used by the design team to reduce the risk of timely convergence of TO methods.
- The potentially high computational cost of TO methods can restrict their application for complex numerical models; for example, in structural contact and non-linear systems. For practical engineering applications it is advisable that TO methods be applied to representative, but simplified, numerical models. Once efficient topologies are identified, they can then be refined by parametric methods. If simple models are applied for TO, there is arguably little advantage to commercial software over open-source code implementations.

Challenges to the direct application of TO outcomes to AM present a significant potential impediment to commercial AM optimization, as well providing a basis for valuable DFAM research. These challenges include:

- TO integrated DFAM tools accommodate AM manufacturability requirements either by the modification of TO outcomes, or by providing additional physical structure as required. These approaches result in fundamentally different geometric outcomes. The former approach modifies the TO geometry and therefore compromises technical function; however, it is advantageous for scenarios where manufacturability is the primary design objective, as well as for design scenarios that are to be implemented algorithmically. For design scenarios where overall structural efficiency is the primary design objective, methods that retain the optimal topology are preferable.

TO methods typically apply some discretization to the design space, and the conversion to continuous geometry as is typically required for manufacture can be challenging. Two approaches exist in the DFAM literature: smoothing of the discretized TO solution and the generation of parametric data. Both approaches have unique advantages and disadvantages. Smoothing methods can be directly applied to the TO outcomes and are therefore directly compatible with generative design methods. Parametrized data provides advantages for detail optimization; however, manual parametrization is typically required to achieve the required design intent.

- Methods to convert discrete digital data to parametric forms have been an active research focus for some time, and yet have not reached the level of robustness required to be integrated in standard commercial software and design workflow. This is obviously a challenging research area, and those intending to contribute to this field should be aware of the potential intractability of this problem. However, there exist many research opportunities for methods that promote Boundary Condition to AM (BC2AM) outcomes, whereby advanced smoothing and parametrization methods are used to enable generative AM design directly from the associated boundary conditions (Chapter 7).

- The current generation of TO integrated DFAM tools are based on assumptions of a homogenous microstructure and simplistic failure modes; neither of these assumptions is acceptable for the robust design of optimized AM structures. The omission of realistic manufacturability and material variability within the current generation of TO integrated DFAM tools results in critical flaws in their commercial applicability. Consequently, significant commercial and research opportunities exist in the development of improved TO integrated DFAM tools that accommodate the specific requirements of AM, specifically:
- There exists a pervasive misrepresentation of AM technologies as enabling constraint-free manufacture. In fact, the manufacturing constraints of relevance to AM systems are as extensive as those of relevance to traditional manufacture but are fundamentally different and are much less well understood and documented. The focus of TO integrated DFAM tools is predominantly on inclination angle as the defining characteristic of TO manufacturability. More sophisticated AM failure modes, such as those presented in detail in Chapter 4 are required for the commercially valuable implementation of these DFAM methods.
- TO methods are inherently expensive in terms of the required computational resources and even solutions for relatively few voxels can be prohibitively time consuming. The topological optimization of thermal AM systems requires that temperature fields be constrained such that desirable surface finish and microstructure is promoted. Current methods for temperature field prediction are themselves computationally expensive; therefore, research outcomes that can reduce the time required to predict thermal fields are required for the robust integration of MAM with TO methods.
- The optimization of engineering structures is of considerable value for AM applications. The current generation of TO tools provide a significant enabling capability; however there exist challenges in their application to commercial AM design. In response to this opportunity, a wave of TO integrated DFAM tools are being released to the market and within the research literature. This first generation of design tools provides enhanced design capability; however, there remain significant research opportunities, especially associated with the multidisciplinary research fields that are necessary to enable breakthrough contributions.

References

Seminal works on stress-concentration factors and closed-form solutions for structural problems:

[1] Young WC, Budnyas RG. Roark's formulas for stress and strain. McGraw-Hill; 2017.
[2] Pilkey WD, Pilkey DF. Peterson's stress concentration factors. John Wiley & Sons; 2008.

Resources that summarise efficient embodiments for machine design and kinematic systems:

[3] Parmley RO. Illustrated sourcebook of mechanical components. McGraw Hill; 2000.
[4] Sclater N, Chironis NP. Mechanisms and mechanical devices sourcebook, vol. 3. New York: McGraw-Hill; 2001.

[5] Jones FD, Horton HL, Newell JA, editors. Ingenious mechanisms for designers and inventors, vols. 1–4. Industrial Press Inc; 1930.

Resources that summarise topological optimisation methods include:

[6] Michelll AGM. LVIII. The limits of economy of material in frame-structures. The London, Edinburgh, and Dublin Philosophical Magazine and Journal of Science 1904;8(47): 589–97.
[7] Dorn W, Gomory R, Greenberg H. Automatic design of optimal structures. Journal of Mecanique 1964;3(1):25–52.
[8] Rozvany GIN, et al. Layout optimization of structures. Applied Mechanics Reviews 1995; 48(2):41–119.
[9] Mirzendehdel AM, Suresh K. Support structure constrained topology optimization for additive manufacturing. Computer-Aided Design 2016;81:1–13.
[10] Xie YM, Steven GP. Basic evolutionary structural optimization. In: Evolutionary structural optimization. London: Springer; 1997. p. 12–29.
[11] Huang X, Xie M. Evolutionary topology optimization of continuum structures: methods and applications. John Wiley & Sons; 2010.
[12] Huang X, Xie YM. Natural frequency optimization of structures using a soft-kill BESO method. In: IOP conference series: materials science and engineering, vol. 10 (1). IOP Publishing; 2010. p. 012191.
[13] Zuo ZH, Xie YM. A simple and compact Python code for complex 3D topology optimization. Advances in Engineering Software 2015;85:1–11.
[14] Sigmund O. A 99 line topology optimization code written in Matlab. Structural and Multidisciplinary Optimization 2001;21(2):120–7.
[15] Bendsøe MP, Sigmund O. Material interpolation schemes in topology optimization. Archive of Applied Mechanics 1999;69(9–10):635–54.

TO integrated DFAM tools that accommodate 2D space field include:

[16] Gaynor AT, Guest JK. Topology optimization considering overhang constraints: eliminating sacrificial support material in additive manufacturing through design. Structural and Multidisciplinary Optimization 2016;54(5):1157–72.
[17] Zhang K, Cheng G, Xu L. Topology optimization considering overhang constraint in additive manufacturing. Computers & Structures 2019;212:86–100.
[18] Guo X, Zhou J, Zhang W, Du Z, Chang L, Liu Y. Self-supporting structure design in additive manufacturing through explicit topology optimization. Computer Methods in Applied Mechanics and Engineering 2017;323:27–63.
[19] Leary M, Merli L, Torti F, Mazur M, Brandt M. Optimal topology for additive manufacture: a method for enabling additive manufacture of support-free optimal structures. Materials & Design 2014;63:678–90.
[20] Zhang W, Zhou L. Topology optimization of self-supporting structures with polygon features for additive manufacturing. Computer Methods in Applied Mechanics and Engineering 2018;334:56–78.

TO integrated DFAM tools that accommodate 3D space field include:

[21] Langelaar M. Topology optimization of 3D self-supporting structures for additive manufacturing. Additive Manufacturing 2016;12:60–70.

[22] Panesar A, Abdi M, Hickman D, Ashcroft I. Strategies for functionally graded lattice structures derived using topology optimisation for additive manufacturing. Additive Manufacturing 2018;19:81–94.

[23] Primo T, Calabrese M, Del Prete A, Anglani A. Additive manufacturing integration with topology optimization methodology for innovative product design. The International Journal of Advanced Manufacturing Technology 2017;93(1–4):467–79.

Resources that provide insight into formal optimisation methods relevant to parametric optimisation include:

[24] Roy R, Hinduja S, Teti R. Recent advances in engineering design optimisation: challenges and future trends. CIRP Annals 2008;57(2):697–715.

[25] Snyman JA. Practical mathematical optimization. Springer Science+ Business Media, Incorporated; 2005. p. 97–148.

The following reference provides an example of how smoothing can be applied to achieve BC2AM outcomes by smoothing discretised data:

[26] Liu S, Li Q, Liu J, Chen W, Zhang Y. A realization method for transforming a topology optimization design into additive manufacturing structures. Engineering 2018;4(2): 277–85.

[27] Liu H, Zhang P, Chien E, Solomon J, Bommes D. Singularity-constrained octahedral fields for hexahedral meshing. ACM Transactions on Graphics (TOG) 2018;37(4):93.

Generative design

7

Additive manufacturing enables the design of high-complexity bespoke structures to achieve custom engineering objectives. The ability of a human designer to manually accommodate such complexity is limited. In such cases, Generative Design (GD) is an enabling DFAM capability whereby the designer formalizes the constraints and objectives required of a satisfactory design in some *expert system*; and defines some *optimization system* to algorithmically satisfy these requirements. GD methods range from fully autonomous implementations that generate absolute solutions, to interactive systems that efficiently generate potential solutions for evaluation by the design team. In addition, the sophistication and scope of the expert system varies between these potential implementations. Consequently, there exists variability in the lexicon and scope of generative design methods; a taxonomy and associated nomenclature is presented to assist in clarification and to provide formal insight into the complexity and risk inherent to the implementation of various GD taxonomies.

Generative design enables optimal results with minimal engineering effort, in particular by enabling the mass-customization of AM design in an economically feasible manner. Furthermore, generative design provides an opportunity to satisfy regulatory and certification requirements for the design of high-complexity bespoke products, such as patient-specific medical implants; whereby the engineering and regulatory team defines the allowable geometric and processing limits for a particular structure, but the design details for each specific structure are algorithmically implemented. Illustrative case studies are presented briefly for implementations of GD systems with various taxonomies, allowing discussion of the inherent challenges and opportunities of these implementations (Fig. 7.1).

7.1 Generative design challenges and opportunities for AM

There exists much uncertainty in the precise meaning of the emerging term *generative design*. In this chapter, Generative Design (GD) refers to methods that enable DFAM outcomes that are practically unachievable by a human designer – whereby the generative design system takes some (perhaps all) of the design effort from the designer, thereby enabling decision-making to progress with the speed and repeatability of a computational system. Such generative design systems inherently enable highly complex design outcomes and the commercial implementation of mass-customized AM systems (Chapter 2). Despite the profound opportunities for generative design methods to add value to AM, there exist numerous inherent challenges to their robust implementation. These challenges are notably associated with the documentation and

Design for Additive Manufacturing. https://doi.org/10.1016/B978-0-12-816721-2.00007-5

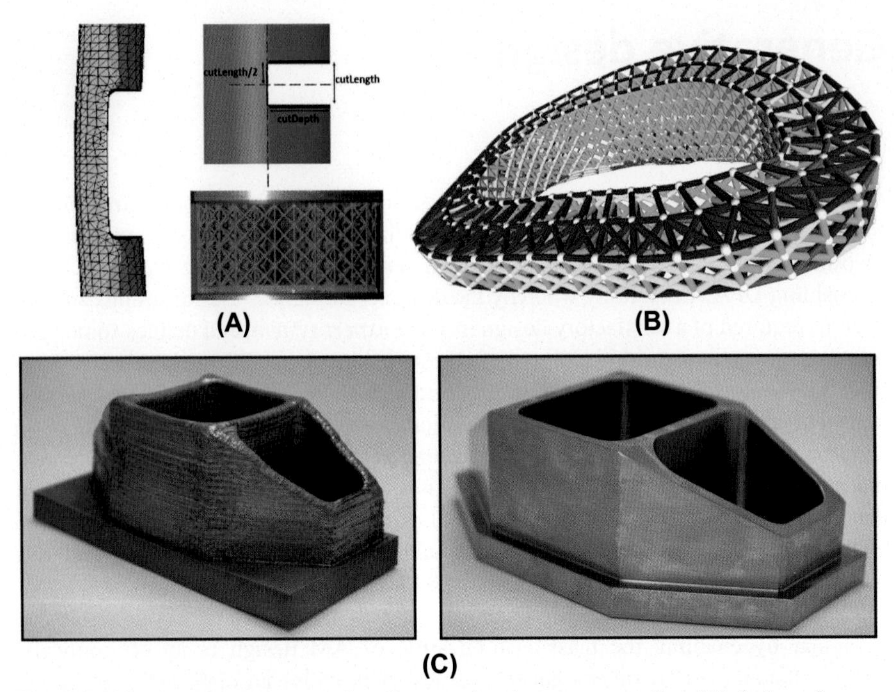

Fig. 7.1 This chapter presents a series of generative design case studies: (A) design of standardized patient-specific medical implants, (B) patient-specific implant generation, (C) design of mass-customized aerospace bracketry.

certification of high-value applications typically associated with mass-customization design scenarios. To illustrate these opportunities and challenges, a series of generative design case studies are presented (Section 7.4).

7.1.1 What is generative design?

The term *generative design* has emerged relatively recently within the design lexicon and is used with varying meaning within different design communities. Generative design can be broadly defined as 'the rules for generating form, rather than the forms themselves' [1]. The intent of this chapter is to present methods that enable DFAM outcomes that are not reasonably achievable by a human designer. This scope of GD is more broad than many literature definitions, specifically that it includes prosaic as well as profound GD implementations; however, it does serve our pragmatic purpose of defining methods that enable enhanced DFAM outcomes.

A simplified, but useful, schematic representation of the formal design process (Chapter 3) as relevant to GD is presented in Fig. 7.2. This schematic includes phases of *problem definition*, where the problem is formally defined in terms of the associated constraints and objectives; *embodiment design*, which consists of the generation and

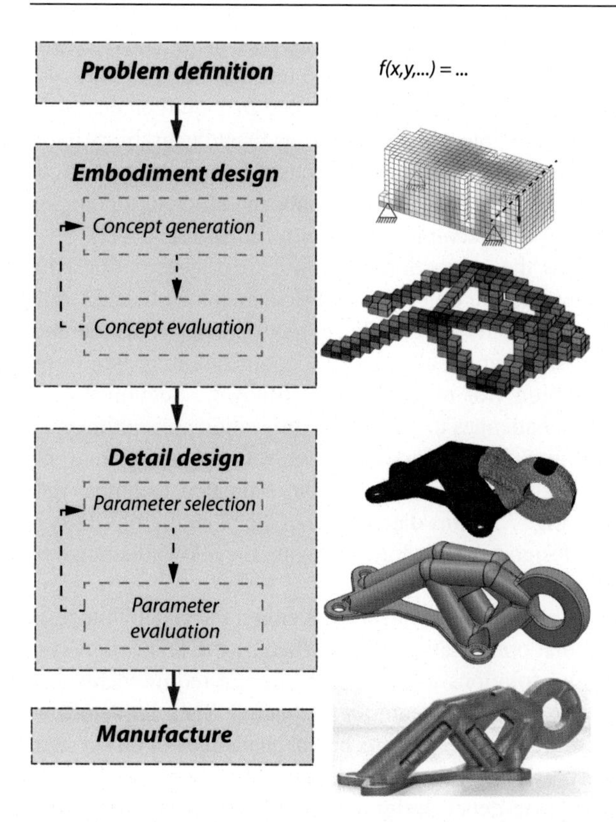

Fig. 7.2 Simplified schematic of the formal design process as relevant to GD.

evaluation of high-level solutions to the problem definition; and *detail design*, where the preferred solution is formally specified to enable manufacture. In order to provide a useful DFAM outcomes, the generative design system must replace one or more of these identified design phases.

7.1.2 Generative design system architecture and implementation

Disparate definitions of the term *generative design* exist within the literature. Here we define generative design to refer to DFAM tools that utilize autonomous systems to aid in the generation of forms that satisfy a specific set of design requirements. This definition is further constrained such that these generative methods be effective, i.e. they enable some output that is not practically feasible for a human designer. A fundamental definition of GD architecture then must include some element that generates feasible solutions, and another element that evaluates the feasibility of these potential solutions. These elements are referred to respectively as the *expert system* and *optimization system*. These GD systems may be applied to either or both of the embodiment and detail design phases of Fig. 7.2; permutations of these implementations are presented in the case studies of Section 7.4.

The *expert system* refers to some database or set of logical rules that intends to replicate the experience and observations of some expert in the relevant field. Expert systems are used to generate potential solutions to specific design requirements. This expert system can vary from a prosaic implementation of some manufacturability database (as is often implemented in detail GD systems, for example, Case Study A, Section 7.4.1); to sophisticated algorithms that generate complex and efficient topologies (as is required for embodiment GD, for example, Case Study B, Section 7.4.2).

The *optimization system* curates the interaction with the expert system, allowing decision-making according to the multiple objectives and constraints associated with specific design requirements. For detail GD systems, the *objective function* that defines system performance is often clearly defined, allowing the optimization system to be implemented in a procedural fashion (for example, Case Study C, Section 7.4.3). For embodiment GD systems, the challenges to effective optimization are significantly higher, as the mathematical definition of the associated objective function may be elusive. A matrix of these GD system taxonomies and the estimated technical risk to implementation is provided in Fig. 7.3. This data assists in the planning of GD system architectures that maximize design benefit while managing the allowable range of technical risk that the design team is willing to accept.

In response to these potentially significant technical challenges, the GD architecture may be implemented with a manual optimization phase, whereby the GD expert system generates a series of potential solutions to the specific design requirements and these solutions are then reported to the design team for evaluation. Such implementations can provide a useful compromise between the technical challenges of GD system implementation, and their ability to generate useful outcomes. Examples of this GD system implementation applied to the detail design phase are presented in Fig. 7.4; including an interesting case where all potential solutions were scheduled for 3D printing, in effect using the manufacturing process as a proxy optimization system.

		Optimisation system	
		Generative	***Manual***
Design phase	***Embodiment***	*Embodiment phase concepts are highly abstract and challenging to algorithmically optimise for complex scenarios.*	*Manual optimisation at embodiment phase can alleviate the complexity of algorithmic methods. (Case Study B)*
	Detail	*Detail phase concepts are more readily characterised and are significantly less challenging to implement with GD optimisation. (Case Study A and B)*	*Manual detail phase optimisation allows the rapidity of an algorithmic expert system while allowing designer control (Figure 4, Case Study C).*

Fig. 7.3 Tabular representation of generative design (GD) system taxonomies. Descriptive text provides further explanation. Colour estimates technical risk to implementation: light green — low, dark green — moderate, yellow — high, red — extreme.

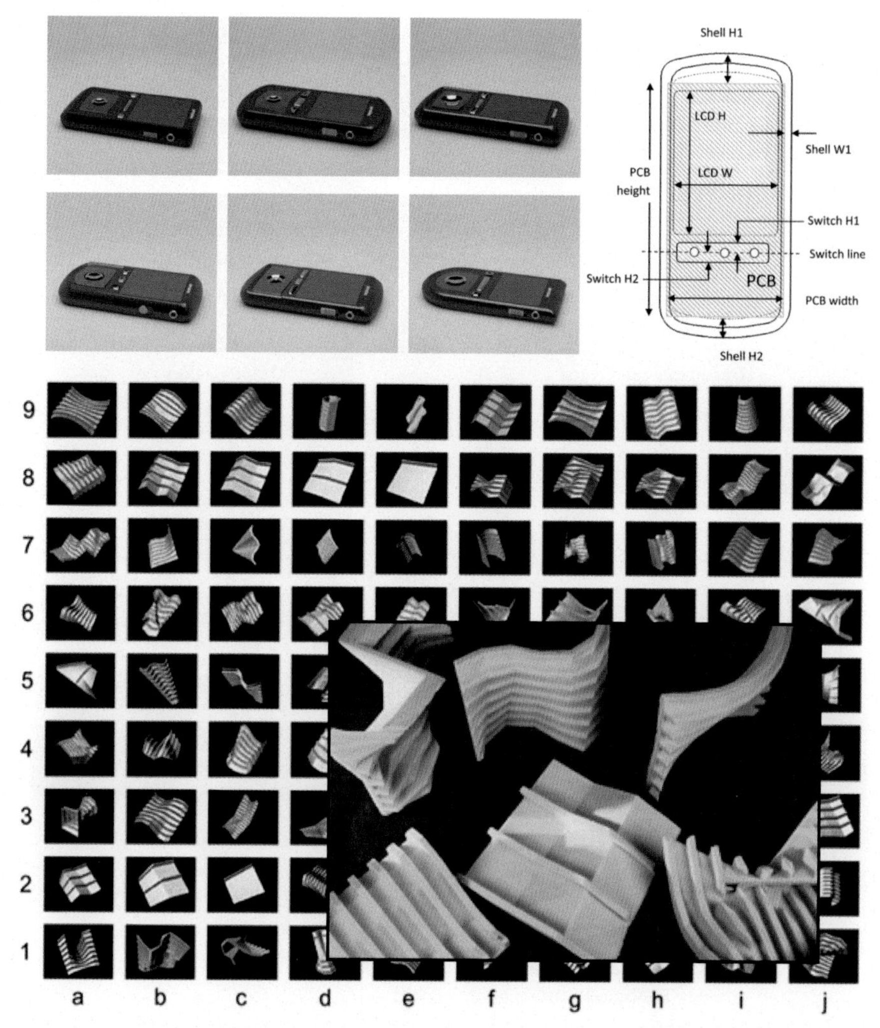

Fig. 7.4 Example of detail GD system implementation with manual optimization applied at the design phase. Upper: In this scenario the expert system generates potential solutions to a commercial product design requirement, these solutions are then manually evaluated for performance [4]. Lower: An expert system generates numerous potential solutions which were all then 3D printed for evaluation (inset) [5].

The specific design phase to which a particular generative design system is applied can be used to characterize the GD itself, as well as the associated opportunities and challenges for implementation (Fig. 7.5).

Embodiment GD systems

These GD systems are applied to the embodiment design phase in order to generate potential solution embodiments. In this scenario, the expert system is required to

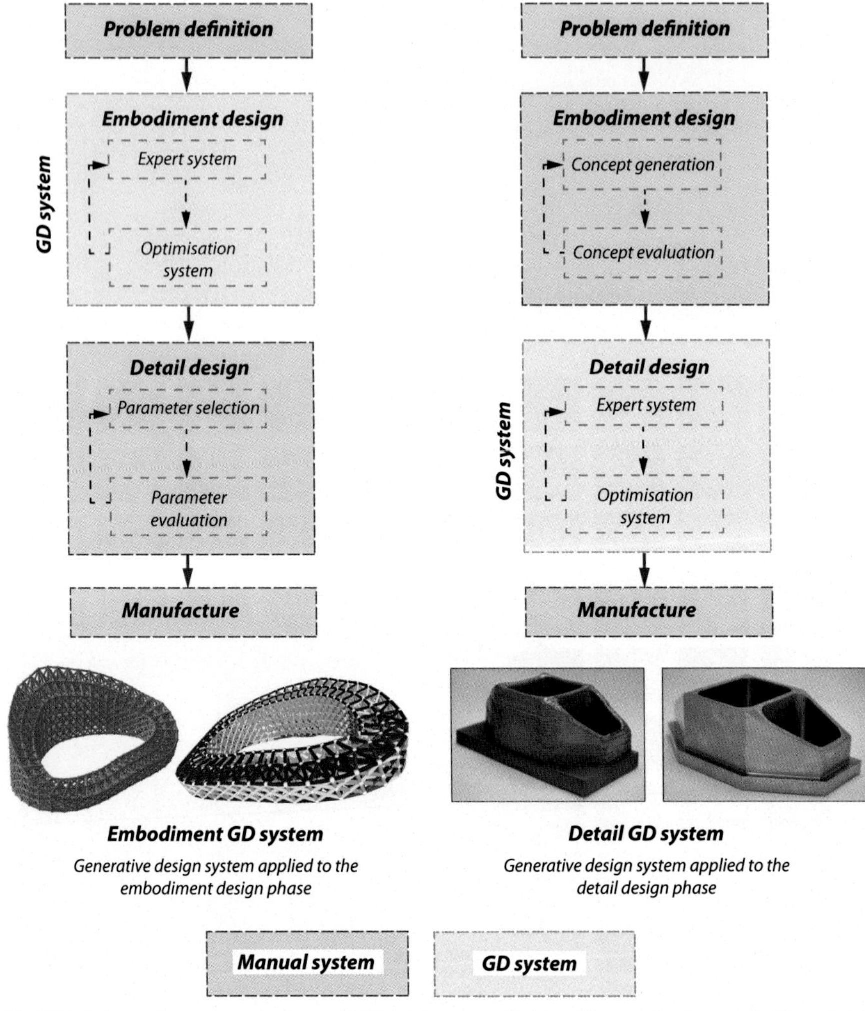

Fig. 7.5 Schematic representations of GD systems applied to either embodiment (left) or detail (right) design phases. Embodiment design is more abstract than detail design, and is therefore more challenging to implement as a GD system (Fig. 7.3).

generate potential solutions to the problem definition, an activity that necessitates a high degree of creative[1] design input. The generation of creative outcomes is technically challenging for autonomous systems, and the implementation and verification of embodiment GD systems may be difficult. Opportunities to imbue a creative response in embodiment GD systems include topological optimization methods, which can be applied to generate multiple high-performing optima (Chapter 6); cellular automata,

[1] For a formal treatment of the elusive definition of the term *creativity*, readers are directed to Ref. [3].

a method to generate complex responses from simple logical rules [6]; as well as genetic algorithms [7] and artificial intelligence systems [8].

Detail GD systems

These GD systems are applied to the detail design phase in order to refine attributes of a defined system embodiment. In these scenarios, the expert system can often be applied in a procedural fashion, and therefore is typically less technically challenging to implement and validate than for embodiment GD systems (Fig. 7.3). Detail GD systems provide a substantial opportunity for increasing DFAM design efficiency while enabling mass-customization and accommodating the requirements for design certification for high-value AM product.

7.1.3 Generative design system complexity

A general consensus in the definition of generative design is that it enables the algorithmic generation of complex outcomes. For this complexity to be useful, it must exceed that which could be reasonably managed by a human designer.[2] This complexity may exist in disperate tangible forms, for example [2]:

Complexity as size

Often described according to the *Kolmogorov complexity* of the *shortest* algorithm that can unambiguously define a specific data set. For the intent of practical DFAM activities it may be useful to define size-complexity in terms of the magnitude of industry standard AM data sets (Chapter 3).

Complexity as coupling

Defined as the degree of connectivity of the problem. This definition is particularly suited to *network* problems such as the design of lattice structures with sophisticated interconnectivity.

Complexity as solvability

May be defined as the number of operations that are required to identify a concept that acceptably satisfies the constraints associated with the problem.

These complexity definitions are provided as a basis for a logical framework to evaluate the intended complexity of implementing a proposed GD system prior to investing design effort. For example, the *solvability* complexity is high for embodiment GD systems as these systems generate unique solutions within a problem definition, and then provide a method for evaluating these systems. Conversely, detail GD systems generate solutions to relatively well-defined problems, but may have high *size*

[2] To be commercially robust, this complexity must be systematic, and implemented in such a way that the outcomes are demonstrably correct and repeatable.

or *coupling* complexity. The capacity for generative design to enable complexity can be utilized for the automated solution of nontrivial engineering problems, for example as is required for mass-customization design scenarios.

7.1.4 Generative design opportunities for mass-customization

The relative engineering economics of Additive Manufacturing (AM) and Traditional Manufacturing (TM) are presented in detail in Chapter 2. One particular aspect of relevance is associated with the intended production volume and associated variation, where *production variation* refers to the existence of distinct permutations of a particular product. These permutations may be associated with distinct material type, processing requirements or specific functional requirements. Traditional manufacture is inherently challenged by the manufacture of production variation, as these variants incur costs associated with custom tooling and custom manufacturing methods.[3] Conversely, as the production variation increases, the production philosophy tends to that of a unit batch-size with mass-customized design; potentially providing a competitive advantage for AM technology, even for nominally high production volumes (Fig. 7.6).

Substantial commercial opportunities exist for the successful implementation of a mass-customized design philosophy. Despite these opportunities, there exist significant technical challenges to be overcome. In particular, as production variation increases, so does the necessity for custom design input to accommodate these variants. This design input typically includes a manual component that incurs cost and lead-time and introduces opportunities for error. Generative design methods provide an enabling opportunity to accommodate the design effort necessary for a mass-customization philosophy without the infeasibly large design cost or opportunity for design error associated with manual design.

The development of robust generative design systems is nontrivial and incurs substantial development effort. This development effort is only commercially justified if it can enable a commensurate reduction in the design effort required to deploy a specific AM design output. This requirement places an economically feasible lower limit to the production volume necessary to commercially justify a particular GD system. As schematically represented in Fig. 7.6, it is apparent that the implementation costs of GD systems are lower when implemented in the design phase than for the embodiment phase. These techno-economic implications must be rationally assessed before engaging in the development of GD systems for a particular commercial AM application; however, significant commercial benefit exists for AM scenarios that are compatible with the implementation requirements for GD systems. A key challenge to the commercial implementation of GD for mass-customization is associated with the certification of these methods for high-value applications.

[3] In some scenarios, traditional manufacturing can be used with flexible manufacturing methods (such as customisable tooling) to be compatible with high variant design. These scenarios are not considered further here.

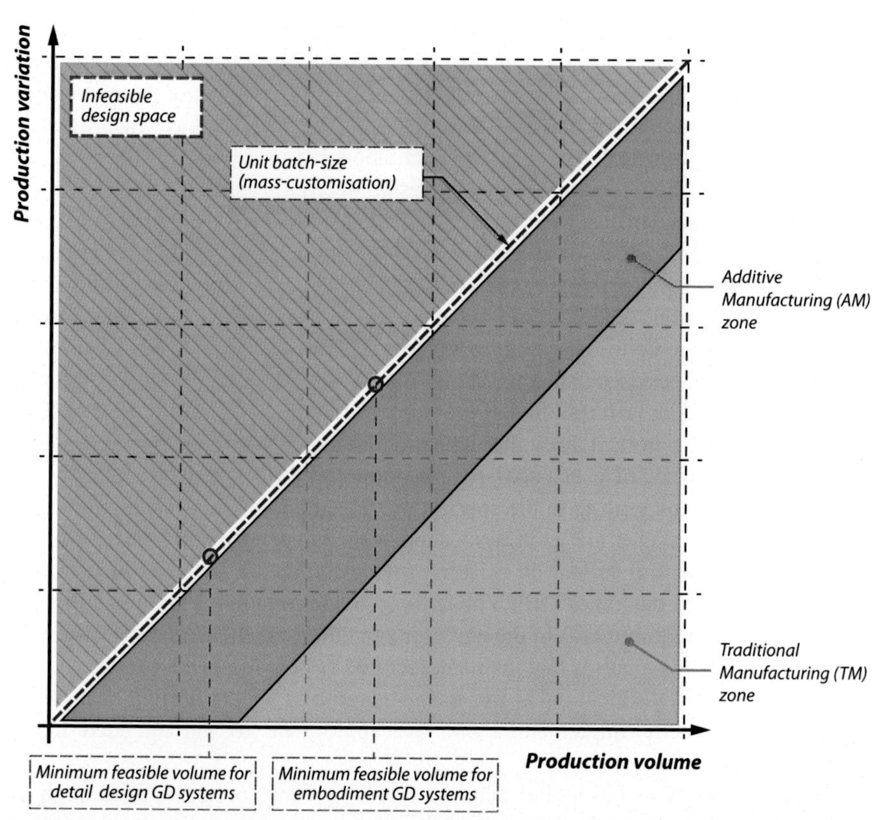

Fig. 7.6 A schematic representation of the optimal production zones for AM and TM with respect to variation in production volume and production variation. Red dashed line indicates a unit batch-size, implying a mass-customized design philosophy. With increasing production variation, the feasibility of AM technology increases. The minimum feasible cost for design GD systems is less than for embodiment GD systems due to their inherent implementation complexity (Fig. 7.3).

7.2 Certification of generative design systems

Generative design provides a significant commercial opportunity to achieve optimized technical results with minimal engineering effort. This outcome is compatible with the digital nature of AM and can enable mass-customization in an economically feasible manner. One particular challenge to this economic opportunity is associated with the certification of the GD methods applied to solve a specific AM design challenge.

Mass-customization by generative design is commercially viable for high-value production.[4] These products are, by their nature, frequently associated with applications that are commercially valuable or safety-critical (or sometimes both). For these scenarios, commercial application is predicated on a process of formal certification to ensure that the designed product meets the intended product definition. For manual design, this certification process includes the documentation of design decisions and reference to relevant codes and formal standards as well as to experimental testing of test-specimens and witness coupons[5] as well as non-destructive testing of manufactured product.

Various formal methodologies exist for the functional certification of manufactured products. The waterfall model is core to the certification of medical devices according to ISO 13485, which defines a quality management system for the design of 'medical devices and related services that consistently meet customer and applicable regulatory requirements' [9]. The waterfall model provides a structured framework for the robust design of high-value product throughout the product life-cycle (Fig. 7.7). The waterfall model includes phases of user needs specification and the associated design inputs, processes and outputs leading to the specific implementation of a manufactured medical device. These stages are each associated with independent review, and nested loops of *verification* and *validation* exist to provide confidence in the appropriateness of the manufactured outcomes [10]. Validation provides assurance that manufactured product meets the requirements of the end user and other stakeholders; for medical device manufacture this requires that the manufactured system meets the patient and surgical requirements. Verification is the technical process of confirming that design methods are correctly implemented and are appropriate to the intended design embodiment.

For clarity, the waterfall model diagram typically shows progression through the process in one direction only. In practice, flow can reverse depending on the outputs of the *review*, *verification* and *validation* stages, which exist to ensure that robust product design decisions are made. GD systems provide a robust mechanism for accommodating the verification phase by providing consistent and exhaustive documentation of decisions made by the GD system (Fig. 7.8).

A practical strategy for the certification of GD systems applied for safety-critical applications is the use of some *experimental validation* technique. Experimental testing adds cost to the AM implementation but can be applied to increase confidence in the validity of the manufactured product for the intended purpose. Examples of destructive and non-destructive experimental validation are presented in the case studies of Section 7.4. The application of experimental validation can be particularly useful when the practical verification of GD systems is stymied by the *curse of dimensionality*.

[4] The use of the term high-value in this context is flexible, and is used here to imply that the costs of establishing the GD system are warranted within the allowable production economics, as described in Chapter 2. The specific threshold associated with the definition of high-value will likely reduce as GD systems become more generally understood and available as DFAM tools.

[5] Witness coupons are fabricated with the production component but are (often destructively) tested to provide confidence in the performance of the production component or system.

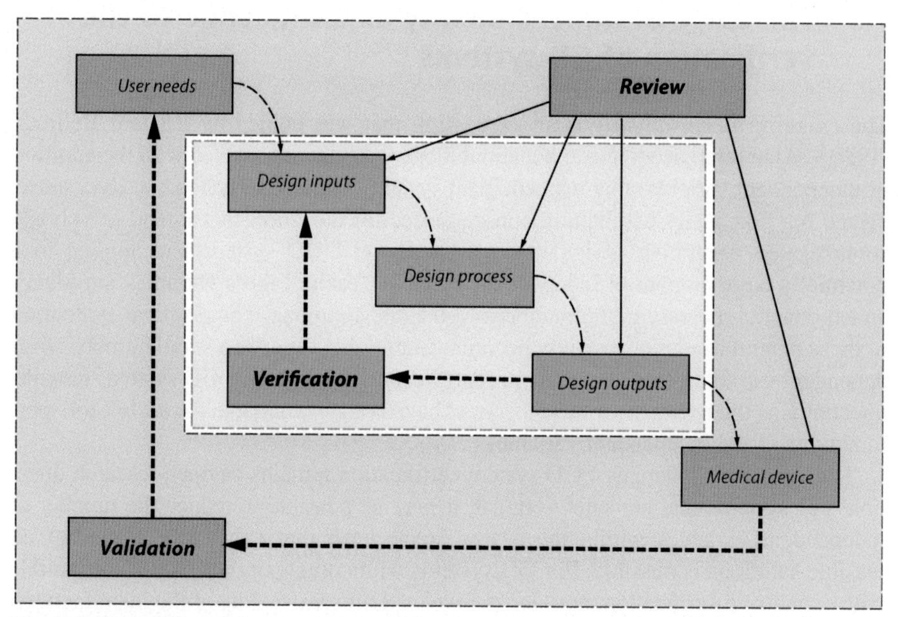

Fig. 7.7 Waterfall model of product design and deployment process including and validation loops, based on ISO 13485 [9]. Highlight indicates scope of embodiment GD (red) and detail GD systems (blue).

FACE CENTERED CIRCUMFERENTIAL									
StartRowsFCCirc2ndNode	StartPerimsFCCirc2ndNode	EndRowsFCCirc2ndNode	EndPerimsFCCirc2ndNode	FCCirc2ndNodeNth	FCCirc2ndNodeSkipInstance	FCCirc2ndNodeShift	ColourFCCirc2	DiaFCCirc2	COMMENTS
4	1	6	1	2	1	0		0.3	PERIM 1
8	1	4	1	2	1	0		0.3	
6	1	8	1	2	1	0		0.3	
8	1	6	1	2	1	0		0.3	
3	1	4	1	1	2	1		0.3	
4	1	3	1	1	2	0		0.3	
9	1	8	1	1	2	1		0.3	
8	1	9	1	1	2	0		0.3	PERIM 3
4	3	6	3	2	1	0		0.3	
6	3	4	3	2	1	0		0.3	
6	3	8	3	2	1	0		0.3	
8	3	6	3	2	1	0		0.3	
4	3	3	3	1	2	0		0.3	
8	3	9	3	1	2	0		0.3	

FACE CENTERED RADIAL						
StartRowsFCRad	StartPerimsFCRad	EndRowsFCRad	EndPerimsFCRad	ColourFCRad	DiaFCRad	COMMENTS
1	1	2	2		0.3	2MM CELLS ROW 1
1	2	2	3		0.3	DIAG DOWN TO RI
1	3	2	4		0.3	
1	4	2	5		0.3	
2	1	3	2		0.3	2MM CELLS ROW 2
2	2	3	3		0.3	DIAG DOWN TO RI
2	3	3	4		0.3	
9	1	10	2		0.3	2MM CELLS ROW 9
9	2	10	3		0.3	DIAG DOWN TO RI
9	3	10	4		0.3	
9	4	10	5		0.3	
10	1	11	2		0.3	2MM CELLS ROW 1
10	2	11	3		0.3	DIAG DOWN TO RI
10	3	11	4		0.3	

FACE CENTERED RADIAL						
StartRowsFCRad2ndNode	StartPerimsFCRad2ndNode	EndRowsFCRad2ndNode	EndPerimsFCRad2ndNode	ColourRad2	DiaFCRad2	COMMENTS
3	2	4	1		0.3	4MM ROW 3->5
3	2	4	3		0.3	
4	1	5	2		0.3	
4	3	5	2		0.3	
3	4	4	3		0.3	
5	2	6	1		0.3	4MM ROW 5->7
5	2	6	3		0.3	
6	1	7	2		0.3	
6	3	7	2		0.3	
7	2	8	1		0.3	4MM ROW 7->9
7	2	8	3		0.3	
8	1	9	2		0.3	
8	3	9	2		0.3	
8	3	9	4		0.3	

Fig. 7.8 The review, verification and validation elements of the waterfall model ensure that design decisions are robust and appropriate. Experimental testing (left) and in-situ documentation by GD systems (right) can be used to provide validation and verification, respectively. These examples are extracted from Case Study B (Section 7.4.2).

7.3 The curse of dimensionality as a challenge to the verification of GD systems

The *curse of dimensionality* is an expression that was coined by Richard Bellman (1920–84) in reference to the exponential increase in data associated with the addition of independent variables to a mathematical system. This phenomenon can have unexpected but practically debilitating consequences for designers that intend to validate sophisticated generative design systems. Non-trivial GD systems are applied to a potentially large number of independent variables, each of these variables introduces an exponential increase in the number of potential solutions. The practical evaluation of these potential solutions rapidly becomes impossible for even a small number of independent variables. Consequently, the certification of non-trivial GD systems remains uncertain, as the response of these is not exhaustively examinable for all feasible permutations of the independent variables.

The practical challenge of GD system certification remains an open research question. For systems that are safety-critical, it may be possible to reduce the number of independent variables within the design scope such that exhaustive search of all feasible solutions is possible. For GD systems with complexity that is not compatible with exhaustive verification, non-destructive experimental testing of the manufactured product provides an incontrovertible method to validate the intended design before use.

Generative methods enable solution permutations to be generated at computational speeds, potentially resulting in a vast number of proposed solutions. *Monte Carlo* methods have the potential to validate such a volume of potential solutions; these methods do not explicitly evaluate every solution permutation, but attempt to provide insight into the overall system response based on a sample of potential design inputs. This process is iterated until the results converge, thereby providing a statistically robust sample of the GD system outputs. The application of Monte Carlo methods for the certification of GD systems remains an open research opportunity.

7.4 Case studies of generative design

To demonstrate the challenges and opportunities associated with the generative design of AM products and systems, a number of practical applications of GD systems for AM design are presented in detail. These applications include GD systems that are applied at both the embodiment and detail design phases, as well as example implementations with both manual and automated optimization systems. The practical challenges and inherent commercial opportunities of these case studies are presented in detail, especially with reference to the research opportunities that are concurrent with the identified commercial challenges.

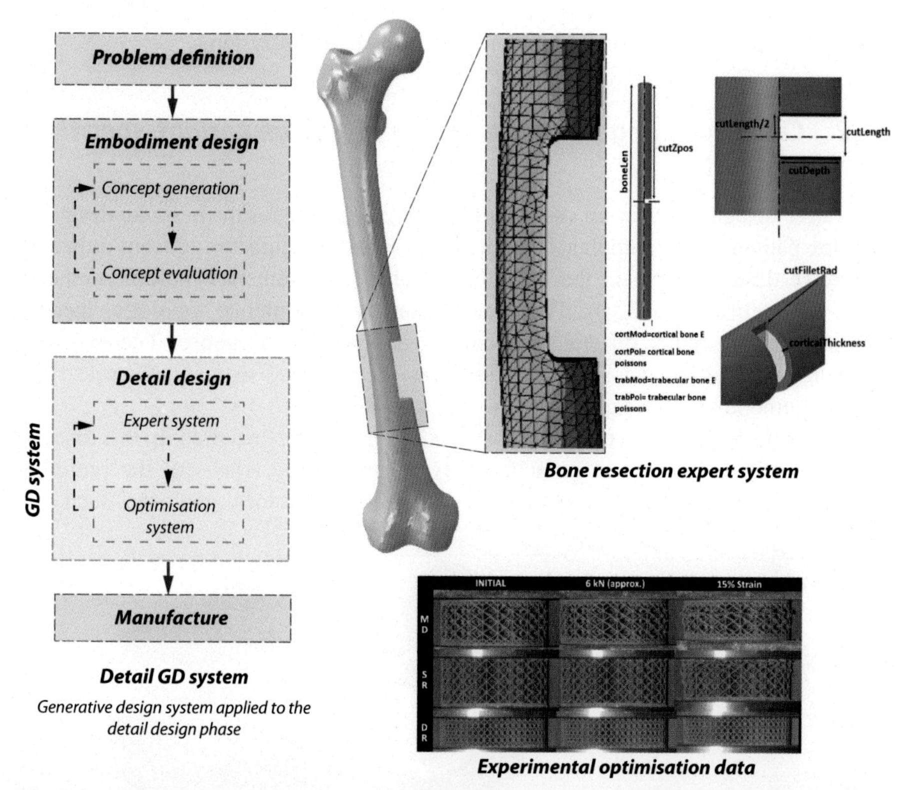

Fig. 7.9 Schematic representation of Case Study A in the context of the GD system workflow. The GD system (blue highlight) provides an enabling mechanism to generate robust patient-specific implants at the detail phase based on predetermined embodiment and a-priori experimental results.

7.4.1 Case study A: generative design of standardized patient-specific implants

Additive manufacture enables a new paradigm of medical implant manufacture; whereby the patient-specific mechanical and physical attributes are accommodated within a mass-customized patient-specific implant. To enable this patient specific design outcome in an economically feasible manner requires that automated generative design systems be implemented. Fig. 7.9 summarizes the generative design of standardized patient-specific implants, whereby the associated expert system represents the femoral bone with a numerical model, enabling the optimization system to select from a series of existing, experimentally validated implant classifications with mechanical and physical response equivalent to the parent bone.

To facilitate validation of this GD system, reference is made to *a priori* experimental validation of candidate implant designs. Although the commercial and clinical implementation of such a mass-customized implant manufacturing system remains largely within the development phase, it is the subject of commercial existing patent literature [11], indicating the commercial value of such GD systems.

7.4.2 Case study B: generative design of unique patient-specific implants

Generative design provides an enabling DFAM technology for the bespoke design of complex patient-specific implants. In this scenario, the manufacturing team utilized a generative design system at the embodiment phase to enumerate multiple embodiments of interest (Fig. 7.10). These embodiments have unique topology, thereby enabling fundamentally different mechanical and physical response. The expertise of the manufacturing and surgical teams was then utilized to manually select the preferred embodiment design.

During the detail design phase, a separate GD system was then used. In this case the expert system defines the manufacturability of geometric variants of the preferred embodiment design. The optimization system then selects local geometry such that

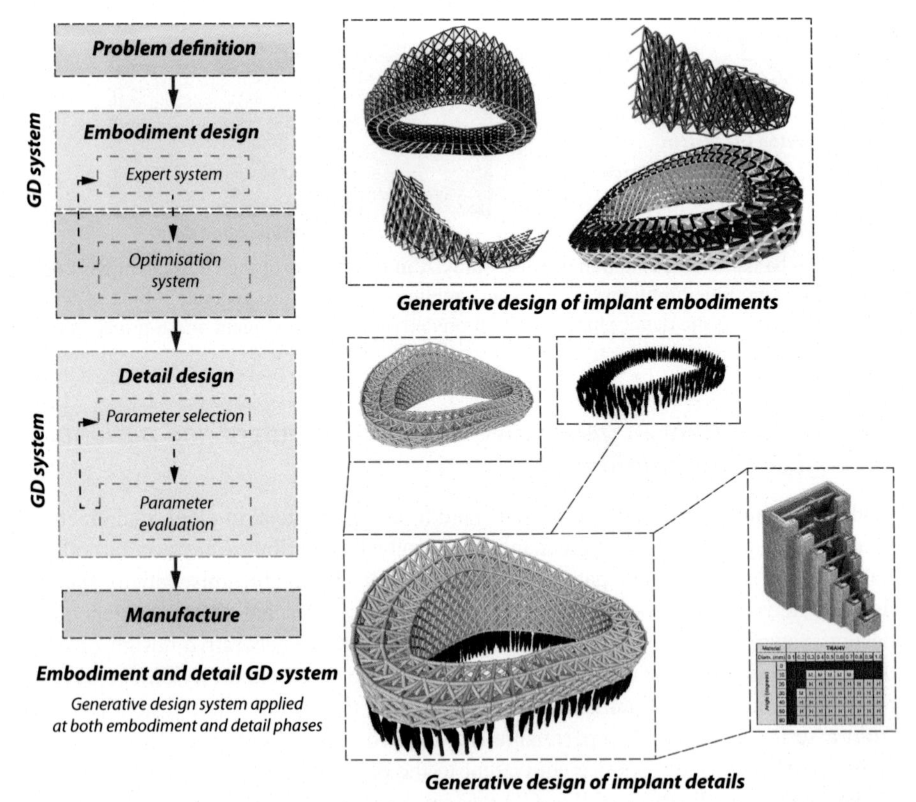

Fig. 7.10 Schematic representation of Case Study B in the context of the GD system workflow. The GD system (blue highlight) provides an enabling capability at both the embodiment and detail design phases.

AM manufacturability requirements are satisfied; thereby enabling a complexity of design that is not reasonably achievable by a human designer, while maintaining record-keeping as is required for medical device certification.

This case study provides insight into the capabilities of generative design to enable efficient design decisions to be made at both the embodiment in detail design levels. The outcomes of this specific case study resulted in surgically significant outcomes in a complex surgical scenario that was non-operable with standard medical implants [12].

7.4.3 Case study C: generative design of space frame systems using topology optimization

Generative design systems can be applied such that manual design effort is replaced by a generative system. These systems are particularly valuable for design scenarios that have low solvability-complexity but are recursively implemented, resulting in high *size-complexity* (Section 7.1.3). This implementation has many commercially relevant benefits including the minimization of design costs and robust documentation outcomes (independent of a single manual designer) as are required for the certification of high-value product. It is perhaps useful to note that many definitions of generative design do not include such parametric systems; however, for the purposes of DFAM, these systems are commercially significant and will be included within the scope of our GD system definition.

A commercially relevant application of such a GD system is implemented within the expert system for the design of space frame systems for structural applications (Fig. 7.11). In this scenario, the expert system generates variants of a particular design based on dimensional parameters including fillet and web dimensions. The outcome of this GD system is then manually optimized for manufacturability based on the expertise of Laser Metal Deposition (LMD, Chapter 12) subject matter experts. This optimization task currently exceeds the capability of existing DFAM tools. However, such GD optimization systems can potentially be implemented as the requisite DFAM tools become algorithmically available, and if a sufficient commercial market exists to support the cost of GD development.

7.5 Research opportunities in generative DFAM

The field of generative DFAM is commercially essential; in particular, generative DFAM provides a mechanism to algorithmically accommodate the design complexity associated with highly customized AM systems. This capability enables the generation of unique commercial systems with a complexity that exceeds the capability of human designers; as well as allowing the economically feasible manufacture of mass-customized commercial products that are inherently infeasible with traditional manufacturing methods. Furthermore, these high-complexity and mass-customized products are typically of high-value and include patient-specific medical implants and structural systems for aerospace and space applications.

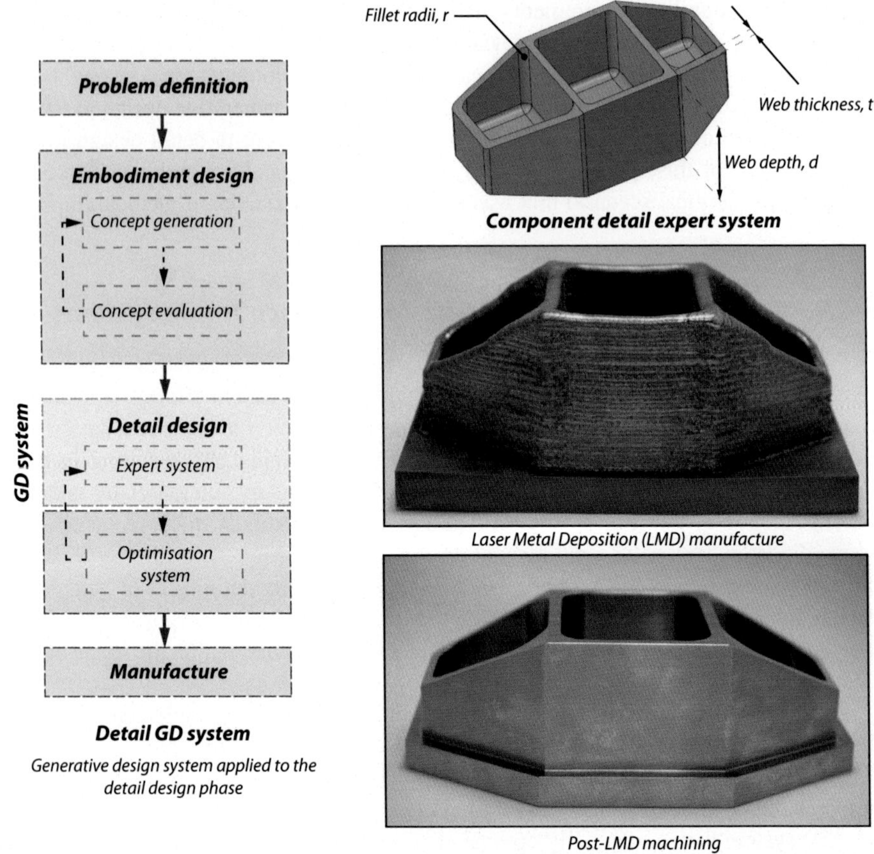

Fig. 7.11 Schematic representation of Case Study C in the context of GD system workflow. The GD system (blue highlight) provides a mechanism to rapidly embody variants of particular geometry. Manual optimization is currently used but could potentially be implemented as a GD system as DFAM technology and economic benefit evolve.

In response to these commercial opportunities and technical imperatives, significant opportunities exist for commercially impactful generative DFAM research contributions. This sophisticated and multidisciplinary research area is rapidly evolving, and specific statements regarding research novelty will quickly become superseded or irrelevant. The following research opportunities are forward-looking and general in nature to minimize the risk of being overly out-dated to future readers:

- The documentation of GD systems for safety-critical applications requires a systematic approach to accommodate the statistical variability inherent to AM structures. For example, statistical models of this variability can be integrated within GD system design utilizing Monte Carlo methods. This research is critical to the commercial deployment of mass-customized AM systems as it enables the inherent statistical variability of these structures to be robustly accommodated.

- Embodiment GD systems with algorithmic optimization remains a technically profound research opportunity. This research space requires that abstract potential design outcomes be generated and then evaluated in a meaningful way. This highly multidisciplinary research activity will benefit from current research findings in artificial intelligence and autonomous systems.
- Practical implementation of embodiment GD systems with automated optimization can be more readily achieved by the integration of both topological and parametric optimization methods. This strategy is briefly presented in Chapter 6 and is currently an underutilized DFAM research opportunity.
- An ongoing challenge for the practical implementation of generative systems is the accommodation of ill-defined constraints and objectives. For these scenarios, the associated optimization system can fail to make robust design decisions; this challenge of optimization in the presence of uncertainty remains an open research question.
- Conversely, implementations of GD systems that are robust to multimodal response[6] and can avoid becoming trapped in local optima are commercially necessary.
- The curse of dimensionality can rapidly overwhelm even well-formed GD methods. Simplified models and heuristics that can increase evaluation speed and intelligently restrict the design space, while still providing the opportunity to identify high performing local optima. These methods provide an important research opportunity for the practical implementation of GD systems; especially for those that require exhaustive validation of the GD system prior to commercial implementation.

7.6 Chapter summary

Generative design is an emerging DFAM philosophy that enables commercial AM outcomes for scenarios that would be otherwise technically or economically intractable. This chapter provides an overview of the challenges and opportunities associated with the implementation of generative DFAM systems as well as a roadmap for commercially relevant research opportunities.

The design lexicon associated with GD systems is in flux and has varying meanings among different design communities. The definition applied in this text encompasses all algorithmic DFAM contributions toward non-trivial commercial outcomes; namely that generative methods algorithmically propose feasible solutions to a design problem based on postulated rules and objectives. Furthermore, to be useful, these generative methods must enable design outcomes that are not feasibly achievable by a human designer – an outcome that is typically measured by the associated problem complexity (Section 7.1.3).

The incorporation of a generalizable definition of generative design methods within the framework of the formal design process enables various taxonomies of GD systems to be articulated. These taxonomies are associated with the design phase to which the

[6] A response is multimodal if it contains multiple local design optima. These local optima can be challenging for optimization systems to avoid.

GD system is applied, as well as the associated GD architecture, which may be implemented with a manual or automated optimization phase. Estimates of the technical risk to implementation are provided to assist in the planning of GD system architectures that maximize design benefit while managing the technical risk that is allowable for the specific design scenario (Section 7.1.2).

For example, by comparing these taxonomies to the waterfall model for medical device certification, it can be seen that GD systems applied at the embodiment phase have higher technical challenges than those applied to the detail design phase. Consequently, the commercial implementation of GD systems for the design of patient-specific implants is typically implemented at the detail design level (for example, Case Study A, Section 7.4.1). However, the design of patient-specific implants at the embodiment level can be clinically implemented, although at this stage with manual optimization of the preferred embodiment based on expert input from surgical and engineering teams (for example, Case Study B, Section 7.4.2).

The implementation of GD systems within the detail design phase can alleviate much manual design and documentation effort, even when combined with manual optimization (for example, Fig. 7.4, Case Study C). Furthermore, this outcome provides an economically viable waypoint in the implementation of fully autonomous detail GA systems.

In particular, generative systems provide an enabling capability for the manufacture of mass-customized systems. These scenarios are inherently incompatible with traditional manufacture (Fig. 7.6); however, they are potentially implementable with AM technologies if a robust digital workflow can be implemented. Despite the associated commercial opportunities of a mass-customized design philosophy there exist significant technical challenges, namely the associated manual design input and the associated opportunities for human-error. Generative design can provide a solution to both the challenges of design effort and design consistency. To aid in the identification of scenarios for which GD systems can enable mass-customized AM design, a schematic representation of optimal AM production zones is presented with reference to product volume and variation, as well as the economically feasible limits of GD system implementation.

Design consistency must be quantified for the certification of high-value, safety-critical applications that are the typical scope of mass-customized AM scenarios. These applications benefit from GD systems in enabling commercially economical design; however, the algorithmic nature of GD systems can also be utilized to provide certification documentation to provide confidence that design decisions are validated and verified according to agreed expert systems and certification protocols. The application of such a certification protocol is especially necessary when the complexity of generative DFAM solutions exceeds the capability of a human designer to characterize and assess.

References

Generative design philosophy

Overview of generative methods

[1] Frazer J. Creative design and the generative evolutionary paradigm. In: Creative evolutionary systems. Morgan Kaufmann; 2002. p. 253–74.

Considerations on the definition of complexity

[2] Adami C. What is complexity? BioEssays 2002;24(12):1085–94.

Overview of notions of creativity

[3] Cropley AJ. Creativity in education & learning: a guide for teachers and educators. Psychology Press; 2001.

Generative design applications

Application of generative design methods with manual optimisation

[4] Krish S. A practical generative design method. Computer-Aided Design 2011;43(1): 88–100.

Application of generative design methods with 3D printing as a proxy optimisation system

[5] Sass L, Oxman R. Materializing design: the implications of rapid prototyping in digital design. Design Studies 2006;27(3):325–55.

Study of cellular automata

[6] Wolfram S. Cellular automata as models of complexity. Nature 1984;311(5985):419.

Summary of genetic algorithms

[7] Gen M, Lin L. Genetic algorithms. Wiley Encyclopedia of Computer Science and Engineering; 2007. p. 1–15.

Discussions of artificial intelligence

[8] Russell SJ, Norvig P. Artificial intelligence: a modern approach. Malaysia: Pearson Education Limited; 2016.

International standard for the development of medical devices

[9] ISO 13485. Medical devices—quality management systems—requirements for regulatory purposes. 2016.

Summary reference for project management

[10] Edition F. IEEE guide—adoption of the Project Management Institute (PMI) standard a guide to the project management Body of Knowledge (PMBOK guide), ISBN 978-0-7381-6788-6.

Application case study for medical generative design

[11] Shidid D, Leary M, Brandt M, Choong P, RMIT University. A method for producing a customised orthopaedic implant. European patent office; 2019. EP31137251.

Application case study for medical implant manufacture

[12] Mobbs RJ, Coughlan M, Thompson R, Sutterlin CE, Phan K. The utility of 3D printing for surgical planning and patient-specific implant design for complex spinal pathologies: case report. Journal of Neurosurgery: Spine 2017;26(4):513—8.

Material extrusion

8

Material Extrusion (MEX) is an ASTM classification of AM technology that is formally defined as ‘an additive manufacturing process in which material is selectively dispensed through a nozzle or orifice’ [1]. MEX technologies include a range of distinct systems; however, in commercial and research sectors, these MEX systems are primarily associated with either the extrusion of thermoplastics or biologically active materials.

Thermoplastic extrusion systems (Section 8.2) are immensely popular due to their compatibililty with a range of polymer systems, and their scalability and robustness of implementation. Whilst these systems are deeply implemented in commercial, research and hobbyist environments, there remain significant opportunities for commercially relevant research contributions in this space, including the development of novel materials, DFAM tools and custom MEX systems.

MEX provides a unique opportunity for the development of AM systems that can process biologically active materials (Section 8.2). These systems include bioactive and cell-laden extrudate that enables the fabrication of high-value clinically functional materials. In response to this valuable opportunity, research and commercial outcomes in this MEX system space are rapidly expanding.

The intent of this chapter is to quantify the attributes of these MEX systems such that practicing engineers can more confidently apply these technologies to commercial projects with a workable understanding of the associated risks and opportunities; and to provide research engineers with a strategic roadmap of potential research contributions of relevance. Commercial MEX systems are almost exclusively associated with the in-situ extrusion of thermoplastic polymers or biological materials (Fig. 8.1). In order to enable a workable understanding of the capabilities of these systems, the underlying processing technology and materials science of these MEX systems is briefly introduced, followed by relevant commercial and clinical outcomes. The chapter concludes with a summary of the technology, DFAM tools and research opportunities associated with the MEX technology classification.

8.1 Thermoplastic filament extrusion systems

MEX technologies are predominantly implemented with a continuous thermoplastic filament input that is locally fluidized in some heating chamber and deposited by an extrusion nozzle as required to generate the required product. Such systems are known according to the ASTM standard designation as Fused Filament Fabrication (FFF), or by the commercial designation Fused Deposition Modelling (FDM). FFF system architecture has converged over time to a typically planar implementation, whereby

Design for Additive Manufacturing. https://doi.org/10.1016/B978-0-12-816721-2.00008-7

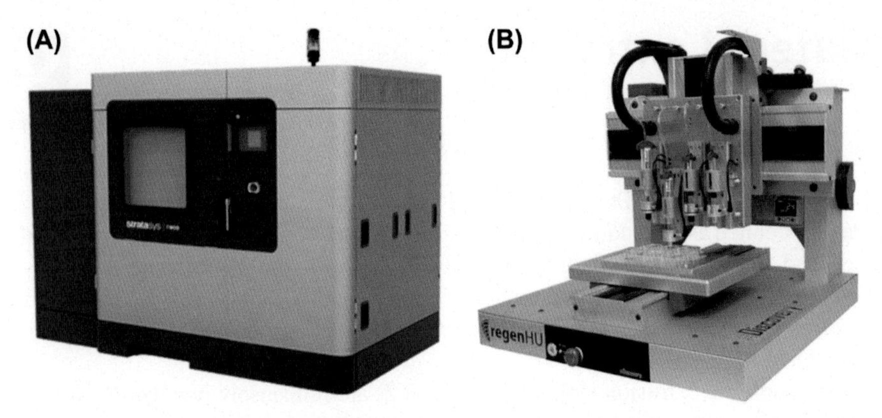

Fig. 8.1 Commercial implementation of MEX technology is predominantly associated with (A) thermoplastic filament extrusion systems (Stratasys, Inc., Fortus 900), and (B) biological polymer extrusion systems (BioFactory, regenHU Ltd).

the deposition head translates on a cartesian (planar) stage that then increments with discrete *layers* in the perpendicular Z-direction. This implementation results in technical challenges and DFAM requirements that are discussed in Chapter 3 and summarized in the following sections.

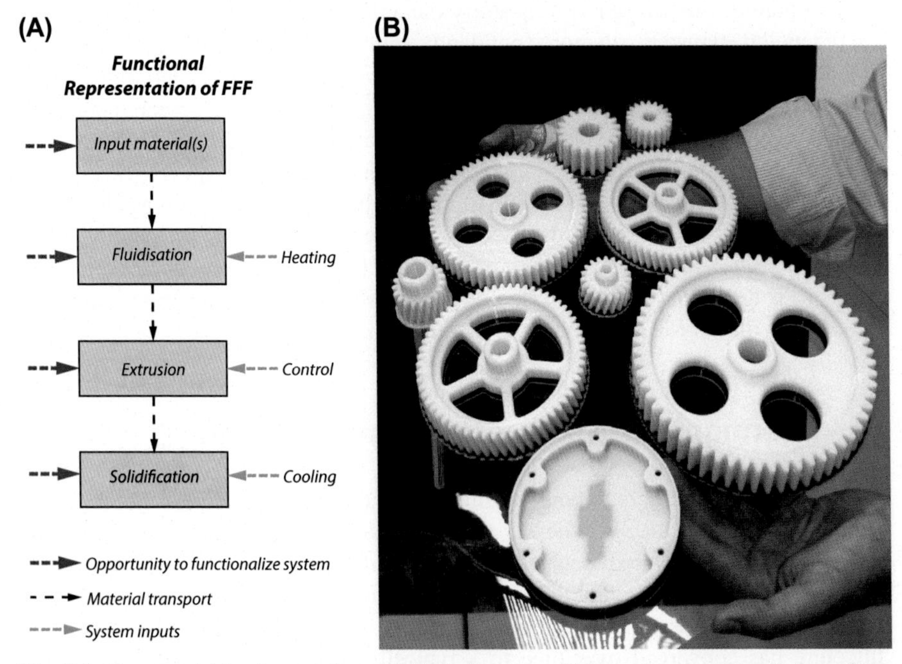

Fig. 8.2 Commercial implementation of FFF technology, (A) functional representation of the FFF process. (B) Typical example of FFF manufactured product.

FFF technologies provide a range of favourable techno-economic attributes. These opportunities include ease of implementation, leading to a low-cost barrier to application; potential for scalability, either with multiple heads or physically large build volumes; and compatibility with readily available and customized input materials, ranging from inexpensive and readily formed materials, to highly technical engineering polymers. Furthermore, FFF technology is highly compatibility with research opportunities in a broad range of fields, including fundamental material science, advanced manufacture, medical research, functional materials and generative DFAM tools.

In response to these positive attributes, FFF technologies are extremely well implemented within commercial, research and hobbyist sectors. As a complement to this technology review, Section 8.2 provides a series of best practice examples of FFF application to a range of commercial applications including the development of functional technical prototypes, the rapid prototyping of potential design solutions, and the manufacture of high-value, mass-customized product (Fig. 8.3).

In addition to these positive attributes, FFF technologies potentially suffer from several technical and economic challenges, in particular (Fig. 8.4):

- the potential for a relatively large surface roughness, especially for orientations that unfavourably align with the base platen
- challenges to the manufacture of unsupported overhangs and internal voids
- complexity in predicting toolpaths that generate defect-free macrostructure
- challenges in the non-destructive characterization of internal defects
- potential for interlaminar structural weaknesses

Despite these technical challenges, the commercial application of FFF technologies is so deeply established that these challenges should be viewed as commercial and research opportunities, rather than any fundamental barrier to the application of FFF technology.

This chapter presents an overview of the pertinent opportunities and potential challenges associated with FFF to assist these technologies to be confidently applied by practicing engineers to commercial applications, and to provide researchers with a strategic roadmap of potential research contributions. Commercial MEX systems are typically associated with the in-situ extrusion of thermally fluidized polymeric materials. In order to enable a workable understanding of the underlying capabilities of these systems, the science of polymeric materials is briefly introduced, especially where relevant to the industrial implementation of the extrusion process.

8.1.1 Polymeric materials

The fundamental polymeric building block is the *monomer*, a single molecule with an innate capacity to form chemical bonds. *Polymerization* generates a sequence of chemically bound monomers that form a *polymer chain* involving potentially thousands of monomer molecules. This *homopolymer* consists of identical units of the fundamental monomer; a *co-polymer* consists of a sequence of varying monomer units.

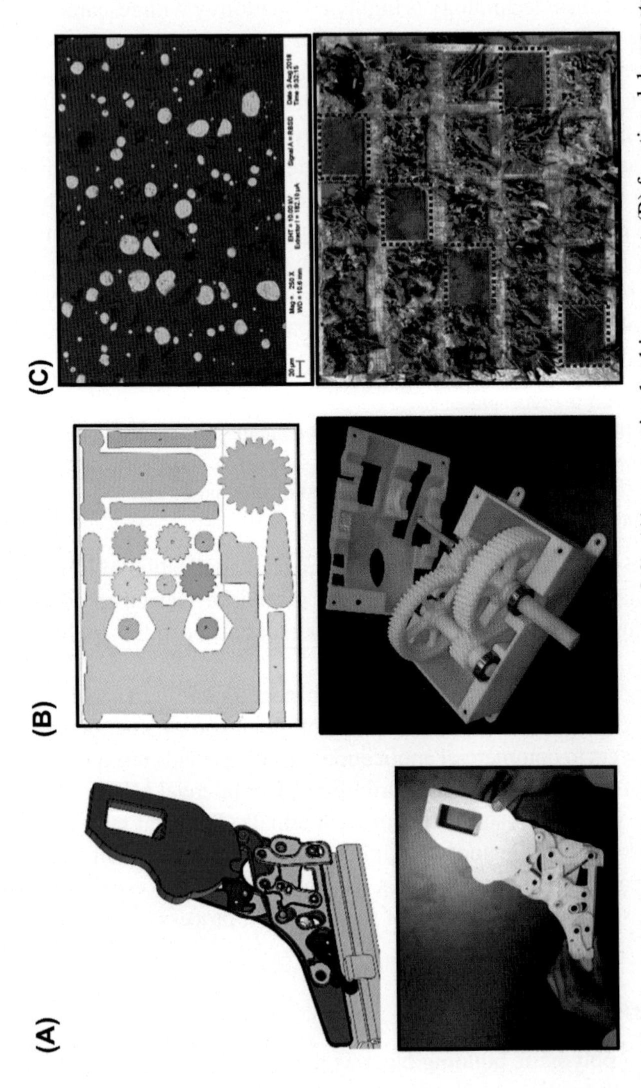

Fig. 8.3 Sample applications of FFF technology application (Section 8.2): (A) automotive latching concept, (B) functional demonstration of proposed commercial gearbox, (C) antibiofouling functional extrudate.

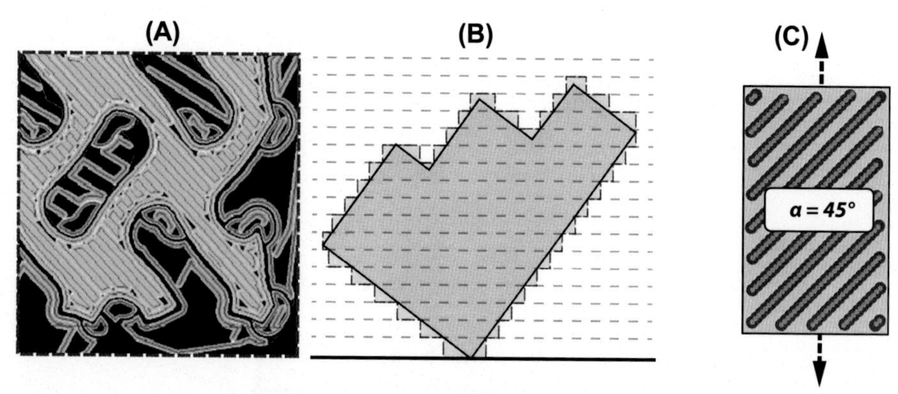

Fig. 8.4 Sample of potential challenges to robust FFF application: (A) porosity due to toolpath selection (see Fig. 8.12), (B) stair-step effects, leading to surface roughness and internal anisotropy (see Fig. 8.15), (C) alignment of filament deposition planes versus loading direction (see Fig. 8.14).

8.1.1.1 Polymer molecular architecture

These polymer chains can be characterized according to their associated *architecture*, referring to their particular molecular arrangement and shape. Relevant measures used to characterize the chain length, and the degree and type of branching include the degree of polymerization, molecular weight, branching type, and associated crystallinity. The relevance of polymer architecture for FFF is, in particular, its influence on as-manufactured mechanical properties and plastic flow during the extrusion process.

Molecular weight refers to the average length of the polymer chains within a particular polymer grade. In general, polymers with higher molecular weight tend to display superior mechanical properties. Conversely high molecular weight polymers increase their resistance to flow at elevated temperatures, thereby imposing a potential technical limit to their practical fabrication in FFF systems. The simplest polymer chain consists of linear combination of repeat units; however, other branching arrangements are also possible, including long, short and mixed combinations (Fig. 8.5). For a similar molecular weight, polymers with different branching arrangements have profoundly different flow characteristics. In general, linear chains tend to align during extrusion with relatively low flow resistance, whereas branched chains can become physically entangled, thereby increasing the resistance to flow.

8.1.1.2 Polymerization

Polymer chains interact to form a bulk polymer. The configuration of this interaction can be either amorphous or crystalline (Fig. 8.6). The *amorphous* (non-crystalline) configuration refers to a random arrangement of neighbouring polymer chains. For interest, all polymers are amorphous at sufficiently high temperature, or when in solution. Regular arrangement where repeat units within the polymer chain forms a *crystalline* structure are also possible. This crystallization can occur both within and

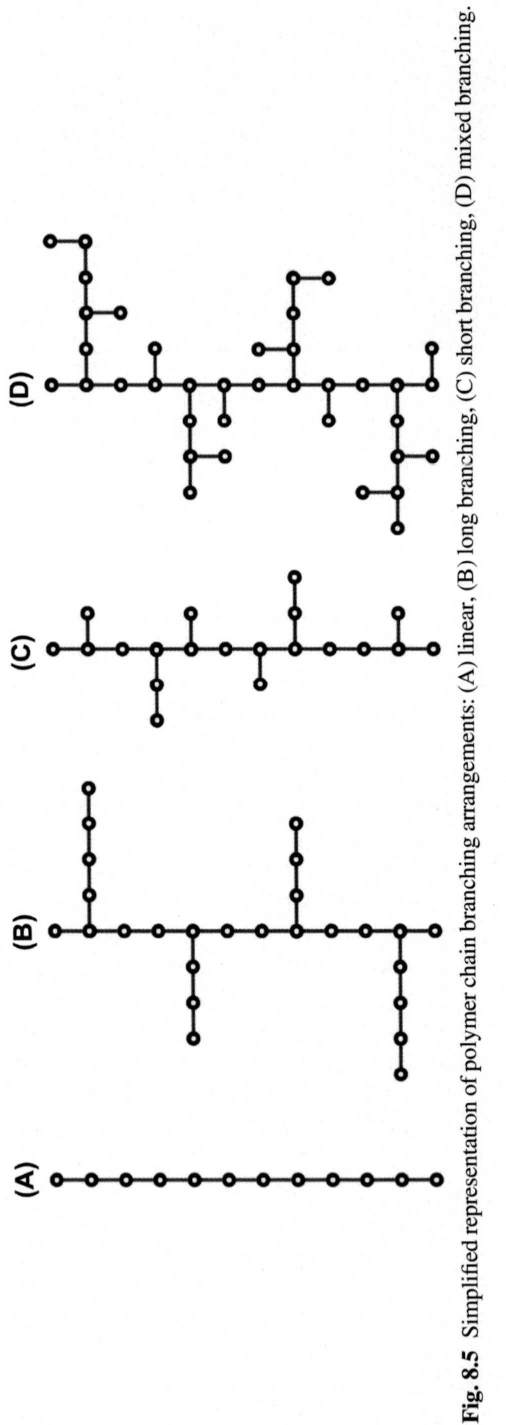

Fig. 8.5 Simplified representation of polymer chain branching arrangements: (A) linear, (B) long branching, (C) short branching, (D) mixed branching.

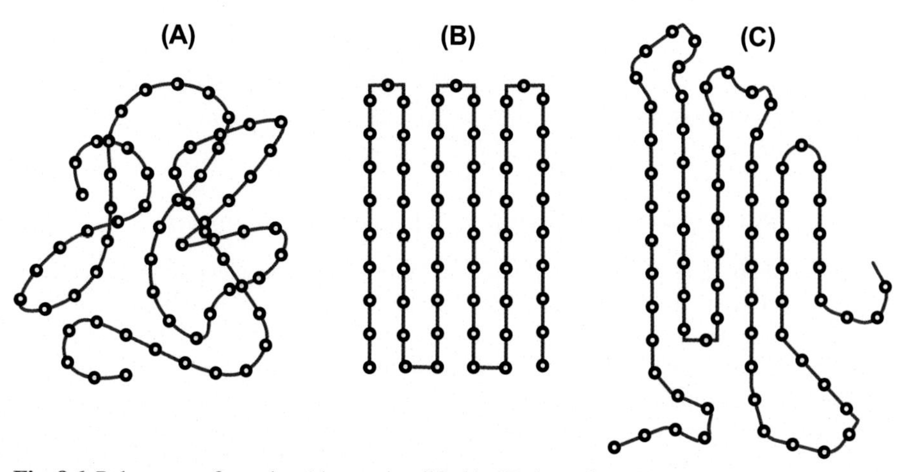

Fig. 8.6 Polymer configuration (shown simplified in 2D, in reality, polymers have 3D structure). (A) amorphous, (B) purely crystalline, (C) semi-crystalline.

across neighbouring polymer chains to imbue some degree of crystallization to the polymer.

The random arrangement of amorphous polymers imparts mechanical flexibility and elasticity. These properties are direction-independent due to this random chain arrangement; however, plastic deformation can align polymer chains, resulting in local behaviour that is more representative of a crystalline structure. The ordered arrangement of crystalline polymers imparts a relatively high strength and rigidity. This response can be direction dependant (anisotropic) depending on the predominant crystal alignment. In practice, these materials are *semi-crystalline*, and includes both regions of crystalline and amorphous arrangement, where the associated mechanical response depends on the relative size of amorphous and crystalline regions.

8.1.1.3 Bonding and polymer classification

Polymer bonding is potentially complex, enabling polymeric materials to manifest a range of physical responses. Monomer units are bound along the polymer chain with an *intra*molecular bond. This bond is of the covalent type meaning that a stable bond is made between shared pairs of electrons across neighbouring atoms. This *primary-bonding* provides the relatively high strength of the polymer backbone.

*Inter*molecular bonds *may* form both across and within these polymer chains. This *secondary-bonding* is relatively weak and provides low stiffness. Secondary-bonding diminishes in strength with increasing temperature, enabling a transition from connected solid state to a fluid state. This phenomenon defines the polymer classification of *thermoplastics* and is necessary for the thermal extrusion required of FFF polymers.

Polymers may also generate primary bonds between neighbouring polymer chains, referred to as network polymers or *thermosets*. These polymers irreversibly harden and do not allow the fluid transition with temperature required for FFF application, and are not considered further in this section.[1] This *cross-linking* occurs by some catalyst including heat, chemical mixing or irradiation, and enables the fabrication of custom biopolymers, discussed in Section 8.1; as well as allowing thermosetting AM processes, including Vat Polymerization (VPX, Chapter 10).

8.1.1.4 Plasticisers, fillers and structural reinforcement

Commercial polymers are typically modified from the fundamental polymer chemistry and architecture by the addition of other polymer chains, chemical or solid inclusions. These provide technical and economic benefits including reduced cost, increased strength, chemical stability and improved flowability.

The role of *plasticisers* is to reduce the stiffness of a polymer as well as (potentially) to reduce the overall material cost. Plasticizing molecules form secondary bonds, but do not cross-link with existing polymer chains; they provide a *lubricating* effect, resulting in reduced stiffness but increasing impact strength and elongation at failure. Plasticisers act similarly to a solvent and reduce flow viscosity and aid in thermomechanical processing and are therefore of significant importance to FFF manufacture.

Relative to other materials, polymers display a relatively low stiffness and strength. In an effort to enhance these properties, reinforcing fillers may be incorporated within the host polymer. These fillers are added as a discrete phase of particulates, or fibres of various lengths. Particle fillers may include various additives for cost or functional benefit; for example, hollow glass beads for reduced density or additives to provide an antimicrobial function. Fibre fillers are typically added to enhance mechanical properties and can include short and long fibres of various classes, including carbon, glass or Kevlar. These fibres provide an as-yet underutilized commercial opportunity for the manufacture of high-value product with exceptional mechanical properties utilizing the commercially robust and inexpensive FFF technology.

8.1.1.5 Polymer thermal response

The stiffness of thermoplastics varies with temperature; much of this observed change corresponds to the thermal influence on chemical bonding (Fig. 8.7). On heating, the mobility of amorphous thermoplastic polymers increases and the state changes reversibly from a relatively hard, *glassy*, state to a more viscous state at the glass-transition temperature, T_g. The response on further heating varies on the specific polymer, particularly with increasing molecular weight. For semi-crystalline thermoplastics, some variation in stiffness is observed at T_g, this is associated with the amorphous regions of the semi-crystalline structure. On further heating the stiffness rapidly, and reversible drops at the melting temperature, T_m, associated with the freeing of secondary bonds. At temperatures above T_m, the polymer acts as a fluid and can be reversibly formed by thermo-mechanical processes such as extrusion.

[1] The development of MEX systems compatible with thermoset polymers presents a nascent commercial opportunity.

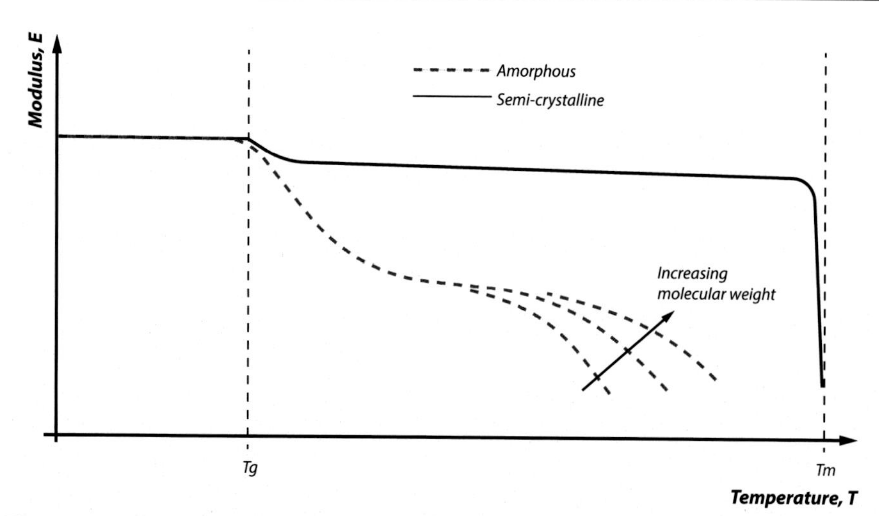

Fig. 8.7 Schematic representation of thermomechanical response of thermoplastic polymers in amorphous and crystalline state. Effect of increasing molecular weight on the modulus of amorphous polymers shown.

Table 8.1 A selection of polymers of relevance to FFF technology.

Material	T_g (°C)	T_m (°C)	*Density (g/cm³)*	*Classification*
ABS	105	NA	0.9–1.5	Amorphous
PEEK	145	335	1.3	Semi-crystalline
PC	145	NA	1.2	Amorphous
PLA	60–65	NA	1.2–1.4	Amorphous
		150–160		Semi-crystalline
PLGA	45–55	NA	1.3	Amorphous
		145–155		Semi-crystalline

Materials are available in a range of grades and physical properties are approximate values.

As discussed, the range of polymers compatible with FFF technologies is extensive: the following grades are of particular relevance to contemporary applications (Table 8.1).

Acrylonitrile-Butadiene-Styrene (ABS)

Acrylonitrile-butadiene-styrene (ABS) is an amorphous copolymer with excellent mechanical properties. The basic function of the ABS repeat units are: acrylonitrile, which provides chemical and temperature resistance; butadiene, which provides impact resistance; and styrene, which reduces cost and enhances processability. Typical proportions of the repeat units vary according to the specific polymer grade but are in the range of acrylonitrile (20–30%), butadiene (20–30%), and stryrene (40–60%).

PolyCarbonate (PC)

PC provides robust mechanical properties for engineering applications, including particularly high toughness. This amorphous polymer has high dimensional stability, can be transparent, has low flammability, and is nontoxic with high biocompatibility – a necessary condition for use in food and medical applications.

PolyLactic Acid (PLA)

PLA is a biodegradable polymer derived from biomass such as corn starch in the form of lactic acid. Many variants exist and architecture varies from amorphous to highly crystalline. PLA biodegrades rapidly (in comparison with hydrocarbon-based polymers) and can be used for the manufacture of resorbable medical implants and devices. PLA has a low glass transition temperature and is therefore technically useful for inexpensive FFF systems.

PolyEtherEtherKetone (PEEK)

PEEK is a semicrystalline polymer with excellent mechanical properties at elevated temperatures, an advantage for many technical applications. However, the elevated processing temperatures for this polymer make manufacture technically challenging; consequently, the associated research opportunities are high in comparison to the more readily processable polymers.

PolyLacticCoGlycolicAcid (PLGA)

A co-polymer of lactic acid and glycolic acid that may exhibit architecture that varies from amorphous to crystalline, PLGA is used in FDA approved devices due to its associated biocompatibility. Furthermore, PLGA degrades by hydrolysis at a rate that can be influenced by the specific polymer chemistry, and the associated by-products are then metabolized through normal bodily functions.

8.1.1.6 Polymer extrusion

The extrusion process is core to the FFF technology and relates to the continuous shaping of a fluidized polymer through an appropriately shaped tool, followed by the extrudate solidification. FFF polymers are typically extruded twice: initially the polymer is extruded in an industrial processing facility to generate the commercial FFF filament, subsequently the filament is extruded through the FFF system to generate the manufactured AM product.[2] The fundamental extrusion process is technically similar for both stages, however the hardware complexity and sophistication

[2] Although less common than filament extrusion FFF systems, commercial FFF systems increasingly utilize injection moulding pellets as the system input material [19]. These systems may be more expensive to setup initially, but do provide commercially attractive opportunities, including the potential for in-situ mixing of input materials and reduced variable costs.

between the industrial filament extrusion process and the in-situ extrusion during FFF processing are radically different. These differences introduce technical challenges that the designer must accommodate as well as DFAM research opportunities for FFF system developers.

Pellet extrusion typically consists of a motor-controlled screw system that transports pellets from a hopper and compresses these through a heated orifice. Temperature feedback control with multiple heating elements is used to precisely control the temperature profile through the system. The screw system is typically categorized by three zones: the transport zone, where solid pellets are loaded in the system; the transition zone, where mechanical working and heating fluidize the polymer; and the melt-conveying zone, where the fluid polymer is driven to the extrusion die. The extrusion die imposes the intended final geometry on the extruded material. An important challenge for extrusion systems is the phenomenon of *die swell*, which refers to the increase in cross-section area observed when the extrudate leaves the restriction of the die. For a given polymer choice, this effect is a function of flow resistance, which in turn is a function of melt temperature. Typical design measures used to stabilize the geometry of the extruded material include the manipulation of the extrusion-die geometry and associated material, as well as active cooling by forced air convection or water quenching. The specific layout of industrial extrusion technology exists in many variants; for example, a venting zone may exist to allow the exhaust of entrapped gasses; multiple counterrotating screws can be used to achieve more thorough mixing and to accommodate shear-sensitive materials; and dimensional feedback can be provided to any of the control inputs to increase dimensional stability and material yield.

The sophistication of commercial extrusion systems is presented here to illustrate the technical disparity between the relatively simple FFF extrusion systems and the highly complex industrial extrusion systems. This disparity provides insight into the complexity of achieving robust polymer extrusion with in-situ FFF extrusion systems and illustrates the opportunities that exist for their enhanced design.

The following section presents the evolution of FFF technology and the typical architecture of FFF systems.

8.1.2 Fused filament fabrication

Fused filament fabrication is a commercially significant AM technology due to its extensive range of beneficial attributes, including the ability to inexpensively fabricate large and geometrically robust geometry; the compatibility with various polymers, ranging from inexpensive and readily fluidized grades to sophisticated high-technology compositions; and the flexibility of system architecture, which allows innovation such as functionally graded materials, chemically active additives, and mechanical strengthening agents. The history of this technology is founded on a sequence of commercial and open-source innovation by two highly innovative engineers, Scott Crump and Adrian Bowyer.

8.1.2.1 Brief history of FFF technology

The FFF technology was invented by Scott Crump who, according to legend, conceived the notion of a novel additive technology while building a toy for his daughter using a hot glue gun. This concept was patented in the early 1990s as an apparatus incorporating a movable dispensing head provided with a supply of material which solidifies at a predetermined temperature [2]. Crump then went on to develop *Stratasys Ltd.* which has since become a globally significant AM technology development company.

Adrian Bowyer from the University of Bath developed a notion that the expiring patents from the original FFF technology could be used to develop an open-source AM system that would be readily manufacturable, and in particular, would be capable of self-replication (Fig. 8.8). This *RepRap* system inspired a movement of makers that has evolved the fabrication of highly functional, inexpensive FFF systems that are affordable for hobbyist use and provide a highly flexible basis for AM research systems, as well as being technically effective for commercial production.

8.1.2.2 FFF system architecture

FFF technology is unique within the AM technology space in that it has benefitted from significant development input at commercial, research and hobbyist level over a number of decades. The core technology developed in Crump's initial patent remains within the design of a modern FFF system. This system architecture (Fig. 8.9) schematically includes a source material (typically filament, but pelletized input is also possible) that is transported in its solid state by some feeder system; a fluidization chamber that allows heating of the source material to a fluid state; an extrusion die, often referred to as a *nozzle*, that imposes some geometric form on the extrudate; and some solidification method, typically fan forced or convective cooling to return the extrudate to the solid state. This material delivery system is then superimposed on some control system and physical architecture that allows the deposition of extrudate to fabricate a 3D structure.

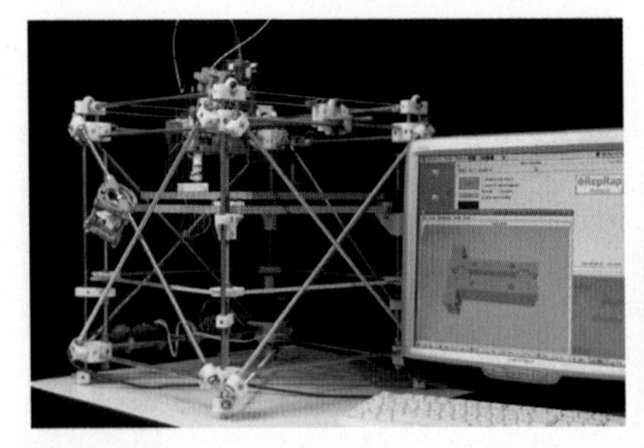

Fig. 8.8 First generation 'Darwin' RepRap Machine with custom parts manufactured with a commercial Stratasys Ltd. FFF system (left). Second generation 'Mendel' RepRap Machine with RepRap manufactured componentry (right) [3].

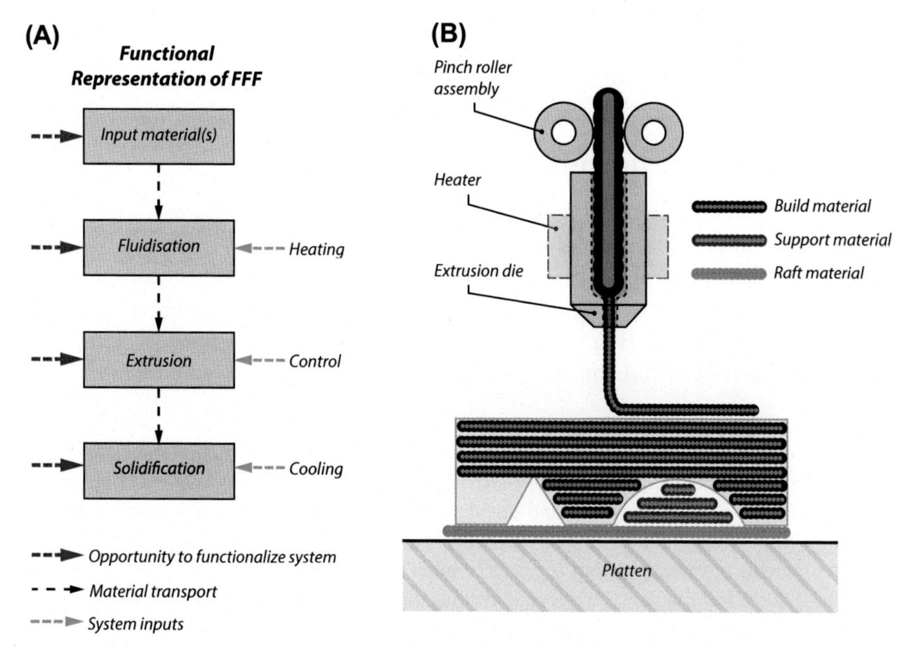

Fig. 8.9 FFF system architecture. (A) Schematic representation of system architecture. (B) Typical FFF implementation.

FFF systems have been subject to decades of intense research and development effort in both commercial and research sectors. These activities have generated a vast range of commercial FFF implementations; however, the most common is associated with linear actuators on a Cartesian (XYZ) positional control system. This implementation typically controls X and Y position by physically translating the nozzle assembly, with the Z-axis positional control being achieved by the independent translation of the support platen.

8.1.3 FFF technology challenges and associated DFAM tools

The FFF technology is arguably the most commercially successful and pervasive of the available AM technologies. Despite this success, the commercial application of FFF technologies potentially suffers from a series of technical and economic challenges that must be addressed by the design team to eliminate or mitigate the associated negative effects. These challenges, associated DFAM tools and commercial best practice, are addressed in detail below, and include:

- complexity in achieving *robust extrudate deposition*
- complexity in generating *robust toolpaths* that avoid potential defects
- difficulties in *non-destructive characterization* of internal defects
- potential for *structural anisotropy* and weaknesses in the Z-plane

- relatively large *surface roughness*, especially for orientations that unfavourably align with the Z-direction layers
- challenges to the manufacture of *unsupported overhangs* and internal voids.

8.1.3.1 Extrudate deposition

The technical complexity and sophistication of commercial extrusion systems is in stark contrast to the austerity and simplicity of typical in-situ FFF extrusion hardware and associated control systems. FFF extrusion systems are based on an integrated fluidization region and heating system and extrudate flow rate is controlled by the external feeding system (Fig. 8.9). As a consequence of the relative simplicity of typical in-situ FFF extrusion systems, technical challenges arise, including the following.

- The relatively simple thermal control systems (typically a single heating unit directly coupled to a convectively cooled zone) for FFF extrusion systems generate challenges for the precise control of the fluid temperature and therefore the flow resistance and associated extrusion rate [4].
- The extrudate positional control is typically implemented with a pinch roller system external to the fluidization chamber; however, in contrast, the fluid flow in industrial extrusion is more precisely controlled by the screw system within the fluidization chamber.
- To further exacerbate this challenge, the FFF extrusion process is subject to stop-start extrusion. Conversely, industrial extrusion processes are allowed to stabilize at constant flow conditions over an extended time period.
- The opportunities for post extrusion cooling are relatively limited in FFF extrusion, as the extruded materials are directly integrated within the manufactured part, rather than being a free filament.

As a consequence of these relative challenges, the implementation of dimensionally robust extrusion systems within FFF systems is a significant technical challenge, especially within inexpensive systems (Fig. 8.10).

8.1.3.2 Toolpath generation

The toolpath geometry is scheduled by some post-processing of the intended CAD geometry (Chapter 3). DFAM tools assist in defining a toolpath and associated process parameters that attempt to define a continuous fill of the desired cross-section; however, this objective is technically challenging (Chapter 4). The typical implementation of toolpath programming algorithms is a perimeter-and-raster[3] strategy, such that a number of self-intersecting continuous contours are defined for the slice perimeter. This remaining area is then filled by some standard pattern, typically a *raster* of of parallel offset toolpaths (Fig. 8.11). The perimeter-and-raster strategy accommodates a range of potential slice geometry and generates continuous perimeter filaments that ensure that the visible surface has relatively low roughness; however, this strategy potentially

[3] The *perimeter-and-raster* terminology is synonymous with *skin-and-core*, the latter being prevalent in thermal metal systems, for example Chapter 11.

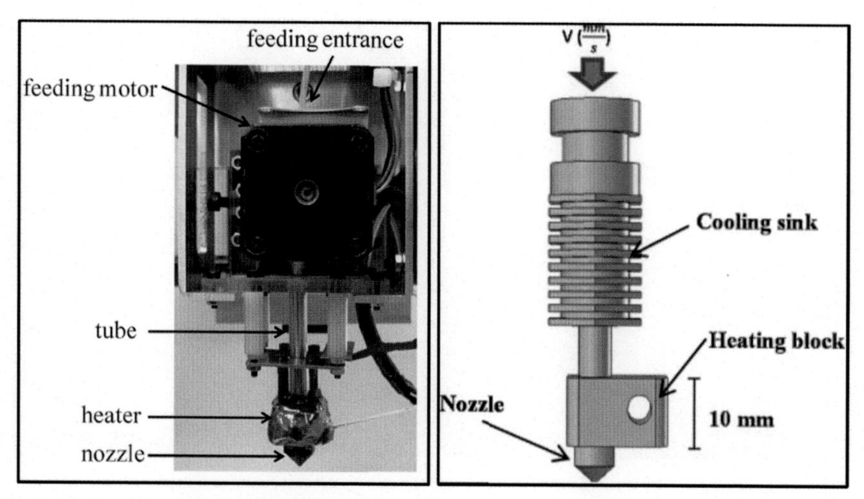

Fig. 8.10 Representation of FFF extrusion head demonstrating simplified design in comparison with more sophisticated industrial extrusion processes. Left: functional representation [5]. Right: schematic implementation [6].

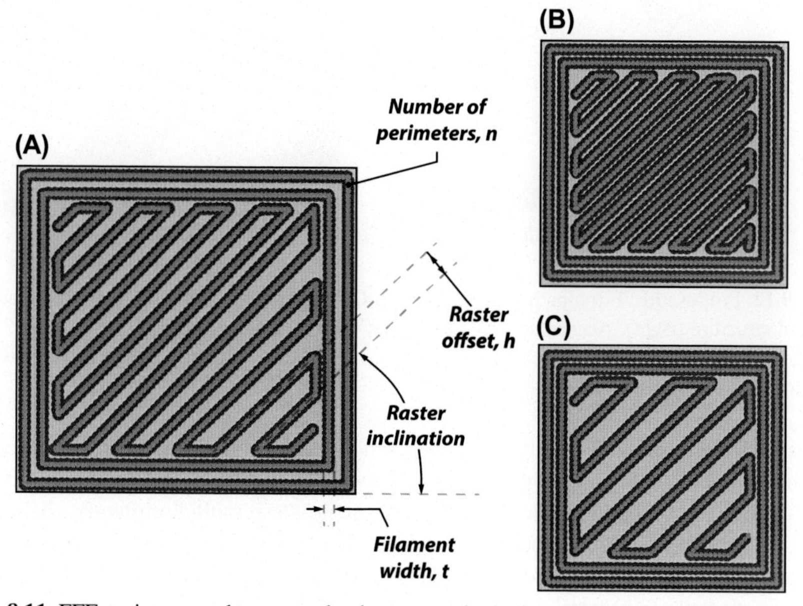

Fig. 8.11 FFF perimeter-and-raster toolpath strategy, including (A) feature definitions, including (B) dense, and (C) sparse infill settings.

introduces local porosity defects, especially for curvilinear geometry, in thin sections and at perimeter/raster intersections (Fig. 8.12).

The DFAM challenges for defect-free FFF toolpath generation are multidisciplinary and involve the mathematics of space-filling; an experimentally valid understanding of

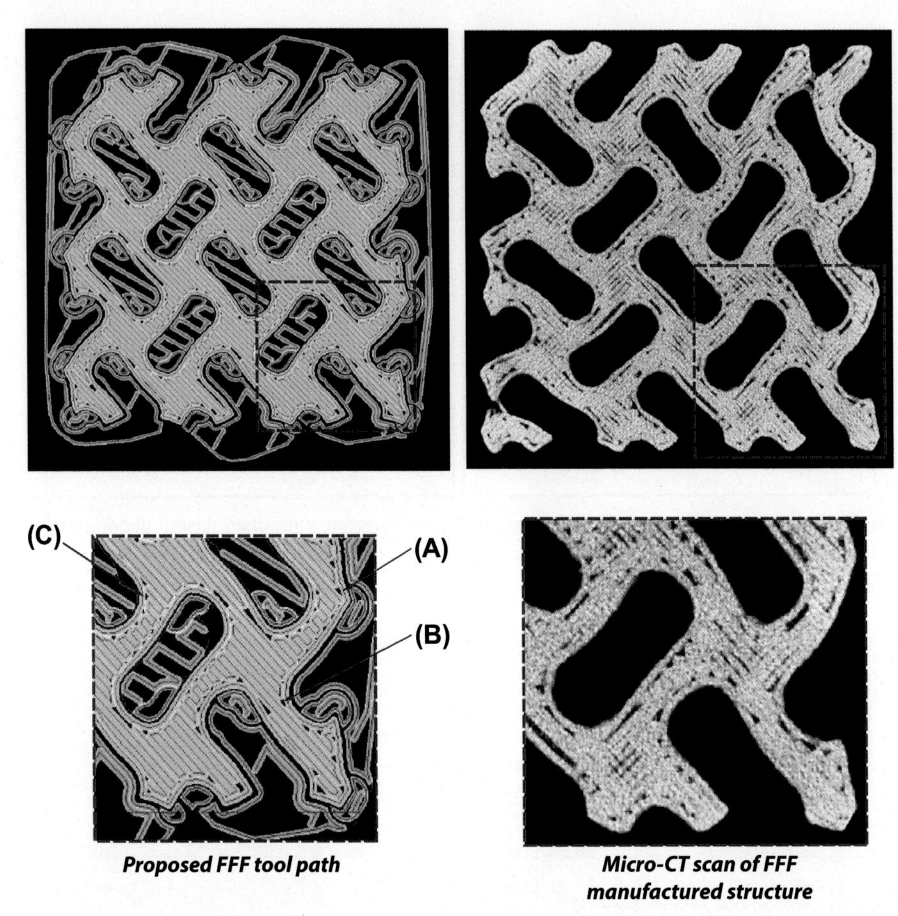

Fig. 8.12 Proposed FFF toolpath (left) and micro-CT reconstruction of FFF manufactured gyroid structure (right). Accommodating toolpaths without unintended porosity may be challenging, especially for: (A) curvilinear geometry, (B) thin sections and, (C) at the intersection between perimeter and raster.

the physical morphology of filament deposition; and a multifunctional optimization methodology that accommodates minimization and structural loading requirements. DFAM tools are emerging that attempt to reconcile these multidisciplinary challenges (Fig. 8.13). However, this opportunity remains an open field for DFAM research, especially for multidisciplinary contributions, including the mathematics of topology, as well as sophisticated representations of as-manufactured extrudate geometry (and polymeric interactions) and the associated DFAM tools required to support this research.

8.1.3.3 Characterization of internal defects

FFF systems are increasingly being utilized for high-value and load-bearing structural applications. These scenarios cannot be commercialized without robust methods for

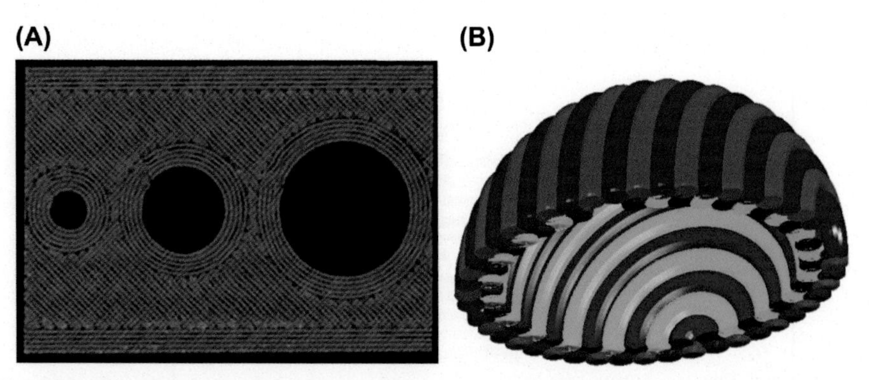

Fig. 8.13 DFAM toolpath optimization. (A) Functionally sensitive filament deployment, in this case including multiple internal filament perimeters to allow tapping of a boss feature. (B) Curved path FFF deposition [7].

understanding and characterizing these internal defects. DFAM responses to these challenges include the fabrication of calibration specimens that allow measurement between the intended toolpath and the non-destructively characterized specimen (as in Fig. 8.12); and the destructive testing of *witness coupons* that intend to replicate the structural features of the as-manufactured product.

8.1.3.4 Structural anisotropy and weaknesses in the Z-plane

The structural application of FFF componentry requires that mechanical response be characterized and predictable. The layerwise fabrication typical of FFF imbues a filament structure that is locally variable within layers parallel to the build platen and potentially introduces interlaminar discontinuities. These characteristics potentially compromise load carrying capability and result in an anisotropic mechanical response (Fig. 8.14).

8.1.3.5 Surface roughness

Surface roughness may be reported by various methods [8,9], and includes qualitative methods that graphically describe the manufactured variability, as well as quantitative methods that characterize either the linear, R, or areal, S, roughness as a single value, typically as either the average deviation from some reference datum, R_a, S_a, or the maximum observed variation between sampled points, R_z, S_z.

FFF technologies must be robustly implemented to avoid undue surface roughness, which can compromise technical or aesthetic function. FFF systems are subject to potentially high surface roughness due to three interdependent effects.

1. **Discrete nature of the deposited filament**

The deposited FFF material is typically simplified in representation as having an elliptical cross-section. This deposition geometry results in local variation in the intended manufactured geometry, whereby the surface of the manufactured geometry exhibits

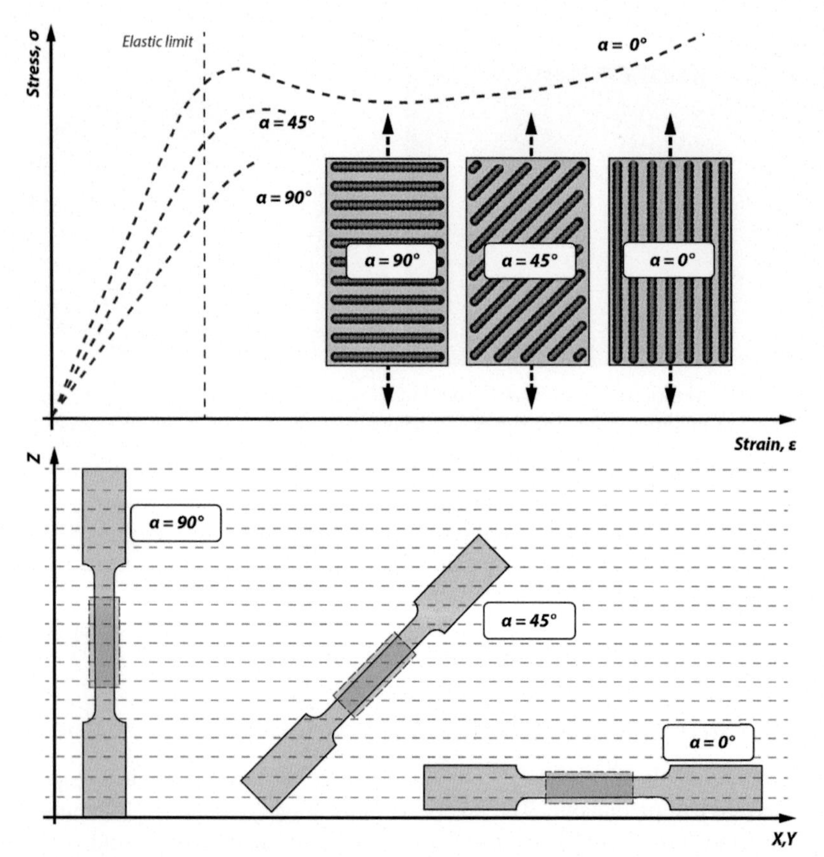

Fig. 8.14 Schematic representation of the effect of specimen orientation on observed mechanical response. Specimens aligned to the build platen ($\alpha = 0°$) have filaments aligned to the load direction and display similar mechanical response to the parent polymer. Specimens with $\alpha = 90°$ have filaments perpendicular to the loading direction and potentially display a brittle response due to the associated interlaminar strength.

local roughness associated with the external volume of this deposited filament. This filament solidifies in-situ and its local geometry is therefore influenced by external factors, including ambient temperature and gravitational and viscous forces, as well as the local toolpath trajectory.

2. Trajectory defined by the associated toolpath

The toolpath trajectory is scheduled by post-processing of the intended CAD geometry, which attempts to define a continuous fill of the desired cross-section, as discussed in Chapters 3 and 4. The perimeter-and-raster strategy is typically applied as it accommodates a range of potential slice geometries and provides a continuous perimeter filament with relatively low roughness on the visible surface. Despite these advantages,

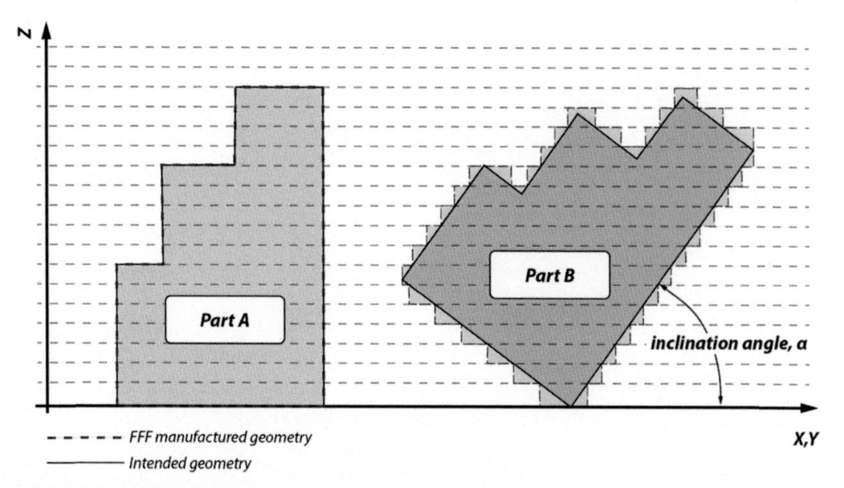

Fig. 8.15 Schematic representation of the stair-step effect introduced by features with some inclination to the manufacturing plane. Geometry orthogonal to the manufacturing plane (Part A) has no stair-step effect, unlike geometry inclined to this plane (Part B).

this toolpath strategy has the potential to introduce porosity defects, especially for curvilinear geometry and thin sections and at perimeter/raster intersections (Fig. 8.12).

3. Fundamental FFF system architecture and its influence on filament deposition

The feasible system architecture for FFF systems is unrestricted, but in practice it is typically of the cartesian (planar) implementation. This implementation results in a technical inability for the system to correctly replicate geometry that is not orthogonal to the manufacturing plane, introducing a stair-step effect in the fabricated geometry (Fig. 8.15).

The experimentally reported consequence of these influential factors is schematically reported for the linear surface roughness, R_a, as a function of inclination angle, α, layer thickness, L_t, and surface orientation, either upward facing or downward facing (Fig. 8.16). For upward facing surfaces aligned to the build platen ($\alpha = 0°$) the observed roughness is that inherent to the cross-sectional geometry of the deposited filament[4] and stair-step effects are nil. This situation persists until the inclination angle increases such that the intended geometry cannot be fabricated within a single layer (Fig. 8.17). A discontinuity then occurs where the local roughness peaks to a maximum as the layerwise manufacturing method attempts (with little success) to implement this intended inclination angle. *Good DFAM practice is to avoid such acute inclination angles.*

[4] Assuming the tool path is scheduled for *fully dense* fill.

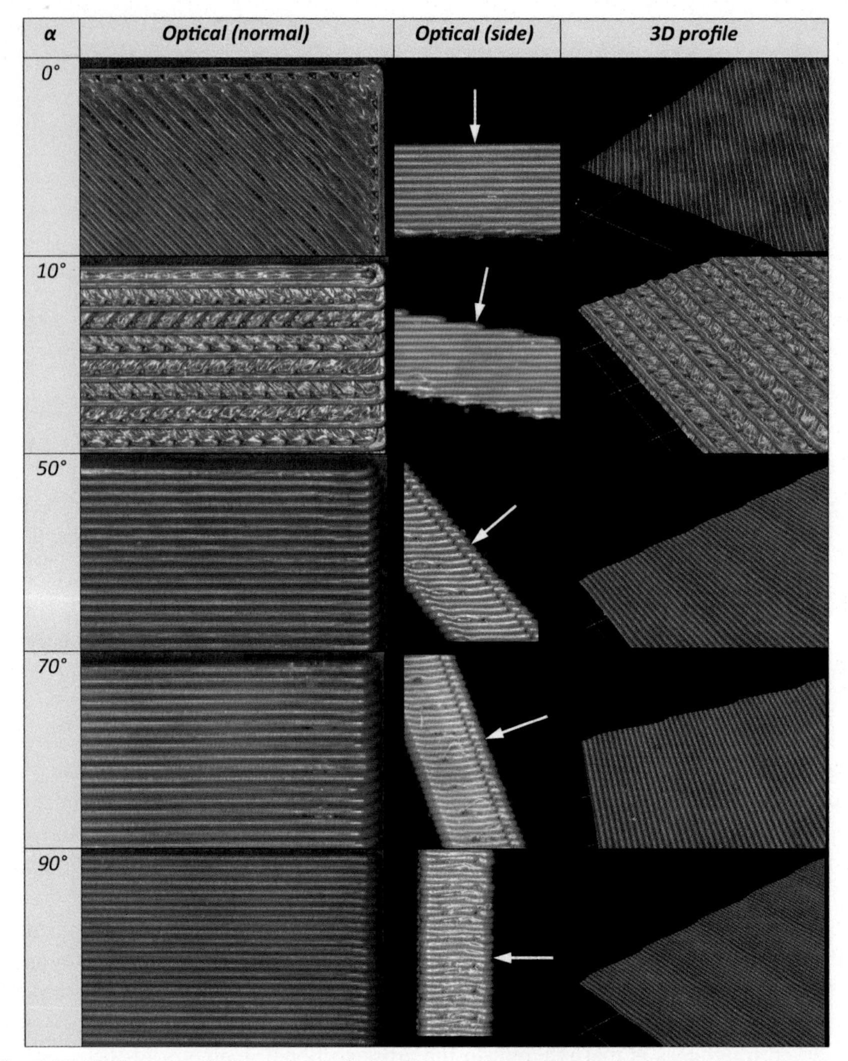

Fig. 8.16 Surface morphology versus inclination angle, α, observed by optical microscopy normal to the upper and side surfaces, combined with reconstructed 3D profile for representative specimens. White arrow indicates viewing direction of 3D profile.

With further increase in inclination, the magnitude of this stair-step effect decreases asymptotically, until for $\alpha = 90°$, the stair-step effect is nil, and the surface roughness again is that incurred by the filament cross-section alone.[5] The stair-step contribution of this observed roughness increases with increasing layer thickness.

[5] Nuances in this (slightly) simplified description exist, particularly that roughness for $\alpha = 90°$ and $\alpha = 0°$ are not necessarily equal and are influenced by layer thickness, local toolpath and extrudate solidification time.

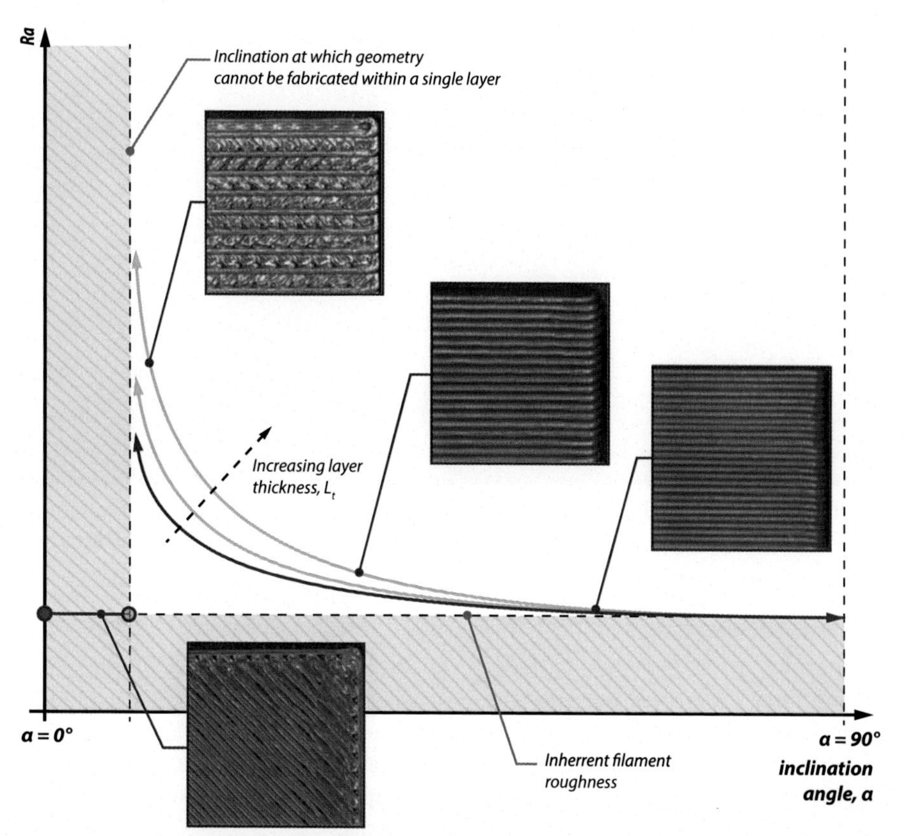

Fig. 8.17 Schematic representation of experimentally observed linear surface roughness, R_a, for upward-facing surfaces of layerwise manufactured FFF specimens.

For unsupported downward facing surfaces the observed roughness response is similar; however, for angles below the critical inclination, $\alpha < \alpha_{crit}$, unsupported manufacture is not possible. These zones can be supported by active support structures if required, although additional costs are incurred due to increased material consumption, manufacturing time and cost of support removal. For increasing inclination angle, the geometry becomes manufacturable; however, it is somewhat compromised until the inclination angle is further increased, and, as for the downward-facing scenario, then tends to the roughness value associated with the filament geometry alone.

8.1.3.6 Unsupported overhangs

Excessively acute inclinations and unsupported islands are accommodated in MEX systems by the use of supporting material. This material may be of the same type as the material used to fabricate the final AM part, or in sophisticated systems, may be of an alternate *support* polymer (Fig. 8.18). This support material is typically frangible or degradable to facilitate removal. Support material increases the set of feasible

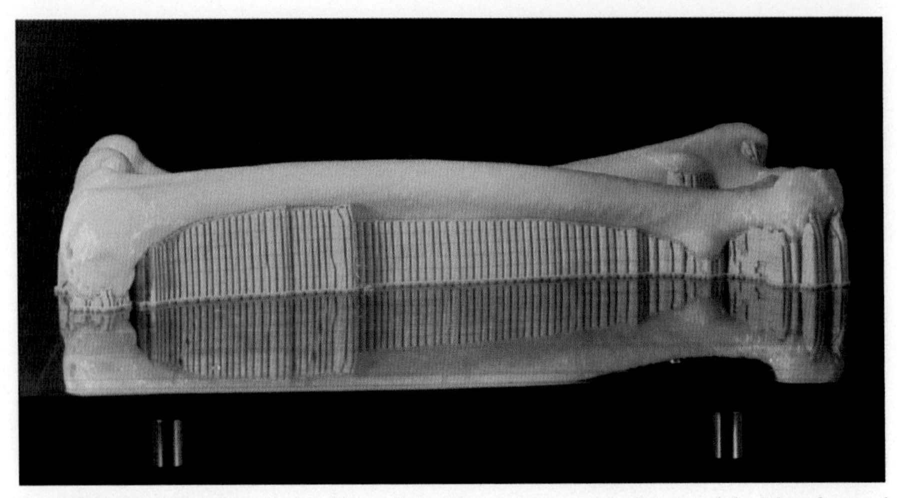

Fig. 8.18 MEX fabricated femoral bone indicating support and build materials. Manufactured with the commercial FFF system of Fig. 8.1 at the RMIT Univeristy Advanced Manufacturing Precinct.

manufactured geometries, but does incur costs due to material consumption, additional fabrication time, and the cost of support structure removal, which for non-trivial geometries can be significant. Emerging DFAM tools are mitigating the costs incurred by support structure generation by the efficient deployment of supporting structures that are fit for their intended functional requirements.

8.1.4 FFF technology summary and associated research opportunities

FFF is a critically important AM technology. The fundamental technology is robust and has been well refined over an extended period of commercial implementation, academic research and hobbyist development. FFF implementations exist in variants that range from high-value commercial installations used in the routine fabrication of commercial product, to inexpensive implementations at the hobbyist level. The broad implementation of FFF systems is in part due to their technical simplicity and compatibility with a range of available polymers.

Despite the broad implementation of FFF systems, there remain significant opportunities for commercially relevant research contributions. These contributions particularly include multidisciplinary contributions that extend the capabilities of current DFAM tools, including: advanced toolpath planning methods, unique and complimentary polymer chemistries and post-processing methods, hybrid materials including filament reinforced and particle impregnated filaments, and the robust fabrication of challenging polymer chemistries with relatively simple extrusion hardware typical of in-situ FFF extrusion implementations.

8.2 Polymer FFF: extended case studies of commercially relevant applications

The ASTM AM technology classification of Fused Filament Fabrication (FFF), commercially referred to as Fused Deposition Modelling (FDM), represents a commercially significant classification of AM technologies. FFF is well developed commercially and provides opportunities for physically large components of high-complexity, as well as being compatible with commercially available thermoplastic polymers ranging from inexpensive materials to high-performing engineering grades. The commercial applicability of polymer FFF extends from the rapid prototyping of functional and aesthetic models, to the manufacture of high-value commercial products. The case studies in this section demonstrate a selection of commercial applications of polymer FFF.

- Rapid prototyping of a proposed *automotive latching concept.* The technical suitability of this concept was demonstrated with a full-scale functional model fabricated with FFF. This functional model provided a high degree of certainty that the intended latching function could be achieved, but without the extensive lead time associated with traditional manufacture, thereby enabling the manufacturer to make informed design decisions early within the design phase.
- Aesthetic and functional demonstration of a proposed *commercial gearbox system*, including functional gearing and integration with commercial bearings as per the intended commercial system. Fasteners were integrated directly within the AM manufacturing process to assist assembly. This functional prototype enabled the manufacturer to effectively pitch the proposed solution to intended end-users and to assist in design finalization and pre-production evaluation.
- Application of novel extrudate to the manufacture of biofouling resistant systems. It has been demonstrated that anti-biofouling particulates can be integrated within FFF extrudate, and that this composite material does not compromise the FFF manufacturability of the host polymer. This capability enables the fabrication of *custom marine structures that resist biofouling*. In this case, the manufactured product is deployed to the end user, and is a manufactured product rather than a prototype. Such high-value application of inexpensive FFF technologies provides a significant commercial opportunity for technically innovative product developers (Fig. 8.19).

8.2.1 Case study: technical evaluation of proposed automotive latching system

Automotive product development is associated with inflexible delivery deadlines and a high-value, high-complexity product. Such product development is therefore typified by short-development timelines with a focus on the de-risking of design outcomes. This design environment is highly compatible with the techno-economic attributes

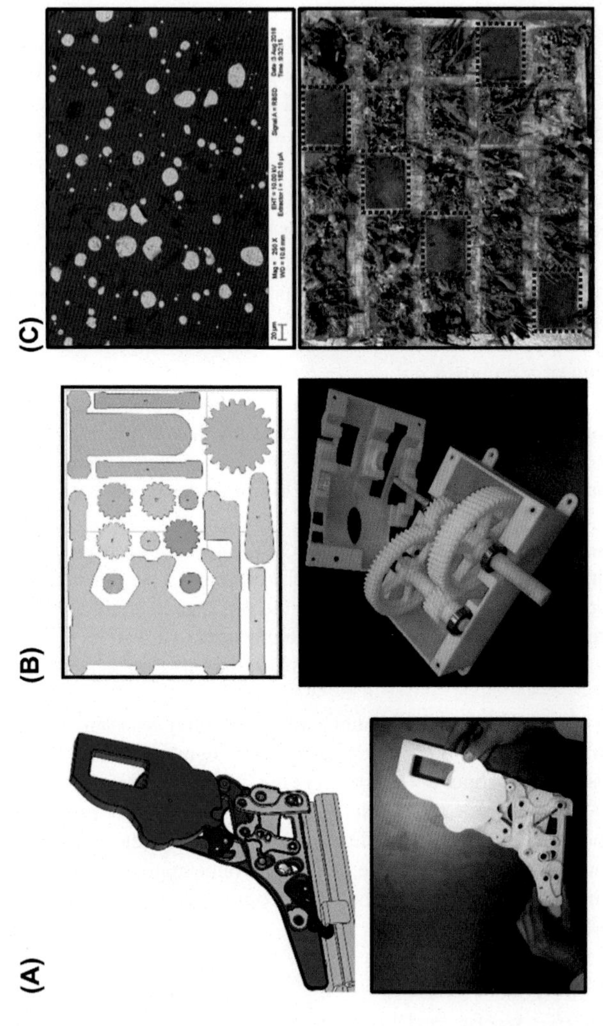

Fig. 8.19 FFF products summarized in this chapter: (A) automotive latching concept, (B) functional demonstration of proposed commercial gearbox, (C) anti-biofouling system.

of AM (Chapter 2); specifically, the ability to rapidly generate high-complexity geometry at low-cost and with functionally robust technical materials.

This case study presents a commercial design project, whereby a traditionally manufactured latching mechanism of high-complexity needed to be technically evaluated according to the original equipment manufacturers technical specification. To fabricate the intended component using traditional methods would incur an unacceptable lead-time. However, the application of FFF technology for the fabrication of a full-scale functional model using high-value engineering polymers enabled the automotive supplier to rapidly evaluate the technical function of the proposed system, but without the extensive lead time associated with traditional manufacture. This outcome enabled the manufacturer to confirm that the proposed latching system functioned as intended, thereby allowing the design team to make informed design decisions early within the design phase.

8.2.1.1 Background

Automotive latching systems enable temporary fastening of enclosures and seating structures and are associated with rigorous and conflicting technical requirements. Automotive fasteners are required to ground impulsive crash loads but must be activated by occupants using relatively low forces and associated displacements. In addition, they are associated with strict requirements for noise and vibration and are to be fabricated at low cost and with low mass. Commercial opportunities exist for automotive suppliers that can develop novel latching systems according to specific customer requirements, while maintaining short lead times and high confidence in the success of the proposed design. The compatibility of these commercial requirements with FFF technology is demonstrated in the following commercial design case study.

8.2.1.2 Design

The initial design of Fig. 8.20 was developed by an experienced automotive design team in response to a commercial latching application with unusually complex design requirements. This latch was required to enable multiple independent latching functions while providing *gating* such that multiple functions could not be concurrently engaged. These technical requirements were to be met while allowing the robust grounding of impulsive crash loading; as well as meeting the noise, vibration and harshness requirements required of this high-value automotive application. Despite the experience of the design team, senior management remained uncertain as to the technical response of the proposed system and were reluctant to fund further development.

In response to a highly constrained delivery deadline, the design team proposed to develop a technically functioning system using AM technology to in order to provide more certainty into the success of the technical function of the proposed design. Based on the outcomes of this rapid prototyping exercise, the proposed latching system could either be rejected or adopted for commercial application using traditional manufacturing methods.

The proposed latching system had been developed to the detail design stage and associated CAD data was therefore available. The requirement for the functional

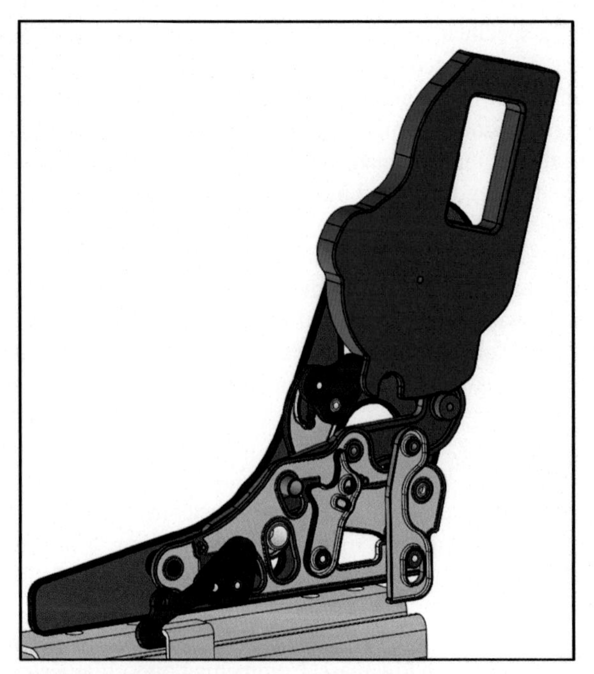

Fig. 8.20 Proposed latching design prototype as developed in response to a complex and unique design brief provided by the original equipment manufacturer.

prototype was therefore to embody the minimum set of functional requirements within a technically equivalent rapidly prototyped system. It was required that the functional prototype be manufactured within a minimal timeline and with minimal risk of failure (the design team had 24 hours to successfully deploy the functional prototype).

The existing CAD data developed by the design team was used as a basis to develop the functional prototype (Fig. 8.21). The DFAM rules of Chapter 4 were applied, specifically those of relevance to FFF manufacture, including:

- Kinematic links were fabricated with a base-plane feature, from which functional surfaces, hinge features, sliders and pins were extruded. This design allowed low-risk functional parts to be readily generated with low risk of manufacturing error. Where a feature was required in the opposing direction from this base-plane feature, it was manufactured as a separate item, and assembled as a net-shape press-fit.[6]
- Overhanging surfaces were avoided by the specific design of AM compatible geometry without compromising component technical function. For example, a chamfer was applied to the latching pin circlip-groove to satisfy the minimum feasible inclination angle, and to enable support-free manufacture[7] (Fig. 8.21).

[6] FFF technology can be readily used to generate functional net-shape press-fit features, particularly when the circularity of these features is within the X-Y plane (Chapter 4).

[7] The removal of support material from this feature was to be avoided as it would add to the delivery time and potentially compromise technical function.

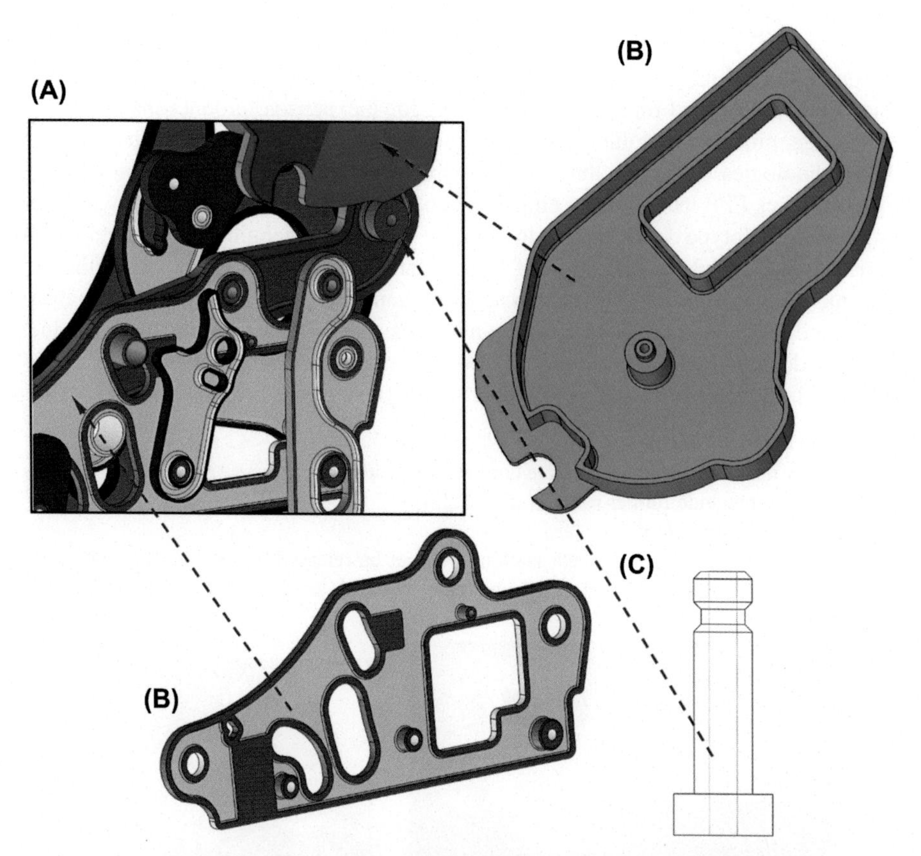

Fig. 8.21 Graphical summary of DFAM methods implemented to achieve net-shape manufacture of functional FFF model: (A) latching components (from Fig. 8.20). (B) To achieve robust manufactured outcome within limited lead-time, the intended product function was replicated with a base-plane feature from which features were extruded to enable rotational and sliding interaction, (C) Where a feature was required in the opposing direction from this base-plane feature, a custom part was fabricated to meet support-free manufacturing requirements and was assembled net-shape by press-fit.

- Standard fasteners and springs were accommodated within the design such that hinge features and spring return was robust and net-shape manufacture was achieved.

These DFAM methods were applied to the design of a functional prototype and enabled minimal pre-processing lead time; FFF file preparation required approximately 4 hours and FFF manufacture was then completed overnight. The following morning, the components were assembled without fault, resulting in net-shape manufacture first-time with an overall lead time of less than 24 hours.

8.2.1.3 Outcome summary

Automotive product development is highly challenging and is associated with inflexible development lead time and requirements for inexpensive product manufactured with large production volume. This techno-economic scenario is challenging for AM technologies. However, the design flexibility and rapidity of production inherent to AM makes FFF technologies eminently compatible with the requirements of pre-production prototype development. In this case study, a challenging automotive design project was enabled by utilization of FFF to fabricate a *works-like* model within an extremely short lead-time. The application of standard DFAM design rules enabled use of pre-production CAD data for the net-shape fabrication of a functional product that was deployed within a 24 hour period. This functional model was then successfully used to demonstrate the following key functions of the proposed prototype:

- ***Seatback-detent function***
  Whereby the seatback is free to rotate until some specified angle, where the seatback detent is captured to preclude further rotation (Fig. 8.22).
- ***Lock-out lever function***
  The lock-out lever precludes seat latching but can be released by rotating the seat to the reclined position (Fig. 8.23), beyond which the seat-back detent can engage (Fig. 8.22).

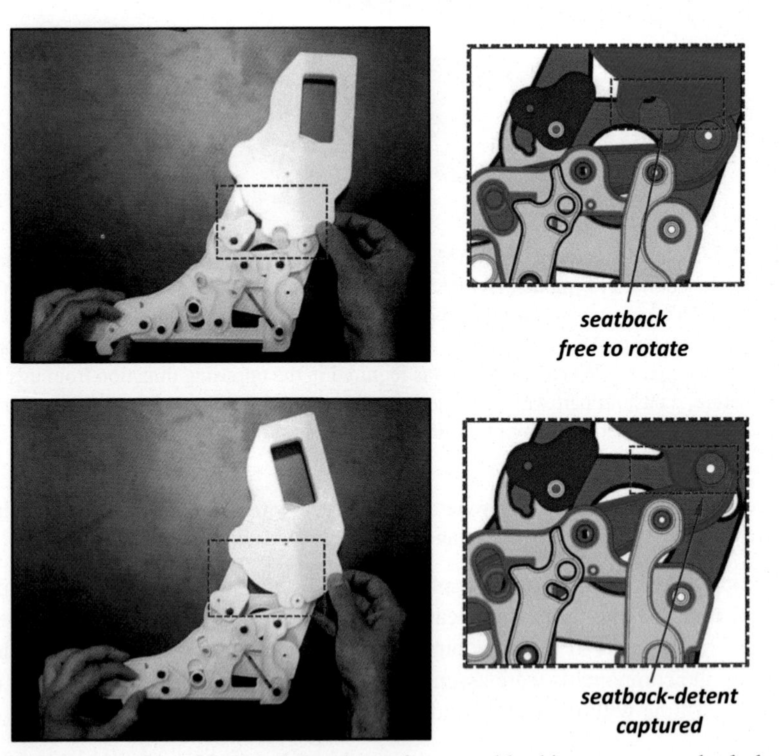

Fig. 8.22 Demonstration of technical functions of proposed latching system: seatback-detent. Upper: seatback-detent disengaged (seatback free to rotate). Lower: seatback-detent engagement (seatback rotation disallowed).

Fig. 8.23 Demonstration of technical functions of proposed latching system: lock-out lever. Upper: lock-out lever precludes seat latching. Lower: seatback rotation successfully disengages lock-out lever.

8.2.1.4 Automotive latching system case study summary

AM technologies provide high flexibility and inexpensive fabrication at low-volumes but are not inherently compatible with the requirements for low-cost high-volume production, as is typical for mass-produced commercial vehicle applications. However, utilization of AM technologies, in this case FFF, can enable the rapid fabrication of functional prototypes such that confidence can be developed for the traditional manufacture of a proposed design. For this urgent application, DFAM design was completed within a few hours and production was scheduled overnight. The rapid prototype was assembled the following morning allowing the proposed prototype to be fabricated net-shape and technically evaluated the following day. For these cases, the DFAM rules of Chapter 3 provide a robust mechanism to minimize the risk of failure to deliver a functioning system first-time.

8.2.2 Case study: commercial demonstration of mechanical gearbox system

Functional prototyping provides an opportunity for the inexpensive deployment of technical models that can assist design decision making, customer engagement and product sales. In particular, FFF technology enables the inexpensive fabrication of physically large components and systems, and for engineering projects with existing CAD data, this strategy incurs very little cost in developing the input design data required for FFF manufacture.

8.2.2.1 Background

This case study documents the application of FFF technology for the aesthetic and functional demonstration of a proposed commercial gearbox system. The proposed gearbox system delivers a power of 50 kW via a two-stage design with helical gearing and has an anticipated sales value of approximately US$50k. Prior to investing in the manufacture of this high-value product, a functional scale-model of the complete system was developed using FFF technology. The scale-model includes functional gearing and integration with commercial bearings systems as per the intended implemented design. It was successfully used to attract commercial attention and evaluate customer response to potential design variants, as well as to assist in design finalization and pre-production evaluation.

8.2.2.2 Design

The DFAM rules of Chapter 3 were applied as follows.

- Automated algorithms for build orientation provide useful insight into local orientation optima. However, the orientation of practical components within the build volume is a multi-objective design problem, and typically benefits from the input of the designer to provide a

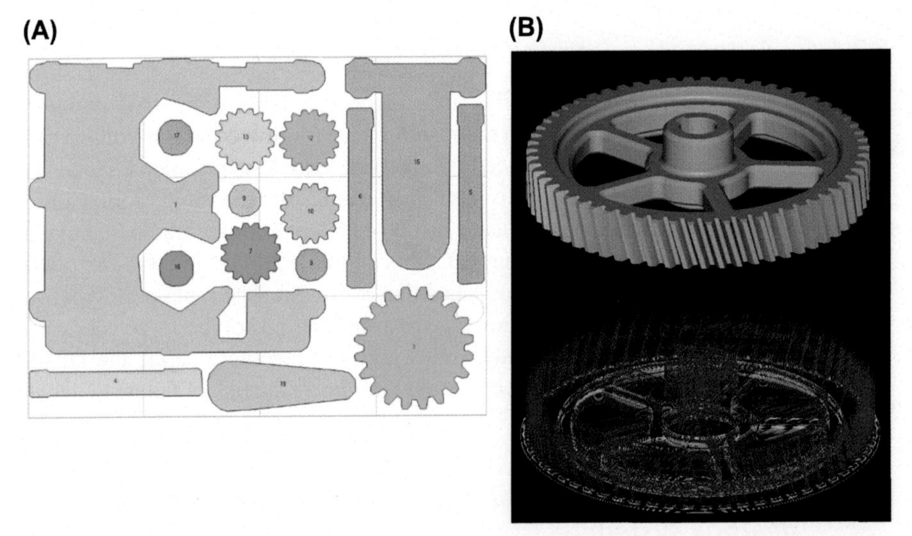

Fig. 8.24 DFAM tools were applied to minimize manufacturing time and cost, while achieving technically robust outcomes. For example: (A) optimized part nesting, including the (B) orientation of technically functional parts such that mating surfaces avoid stair-step effects.

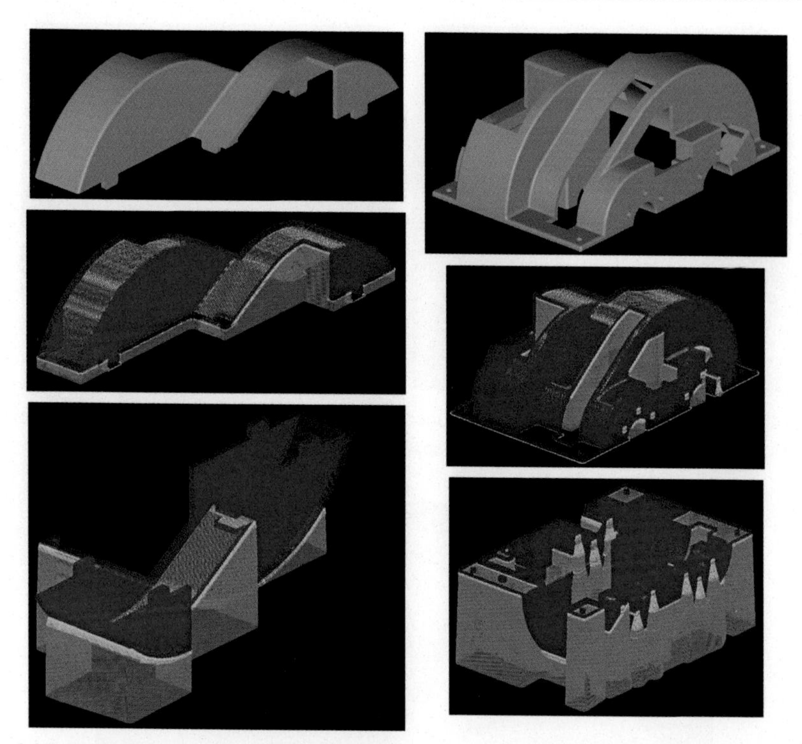

Fig. 8.25 Optimal orientation within the FFF build volume is a multivariable design problem. DFAM tools exist to assist in orientation, but the designer's expertise is typically required minimize support structures while avoiding functional or aesthetic compromise due to stair-step effects.

workable compromise between support consumption and the practical effects of stair-steps on the manufactured components (Fig. 8.24).

- FFF manufacturing costs are a function of overall build-time, material consumption and batch utilization costs (Chapter 2). To enable robust manufacturing outcomes without excessive cost, the nesting of individual components within the build volume should be optimized (Fig. 8.24). This optimization is subject to the requirement that functional geometry (such as mating gear surfaces and functional gear bores in this case) should be oriented such that stair-step effects do not compromise this intended function (Fig. 8.25).
- Functional press-fits (for example, between shafts and gear bore) were implemented such that the system was assembled net-shape, with no post-manufacture processing required other than support structure removal.
- Fasteners were encapsulated directly during manufacture. This was achieved by pre-planning internal voids into which fastener nuts were inserted during manufacture (Fig. 8.26).
- The design effort required to implement product variants (Figs 8.26 and 8.27) was minimized by the re-use of CAD data that was parametrically varied according to the particular application including shafts, spigots and bushings.

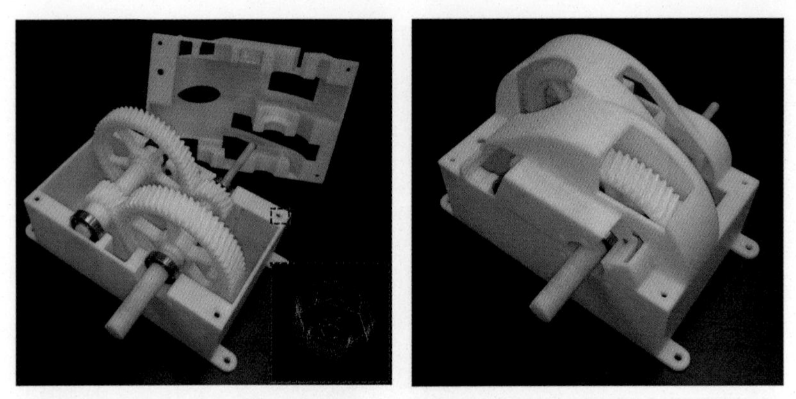

Fig. 8.26 Gearbox variant A. This system incorporates a traditional sump and helical gearing. Inset identifies geometry used to allow encapsulated fasteners.

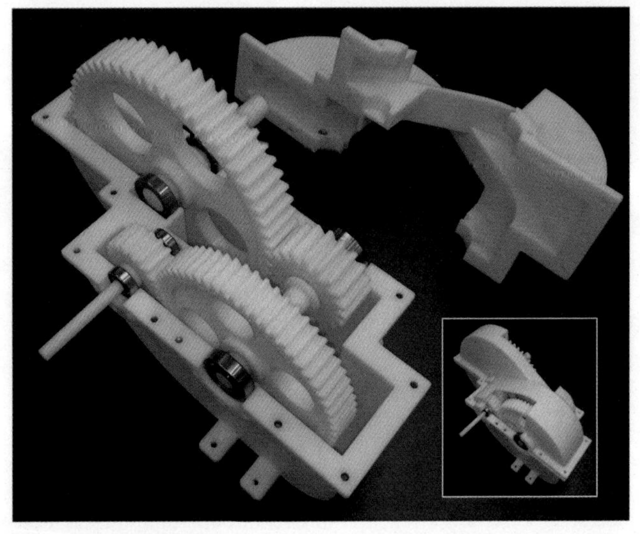

Fig. 8.27 Gearbox variant (B) This system incorporates a contoured sump and spur gearing.

These DFAM rules were utilized to enable the fabrication of multiple permutations of the proposed 50 kW system at a scale of 5:1. The model was fully implemented with a budget of approximately US$1k and (by re-using existing CAD data) required minimal additional file-preparation.

8.2.2.3 Mechanical gearbox latching system case study summary

Functional technical models add-value by assisting design decision-making and by enhancing customer engagement, especially as a prelude to the fabrication of expensive systems. In this case study, DFAM rules were applied to enable prototyping of a functional scale-model of a 50 kW power transmission system for less than 1% of the anticipated sales cost.

8.2.3 Case study: antibiofouling FFF extrudate

It has been demonstrated that antibiofouling particulates can be integrated with the typical FFF extrudate, and that this composite material does not compromise the standard manufacturing processes of the host FFF polymer. This capability enables the fabrication of custom structures that resist fouling of marine systems. In this case, the manufactured product is deployed to the end user, and is therefore a manufactured component rather than a prototype. Such high-value applications of inexpensive FFF technologies provide an example of the significant commercial application opportunity for technically aware product developers.

8.2.3.1 Background

Biofouling refers to the accumulation of microorganisms on exposed marine surfaces. Biofouling presents a significant problem for marine vessels as it significantly increases hydrodynamic friction, leading to increased propulsion drag and associated fuel consumption and running costs. Perhaps more importantly, the biofouling load presents a bio-security risk by the potential for the transport of marine pests. The avoidance of such bio-security risks incurs substantial cost, and existing antibiofouling technologies are sub-optimal for many applications in terms of the inherent toxicity and associated application costs.

8.2.3.2 Design

A particular challenge for biofouling is associated with sea chest and manifold systems; these systems intake and vent water across the marine vessel hull for various technical applications, including the cooling of engines and high temperature equipment. Sea chests and manifold systems are of complex geometry (according to the requirements of the local ship architecture) and therefore inspection and cleaning challenges are exacerbated. Furthermore, these structures are typically associated with elevated temperatures and circulating flow, making for an enhanced biofouling risk. FFF technology presents a robust opportunity for the fabrication of inexpensive, high-value components with locally deployed biofouling formulations that are manufactured on-demand and to the specific geometry required for the particular sea chest system.

The additive manufacture of antibiofouling systems remains an under-researched commercial opportunity; however, research is emerging within the field, for example in antibiofouling for reverse-osmosis and ultrafiltration applications [10], as well as for the manufacture of microbe resistant polymers for medical applications [11]. This case study presents current research in the manufacture of custom structures that mitigate fouling of sea chest applications by the control of streamlines and the localized application of antibiofouling formulations [12].

FFF technology enables the inexpensive fabrication of on-demand structures using engineering grade polymers. This capability allows the fabrication of sea chest insert structures that manipulate streamlines such that circulating flow is eliminated. This reduces biofouling by reducing the physical volume of the sea chest, increasing the shear

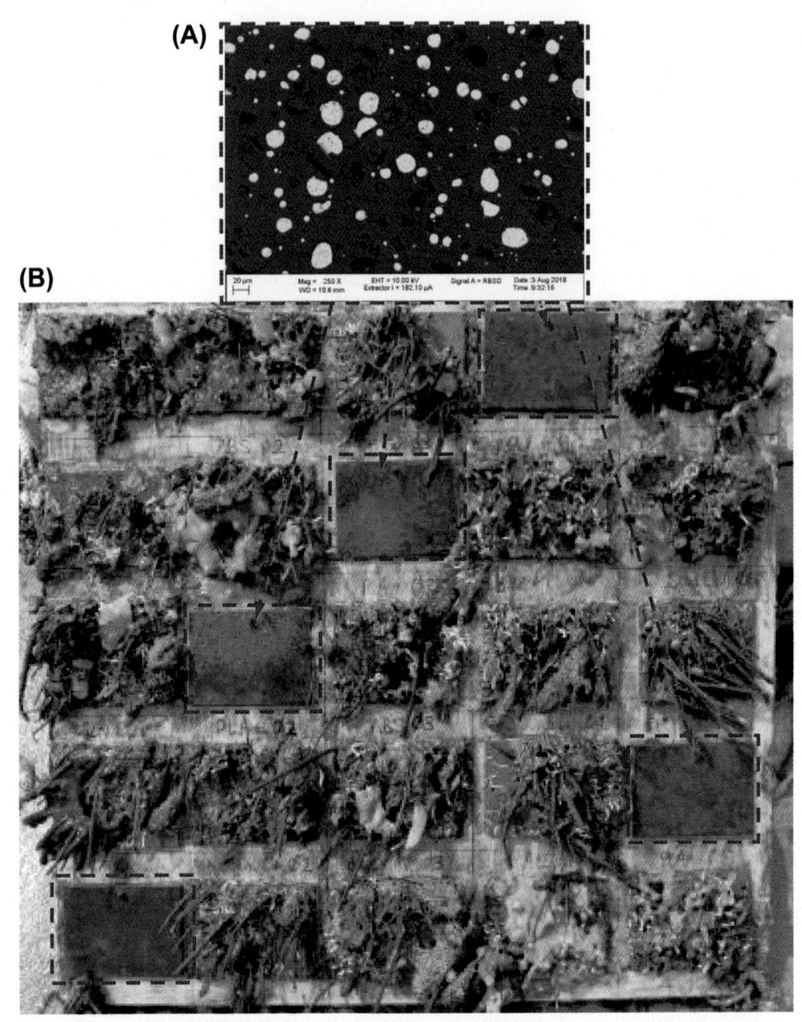

Fig. 8.28 Antibiofouling FFF systems. (A) Electron microscope image of entrained antibiofouling additives (white particles) in FFF filament, (B) the antibiofouling performance of FFF-printed test coupons (red boxes) following 12 months marine exposure. Dr. Richard Piola, Defence Science Technology Group, Melbourne, Australia.

stress observed on the wetted surface, as well as by reducing nutrient access for fouling organisms (Fig. 8.28).

Custom sea chest and manifold geometries provide an inexpensive opportunity to passively reduce fouling load, but do not provide any active antibiofouling mechanism. The post-manufacture application of antibiofouling formulations has been demonstrated [20]; however, this involves the use of solvents and does not accommodate the flexibility of the FFF process for the local customization of antibiofouling formulations. Ongoing research has demonstrated that various antibiofouling

Fig. 8.29 Antibiofouling FFF systems, (A) fabrication of custom geometry, including the local application of antibiofouling materials over a structural substrate (inset), (B) custom antibiofouling FFF housing for marine sensor housing, (C) demonstrating of antibiofouling efficacy over an extended emersion time in marine environment. Dr. Richard Piola, Defence Science Technology Group, Melbourne, Australia.

formulations, including particulates and active chemicals can, under specific conditions, be entrained within FFF extrudate and be robustly deposited with standard manufacturing processes (Fig. 8.29). This capability allows the fabrication of highly customized, high-value antibiofouling applications. For example, local geometry can be manufactured with antibiofouling properties, including biodegradable polymers that deliver active formulations over an extended period and in a controlled fashion. These active materials may be deposited over structural engineering polymers such that the system costs are minimized without compromising biofouling performance or structural integrity. These systems also reduce the use of toxic chemicals by allowing the targeted application of antibiofouling material (Fig. 8.29).

8.2.3.3 Biofouling resistant systems summary

Marine biofouling results in substantial environmental and economic consequence. The application of FFF technologies provides a significant opportunity for innovative, inexpensive and effective response to effects of marine biofouling. The flexibility of FFF systems allow the passive and active optimization of these systems by the elimination of fluid circulation, as well as the application of active antibiofouling formulations. In this application, overall production volume is high. However, each component is unique: a scenario that is not compatible with traditional manufacturing methods but can be commercially implanted with AM by the utilization of generative design methods (Chapter 7). Such high-value, mass-customized applications of FFF technologies provide significant commercial application opportunities for innovative AM product developers.

8.2.4 FFF case study summary

Fused Filament Fabrication (FFF), commonly referred to as Fused Deposition Modelling (FDM), is a commercially significant ASTM AM technology classification. In particular, FFF can accommodate physically large components with high geometric complexity and is compatible with commercially available thermoplastic polymers including inexpensive and high-performance polymers. Consequently, FFF technology is well developed commercially and is available in systems that range from hobbyist machines to commercial implementations. The commercial opportunities enabled by FFF technology are significant as are the opportunities for associated DFAM research tools.

In particular, FFF technologies enable the potential for the manufacture of products and systems that support traditional manufacture. For example, the functional prototype and scale-models presented in this section could not be deployed within an acceptable cost or time budget using traditional methods; however, for such scenarios, the techno-economic attributes of AM technology enable de-risking of traditional manufacture. Furthermore, FFF is commercially successful for commercial production of high volumes when each component has unique, customized geometry. These mass-customized applications can become commercially significant when combined with novel functional materials that can be locally deposited as required by utilizing the mass-customization opportunities of AM.

8.3 Bioplotting

Bioprinting refers to the utilization of AM technologies for the fabrication of engineered constructs that combine biological materials including cells, growth factors, and other biomaterials. By allowing unique combinations of functional biomaterials to be integrated with the flexibility and complexity of AM methods, bioprinting allows significant innovation in biological research and clinical practice, including the development of models for various diseases; the manufacture of simulated tissue for transplantation and biological testing; and the future potential for bespoke organ engineering for transplantation.

Bioprinting is achieved by a range of AM technologies, but is particularly suited to Material Extrusion (MEX) technologies, where it is referred to as *bioplotting*. MEX-bioplotting enables a range of clinically relevant outcomes including control of the architecture and pore size of the biological construct, as well as flexibility in printing a range of biological materials, including viable living cells. Furthermore, bioplotting development benefits from the extensive DFAM outcomes available from FFF research and commercial applications.

The emerging discipline of bioplotting is highly multidisciplinary and includes contributions from polymer chemists, molecular biologists, medical practitioners, bioengineering specialists and manufacturing engineers. The contributions of these specialists have enabled significant biomedical innovation in the development of novel bioplotting systems and materials. These capabilities are being applied in the manufacture of unique 3D constructs that enable novel methods of drug delivery and tissue engineering.

The functional system diagram for bioplotting (Fig. 8.30) attempts to capture the primary opportunities for innovation in bioplotter design. In typical bioplotting architectures, input material (or materials) are extruded as required according to the prescribed toolpath and process settings. Prior to this extrusion phase, there is an

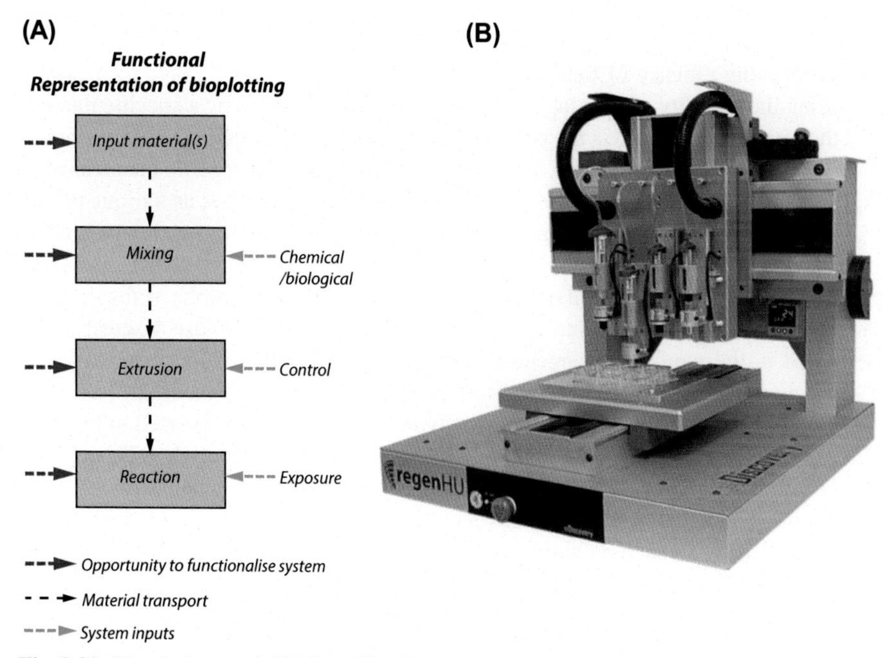

Fig. 8.30 Bioprinting is a MEX-based implementation of bioplotting, with (A) high flexibility in implementation leading to (B) robust commercial applications (see Fig. 8.1).

opportunity for the mixing of additional material, including biological agents, chemical catalysts or other additives that enhance biological function, increase manufacturability by the temporary reduction of flow resistance, or improve mechanical properties of the manufactured construct. A similar opportunity for material interactions exists after the extrusion process, for example, methods to allow extruded materials to become more rigid, including chemical crosslinking or other methods to control *gelation*.[8]

8.3.1 Bioplotting materials

Bioplotting materials range from standard industrial polymers, including grades that are certified for medical applications, to experimental compounds that attempt to achieve biologically useful function. When these materials are functionalized or consist of living cells they may be referred to as bio-ink. Bioplotting materials include both bio-inks and the supporting materials that are used to enable a 3D structure to be fabricated.

Bio-inks and their supporting structure are subject to various manufacturability, mechanical and biological requirements. These requirements include compatibility with AM bioplotting hardware, biocompatibility, resistance to mechanical loading and may require some degree of controlled biodegradability response. Concurrently these inks must be compatible with the associated AM fabrication system and must be processed in a manner that does not excessively compromise their biological function, thereby limiting the allowable shear stress, and temperature to which living cells can be exposed.

The printing efficacy of these bio-inks may be quantified by a range of measures including the viability (percentage living) of cells manufactured by a specific method, and the resistance to flow via the proposed extrusion method. Generic measures of viscosity exist, but the resistance to flow should be known for the specific characteristics (specifically extruder diameter) of the intended bioplotting process, as various beneficial biomaterials can display *shear thinning*, a non-Newtonian response whereby the material viscosity decreases when subject to shear stress (as occurs during extrusion). Shear thinning materials are advantageous for bioplotting applications as they allow a bio-ink to reduce viscosity during extrusion and reversibly increase viscosity once deposited in the required arrangement. Hydrogels are an important classification of bio-ink material that can be designed to exhibit shear thinning behaviour.

Hydrogels refer to three-dimensional polymers with the ability to swell in the presence of aqueous media while maintaining their *overall* shape (Fig. 8.31). This characteristic is useful for MEX-based bioplotting, especially associated with tissue engineering and therapeutic drug delivery. Bioplotting applications include naturally occurring hydrogels such as alginic acid, collagen, agarose and chitosan, as well as synthetic hydrogels such as hyalouric (HA) acid, polymethacrylic acid (PMMA)

[8] Gelation refers to the formation of molecular bonds between the chains of branched polymers such that a single connected molecule is formed. This molecule then loses fluidity and is known as a *gel*.

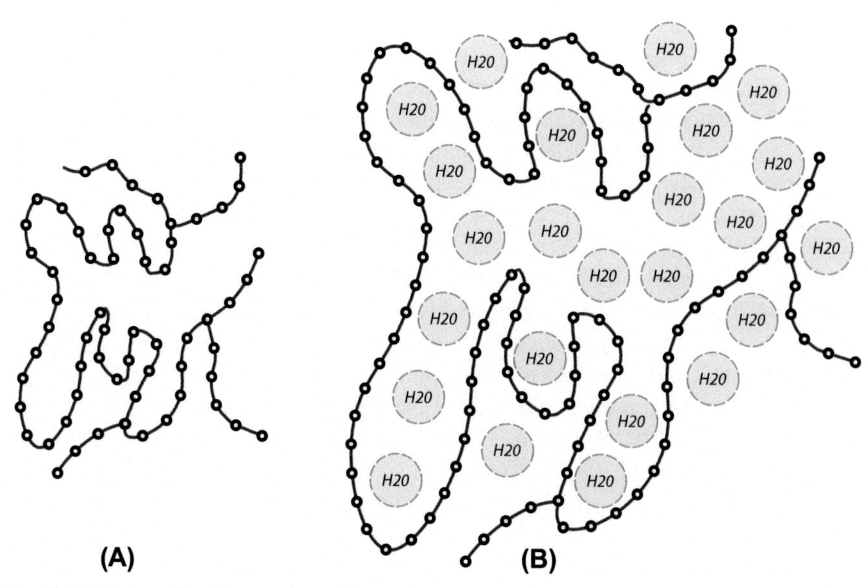

Fig. 8.31 Hydrogels refer to (A) polymers that (B) swell in the presence of aqueous media.

and polyhydroxyethyl methacrylate (PHEMA). The hydrogel may be a homopolymer, copolymer or an Interpenetrating Polymer Network (IPN) consisting of multiple independent polymer components crosslinked in network form.

Hydrogels enable the immobilization of swelling aqueous media, which can include cellular and chemical material, consequently hydrogels are an important classification of bio-ink materials. Recent advances in hydrogel research has enabled clinical innovation in this emerging research space, including advanced polymer architectures, complex IPN materials and stimuli-responsive hydrogels that allow the controlled therapeutic delivery [13].

Bioplotting materials also include a range of commercial and experimental polymer grades that can provide supporting structure for the bio-ink. Of particular interest are materials that biodegrade in a benign fashion within the body (Table 8.1). These polymers provide an opportunity to enable the manufacture of supporting scaffolds that intentionally degrade post implantation, allowing the body's natural processes, such as vascularization, to progressively integrate within these degrading structures.

8.3.2 Clinically relevant research applications

To achieve clinically relevant outcomes, bioplotted constructs must satisfy commercial, technical and regulatory requirements. Commercial requirements include cost-effective manufacture, which is typically concurrent with a need for generative DFAM tools, such that manual design effort is replaced with autonomous methods (Chapter 7). Technical requirements include the necessity for a robust understanding of the influence of bioplotting system design and process parameters on cell viability and geometric fidelity of the manufactured product. Clinical outcomes require that bio-inks and supporting structures be deposited at a resolution that mimics that of the naturally occurring biological constructs that are being replicated. Regulatory requirements include the certification of process repeatability and that risks of biological contamination are effectively mitigated. Research outcomes that contribute to these clinical requirements include the development of novel bio-inks and compatible AM bioplotting systems, a selection of these outcomes is presented below.

The controlled deposition of repeatable scaffold structures with process-controlled porosity is achievable with bioresorbable polymers such as polycaprolactone (PCL) [14]. These structures provide a predictable mechanical response according to their porosity, thereby providing a robust basis for the manufacture of clinically useful biological constructs with engineered biological and mechanical response (Fig. 8.32A).

Tissue engineering applications require the replication of the naturally occurring Extracellular Matrix (ECM), the non-cellular material present in living tissues. The ECM provides structural support to the cellular material as well as enabling biomechanical and biochemical cues that are necessary for normal cell function. The development of bio-inks that are engineered to replicate ECM functionality is of critical clinical importance. For example, the fabrication of specialized hydrogels including gelatin-methacrylamide (Gel-MA) [15]. These Gel-MA hydrogels are comparable to naturally occurring ECM and can be further enhanced, for example by the integration of photocrosslinkable hyaluronic acid methacrylate (HA-MA) to aid the formation of mineralized cartilage and bone, a process known as *chondrogenesis* [16]. These specialized bio-inks enable replication of the precise biological, chemical and structural roles necessary for the replication of functional tissues (Fig. 8.32B).

The fabrication of customized bioprinting hardware enables the fabrication of clinically useful constructs, but requires that chemical, biological, mechanical and manufacturing constraints be satisfied. In response to these complex design requirements, highly innovative custom bioplotting technologies are emerging. For example, custom bioplotting systems with coaxial extrusion capabilities allows the printing of bio-inks with core-shell materials that are variable according to prescribed process parameters. This technology can generate biological constructs that enable controlled structure porosity within both micro- and macro-structures (Fig. 8.32C).

To satisfy the design requirements of sophisticated biological constructs may require multiple distinct biological and mechanical responses. Exemplar bioplotting applications achieve these distinct functions with multiple materials and customized deposition methods. For example, research toward the manufacture of replacement cartilaginous tissues, such as in the human ear, included hybrid bioplotting applications whereby multiple engineered materials are used according to their unique properties [17]. These materials include *sacrificial* materials to aid in the manufacture of the intended geometry; *structural* materials to provide the requisite mechanical response to the implant; and cell supporting *bio-inks*, consisting of cell-laden hydrogels (Fig. 8.32D). These emerging methods provide significant clinical advantages in the fabrication of complex functional tissues and organs [18].

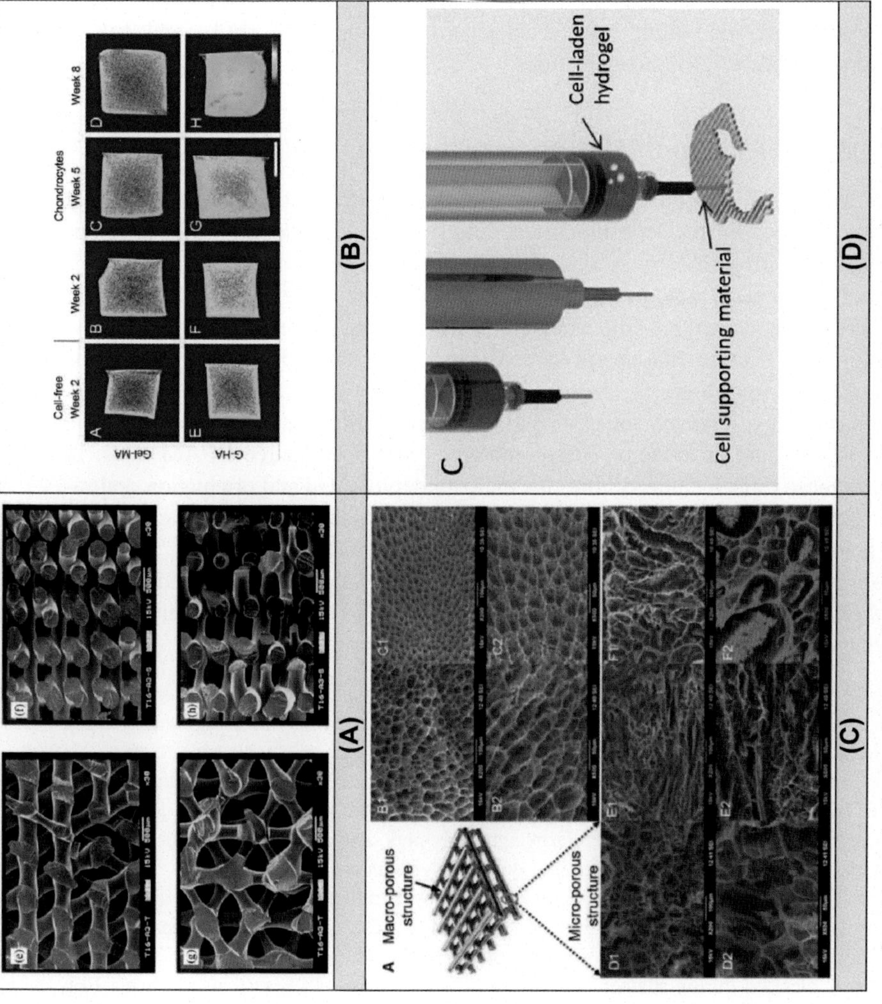

Fig. 8.32 Relevant research outcomes in the development of novel bio-inks and compatible AM systems, including: (A) FFF fabrication of candidate biodegradable PCL scaffolds [14], (B) novel Gel-MA and HA-MA based hydrogels with enhanced chondrogenesis [16], (C) coaxially extruded bio-inks with bespoke macro- and micro-porosity [17], (D) cell-laden hydrogel printed into pre-deposited supporting materials [18].

8.3.3 MEX technology review, DFAM tools and associated research opportunities

The Material Extrusion (MEX) classification of AM technology is highly versatile and commercially relevant. The capability of these systems is enabled by their extensive research and commercial pedigree, compatibility with extensively developed commercial materials and their flexibility in implementation for innovative and sophisticated specialist applications. The majority of research and commercial interests for MEX systems are observed within the Fused Filament Fabrication (FFF) and MEX-bioprinting (bioplotting) implementations.

8.3.4 Fused Filament Fabrication technology review

Initially developed by Scott Crump (founder Stratasys Ltd.) as a patented commercial technology termed Fused Deposition modelling (FDM), this innovative AM development was later promoted by Adrian Bowyer in the *Darwin* and subsequent open-source self-replicating rapid prototyping (*RepRap*) systems. The intense development of the FFF AM technology implementation has resulted in such a ubiquitous AM application that many casual observers consider FFF technologies to be synonymous with the term 3D-printing.

The technology development applied to FFF implementations means that extensive support is available in terms of both commercial and open-source DFAM tools. In particular these tools include systems for toolpath deployment and orientation optimization. Standard commercial polymers are routinely available for these systems as are novel grades developed for specific techno-economic benefit.

Despite (or perhaps due to) this intense technology development, there remain open DFAM research questions of commercial significance. These opportunities include the following:

- The development of advanced polymeric input materials, including AM systems that allow the repeatable manufacture of materials with challenging manufacturing requirements including high melting temperature.
- In general, polymers with higher molecular weight tend to have superior mechanical properties but display relatively high flow resistance, thereby imposing a technical limit to the practical fabrication of high molecular weight polymers. Advanced FFF extrusion systems that can accommodate these challenging polymer systems are required.
- FFF systems typically use filament extrudate as the system input. An alternative architecture is the use of pellet extrusion systems at the FFF head. These systems enable access to lower cost input materials and allow in-situ blending of materials.
- Whether blended during filament extrusion, or in-situ in the FFF system, extensive FFF research opportunities exist for the development of enhanced extrudate, including:
  - the blending of extrusion of chemically active inclusions, for example the antibiofouling additives presented in detail in Section 8.2.3.
  - the entrained extrusion of strengthening filaments, for example the extrusion of composite fibre materials.
- FFF systems are almost exclusively of planar architecture. Such systems have many advantages, but are limited to layerwise fabrication within sequential planes, thereby disallowing

strengthening structures perpendicular to this plane that reduce delamination strength. Alternative architectures have the potential to overcome these limitations; for example, series-robotic systems.
- Continuous build systems are required for efficient commercial applications such that product removal and material re-loading methods are automated.
- Industrial polymer manufacture typically involves the fabrication of molecules and their polymerization as two distinct stages - the in-situ polymerization of polymer materials in the FFF head remains unresearched opportunity. This opportunity would allow the manufacture of thermosetting materials.

8.3.5 Bioplotting technology summary and associated research opportunities

The manufacture of biological constructs provides a significant clinical and research capability. For example, in the fabrication of systems for disease modelling, functional tissues and even the future potential for the manufacture of functional organs for human transplantation. The inherent techno-economic advantages of MEX systems allow for these outcomes to be economically mass-customized to patient-specific medical requirements. In response to these clinical opportunities, a significant research output has emerged. This research effort includes the scientific research associated with fundamental biological response and useful polymer chemistries, as well as the translational aspects required to enable the clinical application of these advanced manufacturing concepts.

MEX-based bioprinting (referred to here as bioplotting) is relatively more recently investigated than equivalent FFF technologies and arguably more sophisticated to commercialise. Consequently, the associated research opportunities for bioplotting are extensive, and relatively varied, including the following:

- The development of novel bio-inks and biologically functional polymer chemistries, as well as the documentation of existing systems in terms of their biological and chemical compatibilities.
- Clinically relevant materials must be characterized in terms of their compatibility with existing and proposed bioplotting systems. This requirement includes the generation of experimental data and associated predictive DFAM models that allow the mechanical and biological design requirements to be mapped to their associated process inputs.
- The manufacturability requirements of bioplotting materials must be quantified such that robust toolpaths can be algorithmically prescribed.
- The routine clinical application of bioplotting systems requires that generative DFAM methods be available to automate system workflows such that manual decision-making bottlenecks are eliminated, and regulatory certification is ensured (Chapter 7).

In addition to these high-level research themes, it is apparent that bioplotting is a clinically enabling technology that is generating unanticipated multi-disciplinary research outcomes. These DFAM outcomes provide significant associated research opportunities for emerging researchers as well as for established research teams in academic and commercial environments.

References

International standard reference for principles and terminology of AM

[1] ISO/ASTM 52900. 2015(en), Additive manufacturing — general principles — terminology. Geneva, Switzerland: International Organization for Standardization (ISO); 2015.

Initial FFF patent document

[2] Scott CS. Apparatus and method for creating three-dimensional objects. 1991. US Patent, 5121329.

RepRap project on self-replicating AM system

[3] Jones R, Haufe P, Sells E, Iravani P, Olliver V, Palmer C, Bowyer A. RepRap—the replicating rapid prototyper. Robotica 2011;29(1):177—91.

Emerging research to enhance FFF extrusion process control

[4] Anderegg DA, Bryant HA, Ruffin DC, Skrip Jr SM, Fallon JJ, Gilmer EL, Bortner MJ. In-situ monitoring of polymer flow temperature and pressure in extrusion based additive manufacturing. Additive Manufacturing 2019.

Advanced FFF system design

[5] Lee WC, Wei CC, Chung SC. Development of a hybrid rapid prototyping system using low-cost fused deposition modeling and five-axis machining. Journal of Materials Processing Technology 2014;214(11):2366—74.

FFF process optimisation for polylactic acid printing

[6] Balani SB, Chabert F, Nassiet V, Cantarel A. Influence of printing parameters on the stability of deposited beads in fused filament fabrication of poly (lactic) acid. Additive Manufacturing 2019;25:112—21.

Generation of three-dimensional FFF toolpaths

[7] Chakraborty D, Reddy BA, Choudhury AR. Extruder path generation for curved layer fused deposition modeling. Computer-Aided Design 2008;40(2):235—43.

International standard for surface texture definitions

[8] International Organization for Standardization. Technical committee ISO/TC 57, 1997. ISO 4287. Geometrical product specifications (GPS). Surface texture: profile method. Terms, definitions and surface texture parameters. International Organization for Standardization.

Review of methods for acquiring AM surface metrology

[9] Townsend A, Senin N, Blunt L, Leach RK, Taylor JS. Surface texture metrology for metal additive manufacturing: a review. Precision Engineering 2016;46:34–47.

Additive manufacture of antibiofouling systems for reverse-osmosis and ultrafiltration applications

[10] Sreedhar N, Thomas N, Al-Ketan O, Rowshan R, Hernandez H, Al-Rub RKA, Arafat HA. 3D printed feed spacers based on triply periodic minimal surfaces for flux enhancement and biofouling mitigation in RO and UF. Desalination 2018;425:12–21.

Additive manufacture of antimicrobial polymers for medical applications

[11] Sreedhar N, González-Henríquez CM, Sarabia-Vallejos MA, Rodríguez Hernandez J. Antimicrobial polymers for additive manufacturing. International Journal of Molecular Sciences 2019;20(5):1210.

Additive manufacture of antibiofouling systems for marine applications

[12] Leary M, Piola R, Shimeta J, Toppi S, Mayson S, McMillan M, Brandt M. Additive manufacture of anti-biofouling inserts for marine applications. Rapid Prototyping Journal 2016;22(2):416–34.

Review of therapeutic hydrogel applications

[13] Buwalda SJ, Vermonden T, Hennink WE. Hydrogels for therapeutic delivery: current developments and future directions. Biomacromolecules 2017;18(2):316–30.

Bioplotting of scaffold structures for tissue engineering

[14] Zein I, Hutmacher DW, Tan KC, Teoh SH. Fused deposition modeling of novel scaffold architectures for tissue engineering applications. Biomaterials 2002;23(4):1169–85.

Hydrogels engineered to replicate ECM functionality

[15] Ansari S, Sarrion P, Hasani-Sadrabadi MM, Aghaloo T, Wu BM, Moshaverinia A. Regulation of the fate of dental-derived mesenchymal stem cells using engineered alginate-GelMA hydrogels. Journal of Biomedical Materials Research Part A 2017;105(11):2957–67.

Hydrogels for the formation of mineralised cartilage and bone

[16] Levett PA, Melchels FP, Schrobback K, Hutmacher DW, Malda J, Klein TJ. A biomimetic extracellular matrix for cartilage tissue engineering centered on photocurable gelatin, hyaluronic acid and chondroitin sulfate. Acta Biomaterialia 2014;10(1):214–23.

Hybrid bioplotting of replacement cartilaginous tissues, such as in the human ear

[17] Liu X, Carter SSD, Renes MJ, Kim J, Rojas-Canales DM, Penko D, Angus C, Beirne S, Drogemuller CJ, Yue Z, Coates PT. Development of a coaxial 3D printing platform for biofabrication of implantable islet-containing constructs. Advanced Healthcare Materials 2019:1801181.

Bioplotting for cartilage regeneration

[18] Chung JH, Kade J, Jeiranikhameneh A, Yue Z, Mukherjee P, Wallace GG. A bioprinting printing approach to regenerate cartilage for microtia treatment. Bioprinting 2018:e00031.

Emerging research into pellet based FFF extrusion systems

[19] Volpato N, Kretschek D, Foggiatto JA, da Silva Cruz CG. Experimental analysis of an extrusion system for additive manufacturing based on polymer pellets. International Journal of Advanced Manufacturing Technology 2015;81(9–12):1519–31.

Application of antibiofouling formulations as a post-manufacture process

[20] Yao J, Chen S, Ma C, Zhang G. Marine anti-biofouling system with poly (ε-caprolactone)/clay composite as carrier of organic antifoulant. Journal of Materials Chemistry B 2014;2(31):5100–6.

Material jetting

9

Material Jetting (MJT) is an ISO/ASTM classification for AM technologies 'in which droplets of feedstock material are selectively deposited' [1]. MJT technologies are enabled by highly complex physical processes associated with multiple unique scientific domains consequently there exist significant technical challenges that must be overcome for successful commercial implementation. As a consequence of this challenging development environment, MJT technologies are typically implemented by large-scale technology development companies and are protected by ongoing patents in relevant technology fields, including material chemistry, curing methods and digital control systems. Even as these patents expire, the economic challenges to the development of competing technologies are high[1]; however, the number of commercial suppliers of MJT hardware is increasing, especially as the fundamental machine elements becomes increasingly available as commodity products.

Concurrently, MJT technologies have been demonstrated to enable unique manufacturing capabilities, that in summary include high dimensional accuracy and achievable feature size, as well as relatively high production rates achievable with a range of unique polymeric materials, including the deposition of functionally graded materials. In response to this robust manufacturing capability, MJT has become a standard industrial technology for a range of commercial applications, including design outcomes that would be infeasible with traditional technologies, including the visualization of complex geometry, including mathematically complex structures and sophisticated medical models. Emerging application opportunities include the fabrication of low-volume tooling for injection molding as well as application in the development of patient specific radiotherapy phantoms.

9.1 Technology history and current methods

Material jetting refers to the controlled deposition of discrete liquid droplets. This technology was intensely developed for the domestic paper printing market and is commercially implemented by either continuous or drop-on-demand jetting systems. Continuous inkjet (CIJ) systems provide a stream of ink that forms discrete droplets due to Plateau-Raleigh instability. These droplets are then deflected as required by a controlled electrostatic field. Drop-on-demand (DOD) jetting systems are implemented by either piezoelectric or thermal systems that, either by piezoelectric shape-change, or heating with thin-film resistors, causes an individual droplet to be ejected. MJT technologies utilize these commercial ink-jetting technologies for the controlled deposition

[1] This state of technical development is in stark contrast to, for example, Fused Filament Fabrication (FFF) technologies that are readily implemented on hobbyist budgets.

Design for Additive Manufacturing. https://doi.org/10.1016/B978-0-12-816721-2.00009-9

of curable polymers for 3D printing. However, these sophisticated technologies require extensive experience to implement commercially and are the subject of ongoing patent protection. Consequently, the development of MJT systems is typically restricted to AM technology development companies with extensive research and development resources.

The commercial implementation of MJT technologies has converged to a typically Cartesian architecture, where the jetting array translates across the build plane depositing material in a two-dimensional voxel array (Fig. 9.1). The build platen descends on each deposited layer, allowing the sequential addition of layers to form a three-dimensional structure. Jetted materials are typically stored in air-excluding storage tanks to reduce unintended polymer curing, which can cause jet blockage, and are heated on transportation to the jetting array to control viscosity. Materials are initially jetted as viscous liquids that have no capability to accommodate overhanging geometry, consequently supporting structures are extensively used. These support structures are removed by exposure to solvents or are manually frangible. Post deposition, the deposited materials are cured by exposure to ultraviolet (UV) light.

Commercial research opportunities are somewhat restricted as MJT systems are typically *closed* commercial systems that do not allow operator access to processing parameters. Consequently, much of the DFAM research outcomes for MJT systems involves the repurposing and extension of existing MJT technologies to achieve increased function, including, for example, the deposition to novel substrate materials as well as to functional applications such as radiation therapy. Fundamental research opportunities exist in the enhancement of deposition rates and achievable material properties, including the development of novel support structure materials and the potential for self-supporting materials.

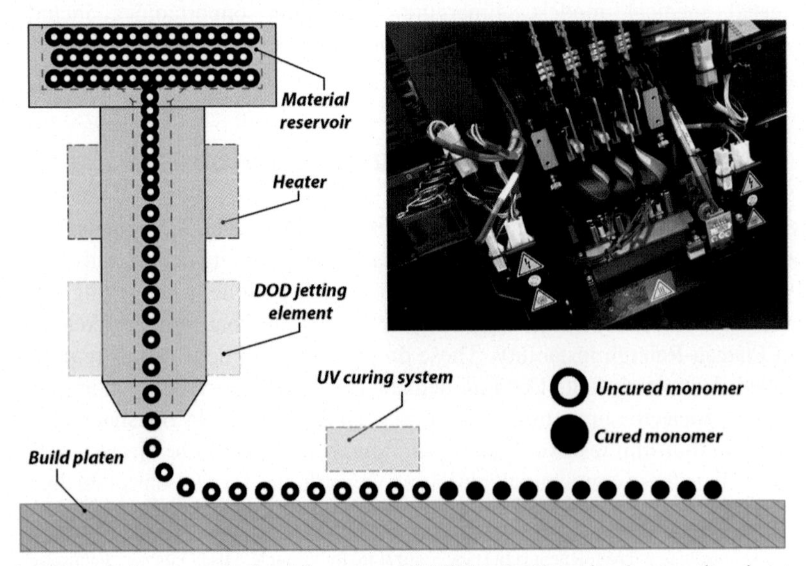

Fig. 9.1 Schematic representation of commercial MJT technology implementation, inset demonstrates the complexity of a commercial jetting assembly.

9.2 Materials of interest

Material jetting technologies fundamentally require that a material be deliverable via jetting technologies and then be cured on demand. Typical commercial implementations use polymeric systems that can be locally cured, typically by UV light exposure. These materials are typically proprietary grades with compositions that are not publicly available, although insight may be obtained informally by reference to mandatory safety data. Specific commercial grades are available for specific MJT technology implementations; however, these grades typically include high strength, low modulus and thermally resistant materials, as well as a range of colour materials being available in advanced systems. Significant research opportunities exist for the development of novel ink chemistries and hardware modifications to allow unique AM outcomes, including: functional additives (such as antibacterial, magnetic and electrically conductive additives); as well as the modification of deposition and curing strategies such that structural reinforcing elements can be incorporated within the manufactured product. Contemporary MJT technologies do not accommodate overhanging structures, therefore a commercial opportunity exists for the design of materials that allow the fabrication of functionally advanced support structures.

9.3 MJT technology applications

Material jetting technologies enables a unique set of manufacturing attributes, including: high-resolution fabrication of relatively large structures; the unrestricted deposition of materials within the three-dimensional voxel field; and, availability of an extensive range of available polymeric materials with variable mechanical and optical properties, including flexible and rigid materials as well as transparent materials and a broad range of colours. This fabrication is made from a digital reference that allows direct control of material attributes for all voxels within the manufactured product and is therefore compatible with generative design methods. These unique characteristics have resulted in the extensive application of MJT in a range of traditional and emerging technology applications, including: visualization of mathematically complex structures (Fig. 9.2), physical representations of numerical simulation data (Fig. 9.3), inexpensive fabrication of lifelike medical models and pre-surgical visualization tools (Fig. 9.4), as well as the generative design of medical dosimetry structures (Section 9.5).

9.4 Technical challenges and research contributions

Material jetting technologies are complex commercial systems, consequently, much of the fundamental technology development occurs within the commercial research and development facilities. Furthermore, these systems are typically not directly compatible with open-source DFAM software, making it difficult to modify existing design algorithms and methodologies. Despite these restrictions, valuable research contributions are being made, for example, associated with application of these technologies to

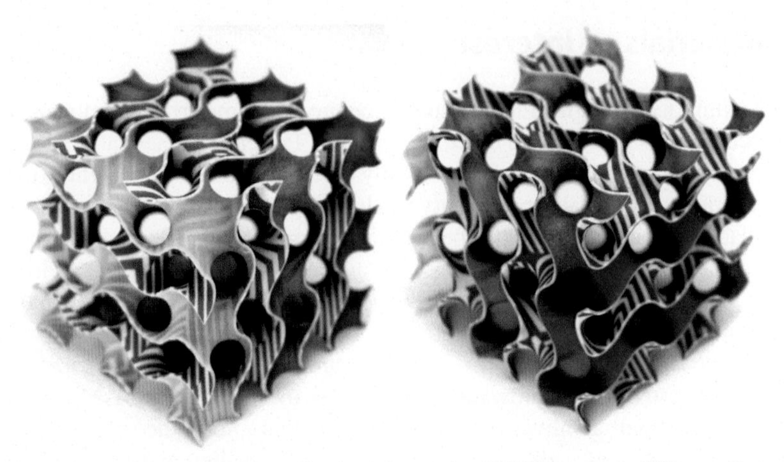

Fig. 9.2 Material Jetting (MJT) application of complex TPMS structure (Chapter 5) designed and fabricated by Phil Pille at RMIT University Advanced Manufacturing Precinct.

allow the fabrication onto novel substrate materials, characterization of the mechanical properties of functional technical products, robust characterization of geometric process capabilities, and the fabrication of short-run injection molding tooling (Fig. 9.5).

MJT deposition onto substrate materials

Contemporary MJT technologies allow direct fabrication onto a build platen (typically incorporating layers of support material such that the manufactured component can be removed without damage), but are not directly compatible with fabrication onto existing structures. Emerging research is demonstrating the capability for commercial MJT systems to achieve robust adhesion of jetted material onto an integrated substrate [4]. This research provides the potential for innovation in MJT application beyond the traditional build chamber; for example, allowing MJT technologies to be integrated within a robotically mounded end-effector for hybrid deposition on large scale marine or aerospace structures for antifouling or aerodynamic benefit.

Geometric process capabilities

The robust engineering design of technically functional products requires a statistically significant understanding of the geometric response of the manufacturing process. This understanding allows Geometric Dimension and Tolerancing (GD&T) to be applied to reliably achieve the intended product function.[2] These data exists for traditional manufacturing processes, but is much less well developed for MJT technologies. Emerging engineering research is providing robust geometric process capabilities that enables the GD&T of MJT products, thereby allowing inexpensive application of this technology in the design of functional technical products [5]. Furthermore,

[2] Or alternately, allows the designer to be aware that the intended function is outside the geometric limits of the intended manufacturing process, such that wasted design and manufacturing effort can be avoided.

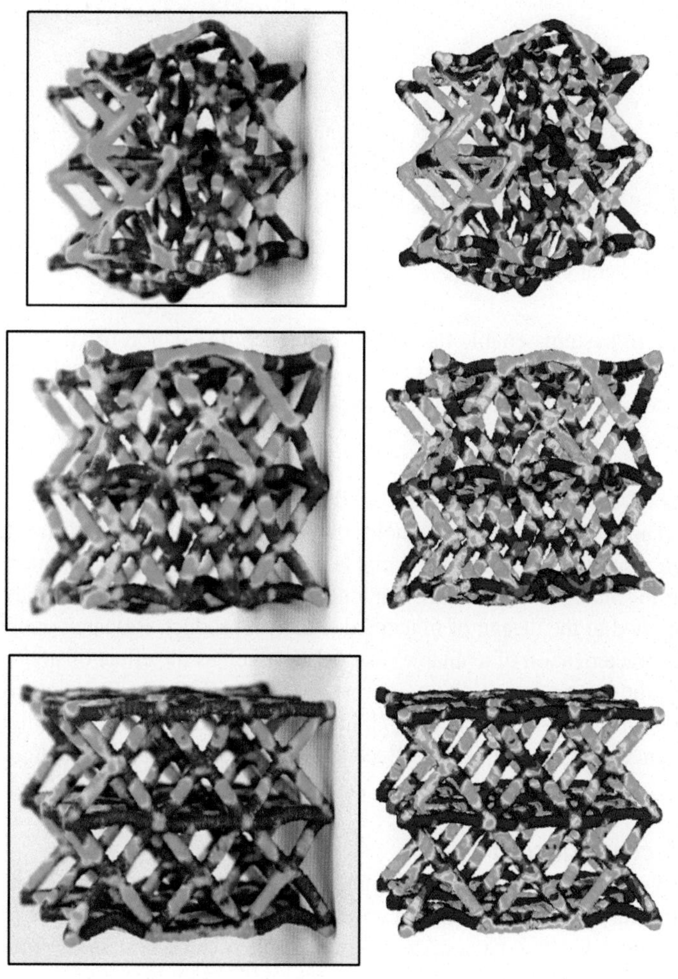

Fig. 9.3 Material Jetting (MJT) technology application to visualize complex stress state and associated deformation (upper) based on the numerical simulation complex lattice structure (lower). Simulation based on the method of [3].

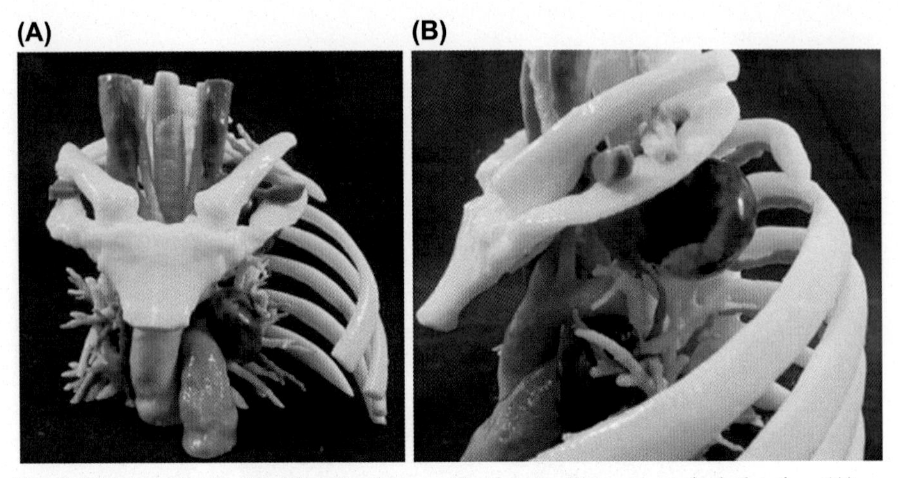

Fig. 9.4 Material Jetting (MJT) technology application to allow pre-surgical planning. (A) Anterior and, (B) lateral view of cancer and lymph nodes [2].

generative design systems enable the algorithmic accommodation of GD&T requirements to enable a specified technical function, and provide a commercially significant DFAM opportunity for the fabrication of mass-customized product [6].

Characterization of mechanical response

The traditional application for MJT technologies is associated with their visual attributes, including high resolution and complex variable color palette. Emerging research is characterizing the mechanical response of material jetting components so that they can be confidently applied to the design of functional technical product. Research opportunities include: the characterization of available materials, as well as the effect of strain concentrations due to the interaction of dissimilar materials, the effect of orientation within the build envelope on mechanical response [7], interlaminar strengths and opportunities for structural enhancement, as well as the effects of thermal and dynamic cycling.

Digital structure design

The extreme flexibility of MJT technologies allows innovative design of mechanical systems, including the manual design of custom structures and systems with increased functionality, as well as the generative design of sophisticated systems and functional gradients based only the defined functional response [8]. Although research outcomes are emerging to characterize the material properties and geometric response achievable with commercial MJT technologies; there are limited DFAM tools that can realize this design opportunity. This lack of available tools provides an opportunity for multidisciplinary design teams to develop systematic and generative DFAM tools to allow the advanced design of MJT systems (Chapter 7).

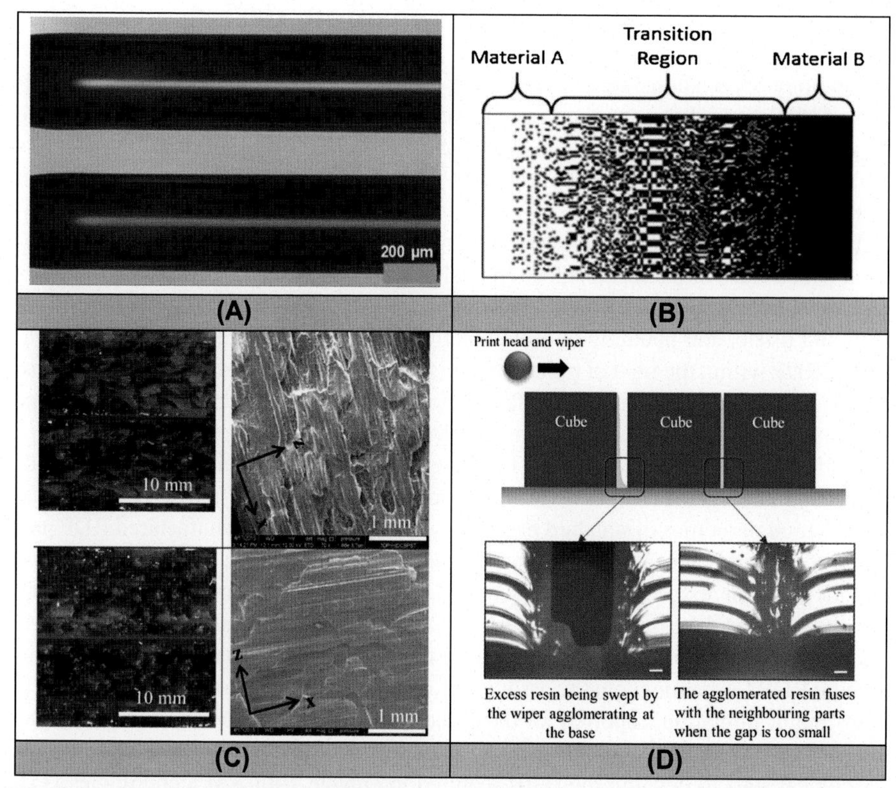

Fig. 9.5 Innovative response to MJT technology challenges, including: (A) deposition onto unique substrate [4], (B) algorithmically designed functional gradient [8], (C) characterization of mechanical response and associated fracture surface morphology [7], (D) characterization of MJT geometric process capabilities [5].

Injection molding tooling

Injection molding tooling is typically fabricated from high-value tool steels that provide a series of technical attributes including high strength, stiffness, elevated temperature wear resistance, and the ability to be polished to a smooth surface. Perhaps surprisingly, MJT technologies are emerging as a potential replacement technology for the manufacture of injection molding tooling. These applications require a particular design methodology, for example, the MJT tool should be protected from shear stresses by being restrained within a bolster, and injection pressures and flow rates may need to be reduced, as well as cycle times increased; however, commercial research has identified that MJT fabricated tools are commercially viable for low-volume production. This opportunity enables the inexpensive manufacture of traditional injection molded components at very short production volumes, as well as providing a mechanism to utilize the present value of knowledge for significant commercial benefit (Chapter 2).

9.5 MJT case study: radiation phantom design

Technology applications are emerging that utilize the unique manufacturing capabilities of MJT technologies to achieve functions that are otherwise technically infeasible or within the domain of traditional manufacturing technologies. In particular, MJT is enabling clinically valuable contributions in the field of patient-specific radiation therapy.

Radiation Dosimetry Phantoms (RDPs) are used to simulate the interaction of radiation with tissues of interest to allow clinical outcomes such as the: design of processes for safe handling of radiation sources, diagnosis of disease, study of radiation effects on living tissue, and planning of radiotherapy to kill cancerous cells [9]. The application of RDP within the field of radiotherapy is to ensure correct radiation dose and geometric precision; this requires the fabrication of *anthropometric* RDP that emulate human tissue geometry and radiographic properties.

Anthropometric RDP are typically fabricated using traditional manufacturing methods (typically molding and casting), whereby proprietary materials are combined to emulate the radiation properties of standard patient tissues. Traditionally manufactured RDP provide a robust and clinically understood radiation response, but are relatively costly and provide data for a specific clinical scenario only. The fabrication of patient-specific RDP by traditional means is possible, but is associated with high cost and long lead times. There exists a significant clinical opportunity for the on-demand manufacture of patient-specific RDP for training, quality assurance, clinical research and radiotherapy [10].

A 'major limitation' reported for the clinical application of AM technologies is the time and cost associated with the generation of AM product, and a potential reluctance for clinical staff to develop the skills required to engage with AM technologies [11]. The medical application of AM-RDP requires that both clinical and engineering requirements are satisfied concurrently, and therefore requires that experts in both domains work together to implement the specific design. This approach is highly flexible, and may be necessary for complex scenarios and for the development of new clinical methods. However, once a preferred clinical method is established, the development of a generative design algorithm (Chapter 7) provides significant opportunities for translating AM-RDP into clinical practice.

Fig. 9.6 presents an implementation of a generative system for clinical AM-RDP design. In this example, the clinician's access to the generative design system is restricted to their field of expert knowledge; specifically, by interacting with the medical image data file (i.e. DICOM file) to characterize the intended density regions of the manufactured phantom. Based on the clinician's specification, the generative design system generates an MJT input file (i.e. an oriented STL data file) that allows direct manufacture of the intended RDP. In clinical practice, these data can be automatically assigned to a specific patient and appended to a digital processing queue for manufacture. In this example, the manufacturability and engineering aspects of the AM-RDP design exist within the generative system, and are modified as required to meet the clinical requirements; associated advantages include: reduced manual effort in RDP design, reduced effort in RDP certification, and improved repeatability between cases.

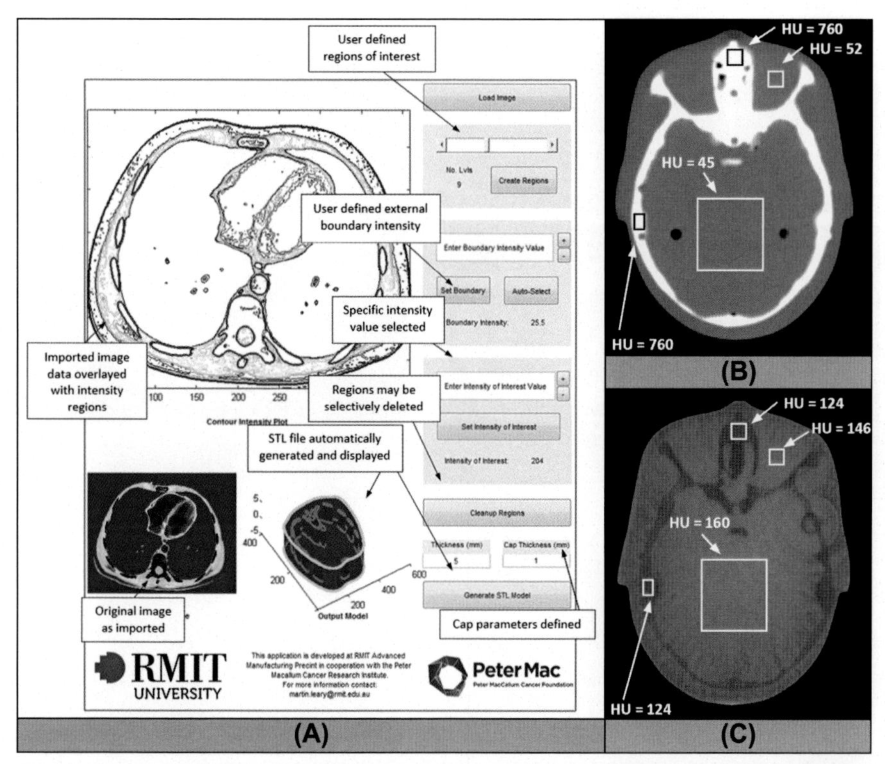

Fig. 9.6 Generative design tool developed for the application of MJT to the fabrication of patient-specific radiation dosimetry phantoms. In this scenario, (A) clinicians provide input on the oncological requirements only, a generative systems then algorithmically converts (B) medical image data to MJT input files required to fabricate the (C) phantom structure [12].

Technical challenges to the clinical application of AM to the fabrication of patient specific RPD include:

- Challenges to the certification of AM manufactured product, including repeatability in geometric deposition, consistency in material properties, and stability of these properties over time and in the presence of radiation exposure.
- Replication of radiation properties, typically characterized by the Hounsfield Unit (HU), a relative measure of the radiation response scaled with respect to the linear attenuation coefficients, μ, of air and water (Eq. 9.1).

$$HU = 1000\left(\frac{\mu - \mu_{water}}{\mu_{water} - \mu_{air}}\right) \tag{9.1}$$

In response to these challenges, a range of AM technology applications have been applied to the development of anthropomorphic RDP. These applications are being developed by various clinical and engineering research groups with varied clinical strategies. These emerging strategies are currently in a state of transition, but are likely

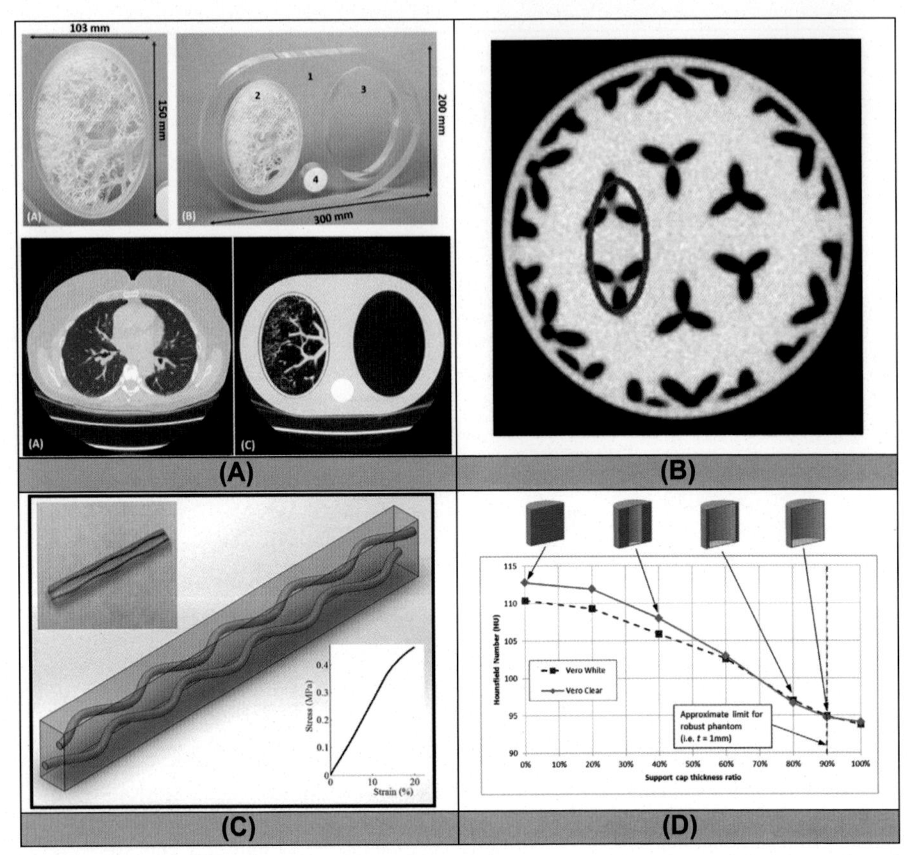

Fig. 9.7 Innovative application of MJT for AM-RDP, including: (A) lung vessel phantom based on an algorithmically generated lung topology and compared with CT scan of original patient CT data [13], (B) tissue-mimicking phantoms to simulate a permeable liver with associated CT scan identifying potential flow challenges [15], (C) dual-material manufacture to emulate mechanical tissue behaviour [14], (D) the application of encapsulated support material to modulate HU [12].

to converge to a preferred standard practice over time. A sample of current research in the application of MJT to the fabrication of clinical AM-RDP includes (Fig. 9.7):

- The generation of patient-specific lung structures based on an algorithmic representation of the lung vessel tree topology; followed by high-resolution manufacture using MJT technologies [13]. These results demonstrate clinically relevant and reproducible data when observed by medical CT.
- The application of multi-material MJT to fabricate structures that emulate the mechanical response of patient-specific tissues [14].
- The replication of fluid flow through simulated liver tissues fabricated with internal porous structures fabricated with MJT technologies [15]. Micro-CT was then used to show the

connectivity of the as-manufactured porous structure and to identify manufacturability issues that must be overcome to control permeability to within the expected values for kidney tissues.

- The encapsulation of support material to allow the modulation of HU within an algorithmically generated RDP [12].

MJT provides a the capability to manufacture a range of clinical RDP types within a single AM technology platform, including: the high resolution replication of complex tissue structures, including fluid flow and simulated vasculature; the simulation of tissue mechanical response; and, the possibility to integrate coloured features for reference, quality control and to represent tissue aesthetics. Furthermore, the 'closed' nature of proprietary MJT systems can provide a clinical advantage for certification and quality assurance, as there is a reduced possibility of unintentional manipulation of process parameters that could result in compromised clinical outcomes.

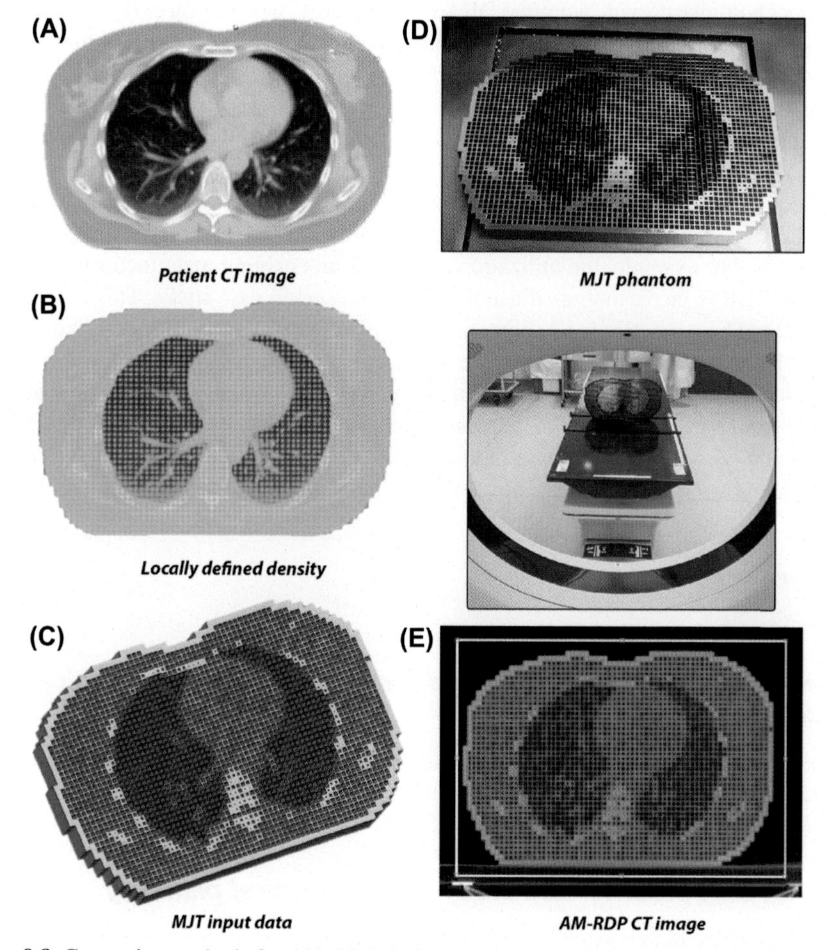

Fig. 9.8 Generative methods for AM-RDP design provide an opportunity to increase clinical engagement, reduce manual design effort and aid in quality assurance and certification. Emerging DFAM methods allow medical image data to be generatively converted to a local density field that when fabricated by MJT generates a patient-specific dosimetry phantom [10].

A persistent challenge to the clinical implementation of AM-RDP is the range of HU achievable with available AM materials. Commercial MJT polymers are chemically similar, and tend to provide a relatively similar radiation response in terms of the measured HU. To provide the required RDP contrast, researchers typically use air or liquid as a contrast media. MJT provides an opportunity to generate high-resolution structures that locally vary the ratio of polymeric material to air voids resulting in a *partial volume effect* that allows the emulation of HU that are in the range required for lung tissues [10]. These methods provide an emerging DFAM opportunity that is compatible with generative design methods and can be used to algorithmically generate patient-specific AM-RDP based directly on medical image data (Fig. 9.8).

9.6 MJT technology summary and DFAM opportunities

MJT technologies provide a unique DFAM opportunity, whereby an extensive array of sophisticated polymers can be deposited on demand within a 3D voxel structure. These design capabilities are extensively applied in the fabrication of high-value product in a range of commercial applications. In particular, the use of MJT as a pre-surgical planning tool has become a clinical standard for complex cases where aesthetic and mechanical function must be visualized. The application of MJT for functional engineering applications is relatively low, and associated DFAM opportunities exist, especially as the characterization of MJT mechanical and geometric properties is increasing. For example, the utilization of MJT as an enabling manufacturing method for AM-RDP is increasing, as illustrated in the presented case study. This translation into clinical best practice is likely to continue, especially as medically certified generative DFAM tools become available to allow direct translation of the clinical requirements to a manufactured artefact. Furthermore, it is likely that as patents expire, and as fundamental hardware elements become commercially available as commodity products, that there will be an increase in low-cost MJT hardware as well as the emergence of the application of these technologies in the hobbyist and research sectors.

References

International standard reference for principles and terminology of AM

[1] ISO/ASTM 52900:2015(en). Additive manufacturing — general principles — terminology. Geneva, Switzerland: International Organization for Standardization (ISO); 2015.

Clinical best practice case study for pre-surgical planning

[2] Gillaspie EA, Matsumoto JS, Morris NE, Downey RJ, Shen KR, Allen MS, Blackmon SH. From 3-dimensional printing to 5-dimensional printing: enhancing thoracic surgical

planning and resection of complex tumors. The Annals of Thoracic Surgery 2016;101(5): 1958–62.

Computational methods for the simulation of as-manufactured AM lattice structures

[3] Lozanovski B, Leary M, Tran P, Shidid D, Qian M, Choong P, Brandt M. Computational modelling of strut defects in SLM manufactured lattice structures. Materials & Design 2019; 171:107671.

Innovative application of MJT to substrate materials

[4] Dilag J, Chen T, Li S, Bateman SA. Design and direct additive manufacturing of three-dimensional surface micro-structures using material jetting technologies. Additive Manufacturing; 2019.

Characterization of MJT geometric process capabilities

[5] Yap YL, Wang C, Sing SL, Dikshit V, Yeong WY, Wei J. Material jetting additive manufacturing: an experimental study using designed metrological benchmarks. Precision Engineering 2017;50:275–85.

Commercial best-practice example of generative integration of GD&T data

[6] Armillotta A. A method for computer-aided specification of geometric tolerances. Computer-Aided Design 2013;45(12):1604–16.

Characterization of the effect of MJT build orientation on mechanical properties

[7] Vu IQ, Bass LB, Williams CB, Dillard DA. Characterizing the effect of print orientation on interface integrity of multi-material jetting additive manufacturing. Additive Manufacturing 2018;22:447–61.

Optimal design by the use of functionally graded materials

[8] Kaweesa DV, Meisel NA. Quantifying fatigue property changes in material jetted parts due to functionally graded material interface design. Additive Manufacturing 2018;21:141–9.

Seminal reference on RDP design

[9] White DR, Booz J, Griffith RV, Spokas JJ, Wilson IJ. Tissue substitutes in radiation dosimetry and measurement. International Commission on Radiation Units and Measurements Report 1989;44.

Review of clinical application of AM technology

[10] Tino R, Yeo A, Leary M, Brandt M, Kron T. A systematic review on 3D-printed imaging and dosimetry phantoms in radiation therapy. Technology in Cancer Research & Treatment 2019;18. https://doi.org/10.1177/1533033819870208.
[11] Rengier F, Mehndiratta A, Von Tengg-Kobligk H, Zechmann CM, Unterhinninghofen R, Kauczor HU, Giesel FL. 3D printing based on imaging data: review of medical applications. International Journal of Computer Assisted Radiology and Surgery 2010;5(4):335−41.

Generative design of AM-RDP

[12] Leary M, Kron T, Keller C, Franich R, Lonski P, Subic A, Brandt M. Additive manufacture of custom radiation dosimetry phantoms: an automated method compatible with commercial polymer 3D printers. Materials & Design 2015;86:487−99.

Algorithmically generated MJT lung phantom

[13] Hernandez-Giron I, et al. Development of a 3D printed anthropomorphic lung phantom for image quality assessment in CT. Physica Medica 2019;57:47−57.

Dual-material MJT manufacture to simulate mechanical tissue response

[14] Wang K, et al. Dual-material 3D printed metamaterials with tunable mechanical properties for patient-specific tissue-mimicking phantoms. Additive Manufacturing 2016;12:31−7.

Fluid permeable liver phantom

[15] Low L, et al. 3D printing complex lattice structures for permeable liver phantom fabrication. Bioprinting 2018;10:e00025.

Vat polymerization

10

Vat polymerization (VPP) is an ISO/ASTM classification for AM technologies in which liquid photopolymer in a vat is selectively cured by light-activated polymerization' [1]. Vat polymerization technologies utilize an extensive existing understanding of polymer chemistry, and the commercial availability of curable polymers, combined with the ready availability of the associated machine elements and technologies. VPP was one of the first AM technologies to be implemented commercially and has received considerable commercial uptake. Consequently, VPP systems are a technically robust turn-key technology.

Advantages of VPP systems include the opportunity to provide precise feature size (determined by the optics of the photo-curing system) combined with a relatively large build volume. Furthermore, the fluid tank provides some buoyancy to the cured polymer[1] and therefore the opportunity for the generation of complex self-supporting geometry is high. In many ways VPP systems provide a *technology complement* to thermoplastic extrusion MEX systems, in that they allow high resolution polymer production, albeit with lower deposition rates and commensurately higher cost per unit volume. VPP systems are cooled by the surrounding fluid polymer, and transient thermal effects are negligible. Overall VPP manufacture is well understood and predictable from a manufacturing standpoint and is therefore highly compatible with process automation and generative design methods (Chapter 7). This commercial capability is illustrated by the extensive application of VPP technology to the manufacture of patient-specific surgical guides and surgical planning models.

Technical challenges exist that are inherent to the typical VPP architecture, curing method, and materials of interest. These challenges include: the requirement for a single polymerization material; potential for movement of this fluid material during manufacture; and, the potential for encapsulated fluid materials. The effect of partially cured monomers is also relevant, and their potential toxicity may be a challenge to MEMS, biological and medical applications, as is the potential for post-manufacture instability of the photo-cured polymers. VPP systems are cured by light-activated polymerization. This light source can be a laser with high energy density and physically small spot-size, which allows very fine feature size but potentially requires extensive processing time.

The commercial technology development of VPP systems has focused on performance optimization in terms of curing rate, including high-power systems, the use of multiple concurrent light sources and increase in physical build volume. Novel research extensions of interest include post-processing to remove uncured monomers,

[1] This buoyancy can be a challenge to cured materials that increase or decrease density on curing, as they either tend to sink or float. Neutral buoyancy of cured polymer is a desirable attribute of VPP systems.

Design for Additive Manufacturing. https://doi.org/10.1016/B978-0-12-816721-2.00010-5

innovative VPP architectures and material systems that allow rapid processing while maintaining high resolution, topology optimization methods that can accommodate process-relevant variables and process-feedback systems that aid in quality control and production certification.

A highly relevant development in VPP technology is the decreasing cost and increasing build rate of commercial systems. This reduced economic barrier-to-entry is enabling a new generation of innovation within the VPP technology space. This innovation may result in an increase in open-source and research outcomes, similar to what has been observed in the Material Extrusion (MEX) technology space.

10.1 Technology history and current methods

The quintessential application of VPP technology is the Stereolithographic (SLA) system. Modern photolithographic AM technologies appeared in the 1970s and stereolithography was commercially developed in the 1980s [2]. Robust techno-economic attributes have enabled SLA to become a mature and widespread industrial technology. The typical SLA technology architecture is implemented with a bath containing a photopolymeric resin. This resin is locally cured by the application of a light-source that causes irreversible polymerization, thereby locally solidifying the intended part geometry. A laser light-source is technically beneficial due to the high local energy density and the dimensional independence of the beam profile to distance.[2] Laser positional control is achieved with a Cartesian hardware setup whereby a liquid permeable bed translates in the build-direction and a galvanometer controls planar positioning of the light source (Fig. 10.1).

Although SLA is possibly the most well-known and extensively applied VPP technology variant, other commercially robust implementations exist, in particular Digital Light Processing (DLP). The distinguishing feature is that SLA uses a galvanometer-controlled light source whereas DLP utilizes a digital light projection across the entire manufacturing plane. Advantages and disadvantages exist for the DLP implementation; processing times can be reduced, as the rate of polymerization is potentially increased, however unique resolution challenges exist due to the inherently voxelated representation of the associated exposure mask (Fig. 10.2).

10.2 Materials of interest

Photo-curable polymers are based on well-developed chemistry and are readily available in commercial grades. Photopolymerization refers to a chemical transformation whereby a one-part uncured thermoset is converted to a cross-linked polymer. This on-demand chain-reaction is initiated by light exposure, which in turn generates

[2] Although the laser light intensity profile is dimensionally coherent, this profile becomes elliptical for planar locations that are not directly below the galvanometer. This non-circular profile results in the potential for geometric dependence on planar location and is exacerbated as the lateral distance increases.

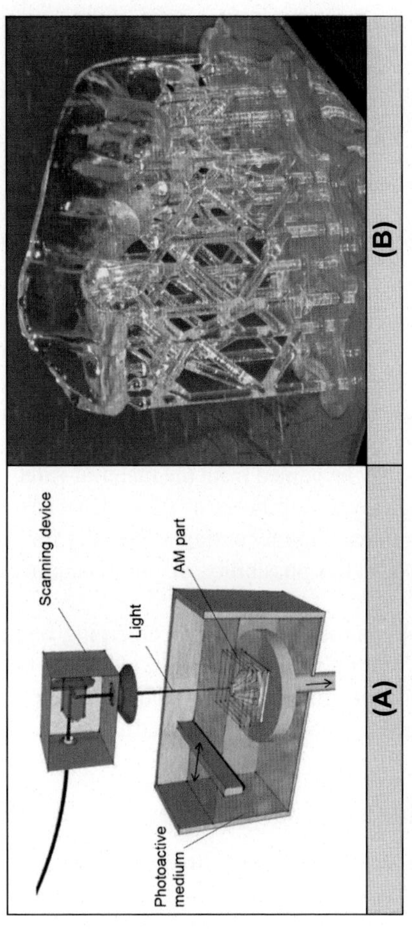

Fig. 10.1 (A) Schematic representation of stereolithography (SLA) system [3], (B) Commercial implementation of SLA product for medical application [4].

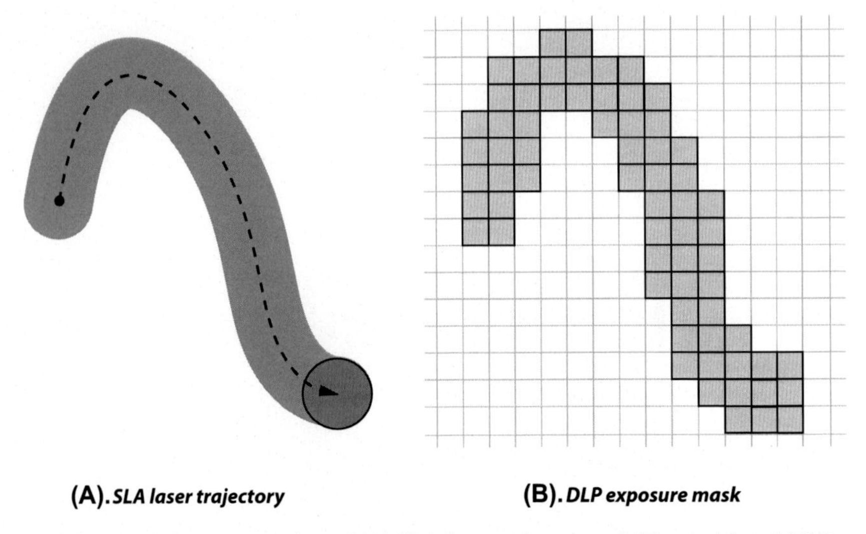

Fig. 10.2 Schematic representation of (A) SLA laser trajectory and (B) voxel-based DLP exposure mask. Relative feature dimensions depend on the specific technology implementation.

ions or free radicals (reactive molecules with an unpaired electron). Many commercial VPP systems use proprietary photopolymeric chemistries that are not publicly documented (although insights may be gained from the material-safety data that is provided for regulatory approval). Relevant attributes of these materials include any potential risks associated with incomplete polymer curing, especially for systems with chemical or biological interaction, and the possibility of post-manufacture instability of the photo-cured polymers.

Further to these typical commercial materials, emerging research identifies opportunities for innovation in VPP polymers of interest, including biologically compatible polymer formulations, polymers that can modify their structure in response to external stimuli, and fibre-reinforced polymers (Section 10.4).

10.2.1 Biocompatible polymer systems

Biocompatible VPP polymer systems have attracted significant research attention for biological applications [5]. In response to this, numerous biocompatible and biodegradable polymer systems have been developed with VPP, including trimethylene carbonate (TMC), ε-caprolactone (CL) and polymer hydrogels.

10.2.2 Shape memory polymers

Shape Memory Polymers (SMP) provide a programmable stimuli-response, the one-way transformation of *heat-shrinking* polymers being the most well-known commercial example. The integration of sophisticated SMP polymers within the VPP technology is an emerging research space that will enable innovative DFAM outcomes.

10.2.3 Composite materials

Although VPP systems are commercially restricted to monolithic photopolymers, emerging research has demonstrated the technical feasibility of fibre-reinforced composite structures. These hybrid polymers still require significant development effort to be commercially robust but are likely to expand the role of commercial VPP applications.

10.3 Commercial application

Vat polymerization technologies are commercially mature, and stable commercial applications have developed, including pre-production validation of aesthetic and functional products, the fabrication of custom surgical tools and the manufacture of custom fluidic and MEMS devices. These applications are described below and opportunities for innovative applications are introduced in the following sections.

10.3.1 Pre-production validation

VPP processes are especially suited to the emulation of plastic injection moulded products. Due to a fine feature size combined with large production envelope, VPP methods can be readily used to assess production aesthetics and optimize product function without the need to manufacture expensive tooling. In these scenarios, VPP may be also be used for commercial mass-customization of for series-production of low-volume product.

10.3.2 Surgical guides, devices and pre-operative planning models

Surgical guides benefit from patient-specific design flexibility and require a relatively fine feature size and high geometric precision. VPP systems are now extensively applied in the fabrication of patient-specific surgical guides, especially in dental applications, where in-clinic guide fabrication is becoming an industry standard. Surgical devices and pre-operative planning models are also routinely fabricated with this technology. The application of VPP for implantation is potentially restricted if residual partially-cured photopolymers could cause harm (Fig. 10.3).

10.3.3 Fluidic devices

Micro-fluidic and microelectromechanical (MEMS) devices are an emerging technology with great innovation potential. These devices control fluid dynamics and utilize the physics of small-scale interactions to achieve functionality that is unique to the micro and nano scales. These devices require complex geometry to be fabricated,

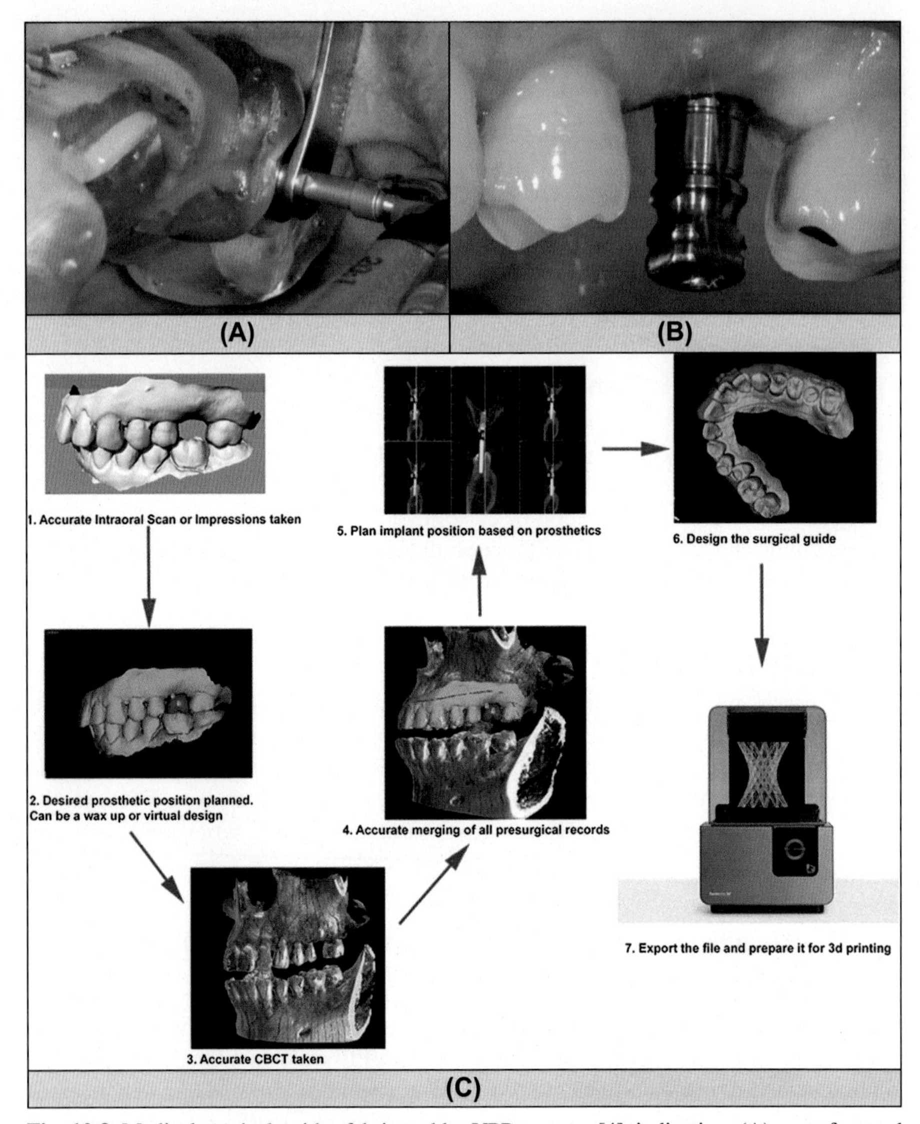

Fig. 10.3 Medical surgical guides fabricated by VPP systems [4], indicating, (A) manufactured guide, (B) in-situ implant, and (C) systematic generative workflow for the implementation of dental surgical guides based on VPP technology [4].

preferably directly from a digital representation and with some generative design system such that highly complex devices can be readily and robustly implemented. VPP systems are compatible with these techno-economic requirements. The potential biological and chemical interaction with partially-cured photopolymers may be a consideration.

10.3.4 Generative methods in VPP applications

The manufacturability and associated response of commercial VPP systems is relatively well understood. This robust understanding has enabled the development of expert systems and algorithmic methods for the generative application of VPP methods to commercial challenges; an opportunity that is routinely commercialized in clinical dental scenarios (Fig. 10.3).

10.4 Technical challenges and emerging research

VPP provides a technologically mature AM opportunity that enables the relatively inexpensive fabrication of high-value products and systems with high resolution, in a manner that is compatible with generative methods. Despite this clear techno-economic opportunity, there remain significant technical challenges that limit the commercial and clinical application of the VPP technology. These challenges are being addressed by emerging DFAM innovations, especially opportunities to challenge traditional notions of VPP as being associated with monolithic structures; including, multi-material systems, innovation in support structure design, generative methods for functional optimization and the incorporation of fibre-reinforced systems (Fig. 10.4).

10.4.1 Multi-materials

A reasonable design assumption seems to be that VPP systems be restricted to a single photopolymeric material. This presumed limitation is being challenged by innovative research groups that have demonstrated, for example, the co-fabrication of multi-material VPP structures, including the deployment of distinct materials within a single layer [6].

10.4.2 Alternate supports

VPP support structures are well implemented for accommodating inherent fabrication limitations, including overhang and island features, such that available DFAM tools can ensure robust manufacture, and are routinely integrated within generative design methods (Fig. 10.3). Emerging research considers alternate methodologies that provide functional advantages. For example, the application of water-based support structures that are self-cleaning and do not consume non-recyclable photopolymers [7].

10.4.3 Generative topology optimization

Topological optimization methods are technically well developed and supported in open-source and commercial DFAM tools. The extension of these optimization methods to the generative design of VPP systems provides a commercially relevant

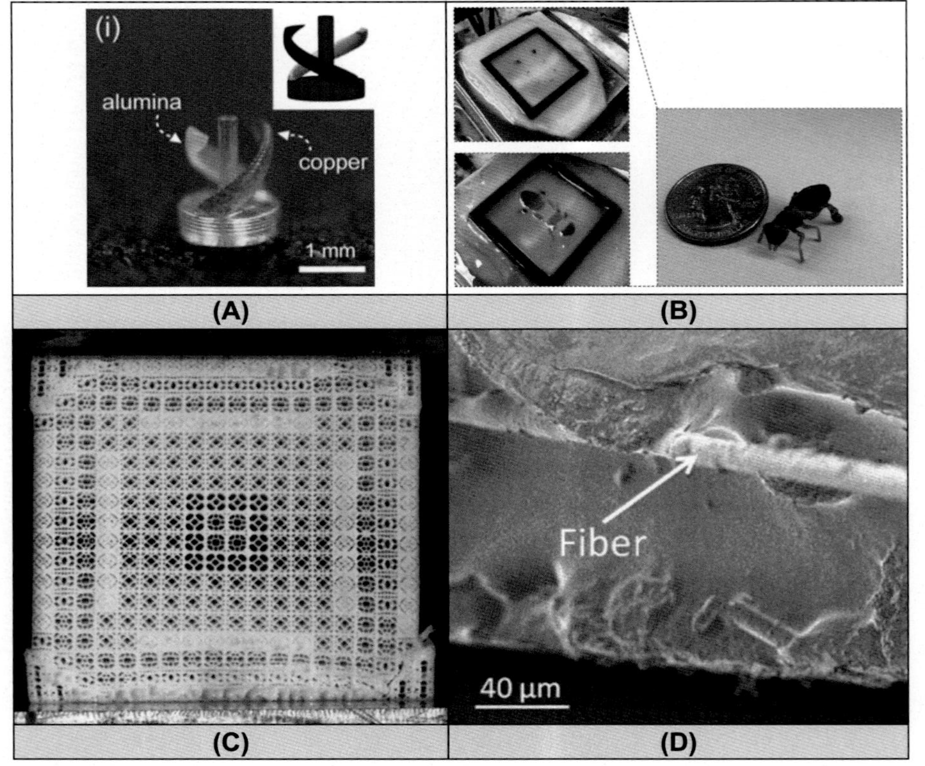

Fig. 10.4 Potential commercial innovation for VPP systems, including (A) innovative approaches to accommodate multi-materials [6], (B) water-based support structures [7], (C) generative topology optimization for the deployment of functionally optimal VPP structures [8], (D). Fibre-reinforced materials [9].

opportunity for the inexpensive fabrication of low-cost structures that are algorithmically optimized for their intended function [8].

10.4.4 Fiber-reinforced materials

Traditional VPP applications are based the fabrication of monolithic structures based on photopolymeric materials. These photopolymers have relatively modest structural properties. Recent innovation has provided insight into the commercial opportunities for structurally-reinforced composites fabricated by VPP methods. This research opportunity has demonstrated increase in mechanical properties of an order of magnitude [9].

These emerging technology innovations are indicative of the potential for DFAM extensions to the traditional application of VPP for monolithic mono-material structures.

10.5 VPP technology summary and DFAM research opportunities

VPP provides a unique AM technology platform that has matured both commercially and clinically over recent decades. The commercial application of VPP systems is consequently well established, especially for the pre-production validation of aesthetic and technical products, surgical guides and micro-fluidic devices. The VPP systems applied in these roles are technically robust, and in contrast to some other AM technology classifications, have well defined manufacturability. In response to this robust understanding, VPP systems are very well suited to generative design methods, and a primary DFAM opportunity is the definition of robust generative methods that integrate with commercial and custom VPP hardware and polymers to meet specific customer needs.

Fundamental research of commercial merit includes the development of unique polymeric systems, particularly biocompatible and biodegradable alloys, SMP and composite materials. These material innovations are in turn compatible with emerging technological developments, such as multi-material and fibre-reinforced structural components that are generatively-optimized and manufactured support-free these innovations are revolutionizing the application of VPP systems, leading to a reassessment of established notions of the technical limits of VPP systems.

The innovative research outcomes currently emerging within VPP technology enable unique DFAM opportunities and are, to an extent, 'solutions waiting for a problem' that will provide a distinct commercial and clinical opportunities for commercial research teams that are sufficiently innovative and agile to utilize these opportunities.

References

International standard reference for principles and terminology of AM

[1] ISO/ASTM 52900:2015(en), Additive manufacturing — general principles — Terminology. Geneva, Switzerland: International Organization for Standardization (ISO); 2015.

Comprehensive review of the SLA process

[2] Bártolo PJ, Gibson I. History of stereolithographic processes. In: Stereolithography. Boston, MA: Springer; 2011. p. 37–56.

Review of laser AM systems

[3] Hagedorn Y. Laser additive manufacturing of ceramic components: materials, processes, and mechanisms. In: Laser additive manufacturing. Woodhead Publishing; 2017. p. 163–80.

Clinical best-practice example of surgical guide manufacture

[4] Whitley III D, Eidson RS, Rudek I, Bencharit S. In-office fabrication of dental implant surgical guides using desktop stereolithographic printing and implant treatment planning software: a clinical report. The Journal of Prosthetic Dentistry 2017;118(3):256–63.

SLA biomedical applications

[5] Melchels FP, Feijen J, Grijpma DW. A review on stereolithography and its applications in biomedical engineering. Biomaterials 2010;31(24):6121–30.

Multi-material SLA printing

[6] Han D, Yang C, Fang NX, Lee H. Rapid multi-material 3D printing with projection micro-stereolithography using dynamic fluidic control. Additive Manufacturing 2019;27:606–15.

Innovative SLA support structures

[7] Jin J, Chen Y. Highly removable water support for Stereolithography. Journal of Manufacturing Processes 2017;28:541–9.

Generative design of optimised SLA structures

[8] Liu T, Guessasma S, Zhu J, Zhang W, Belhabib S. Functionally graded materials from topology optimisation and stereolithography. European Polymer Journal 2018;108:199–211.

Fibre reinforced SLA

[9] Sano Y, Matsuzaki R, Ueda M, Todoroki A, Hirano Y. 3D printing of discontinuous and continuous fibre composites using stereolithography. Additive Manufacturing 2018;24: 521–7.

Powder bed fusion

11

Powder Bed Fusion (PBF) is an ISO/ASTM classification for AM technologies 'in which focused thermal energy is used to fuse materials by melting as they are being deposited' [1]. The PBF classification is commercially significant, especially for high-value technical products, including the mass-customisation of medical implants with patient-specific mechanical properties and associated geometry; and the series production of highly efficient functional geometry, such as for aerospace componentry. Powder bed fusion technologies are typically implemented with a thermal energy source generated by either a laser or electron-beam. Both implementations are commercially viable, and the selection of the optimal system is a function of the techno-economic attributes of the intended application. Since the inception of PBF methods, the associated technology development has focused on the optimization of system performance in terms of available heating power, number of concurrent power sources, physical build volume, in-situ inspection systems, and process automation.

Despite the significant technology development contributions made for PBF, commercially relevant DFAM research opportunities exist. These include process optimization for known engineering alloys; the specification of material systems that are optimized for the solidification attributes inherent to PBF; process parameter optimization as a function of intended local geometry and associated temperature profiles; toolpath optimization to mitigate thermal defects and local porosity; topology optimization methods that can accommodate process-relevant variables, such as temperature field, as an optimization variable; and process feedback systems that can provide in-situ quality control and aid in production certification.

11.1 Technology history and current methods

Powder bed fusion systems enable commercially significant outcomes and have benefitted from significant technology development by both commercial and government funded research organisations [2]. Early technologies provided only partial fusion, or sintering, whereby further thermal processing was required to allow fusion of the manufactured component. Contemporary methods, for example as developed by Fraunhofer Institute for Laser Technology, allowed sufficient energy density such that the 'metallic powder is fully molten throughout' [3]. These industrial PBF systems are now very much commercially robust turn-key production systems and are typically implemented as a Cartesian hardware system with the following sequential operating phases.

Design for Additive Manufacturing. https://doi.org/10.1016/B978-0-12-816721-2.00011-7

1. System setup, including charging of the powder source, stabilization of thermal heating and establishment of the required operating environment, typically either a vacuum or shielding gas.
2. Powder distribution, whereby a layer of powder is screed across the build platen using various methods including roller, blade or wiper.[1]
3. Selective powder fusion, whereby a thermal heat source selectively fuses material according to the specified process parameters and associated toolpath.
4. Build-platen descent, whereby the build-platen descends to the required layer thickness. Phases 2–4 are then repeated for all required manufactured layers.
5. Manufactured components are removed from the build platen. This process may involve in-situ heat treatment and the controlled restoration of the ambient environment.
6. As-manufactured components are then subject to support material removal processes and heat treatment as required.

PBF systems are available with either laser or electron thermal energy sources: both variants have unique attributes that are variously beneficial for specific commercial applications. The following summary attempts to provide an independent review of the relative merits of these PBF implementations.

Electron Beam Melting (EBM) systems utilize an electron-beam as the thermal energy source. The electron-beam is directed by a series of magnetic lenses with sufficient agility to maintain multiple melt-pools concurrently. EBM-PBF is typically not compatible with ferrous metals due to magnetic interference. The electron beam energy source requires the establishment of a vacuum within the system prior to manufacture. It is relatively time-consuming to establish this vacuum (in comparison with shielding gas environments), but once established it allows multiple technical advantages. For example, the vacuum environment can be readily heated, and does not include convective heat transfer to a shielding gas, thereby allowing elevated operating temperatures and a thermally stable operating environment. Consequently, EBM systems are typically associated with relatively low thermal distortion of the as-manufactured component. EBM systems allow for in-situ support structures to be anchored directly within the powder bed, a capability that is not yet demonstrated in laser-based PBF systems (Fig. 11.1).

Selective Laser Melting (SLM) systems utilize a laser-beam as the thermal energy source. These systems do not require a vacuum environment, but instead utilize inert gas shielding (typically of either argon or nitrogen) to mitigate in-situ oxidization. This scenario allows for potentially faster initialization of the SLM system but does introduce a convective cooling effect, which makes system heating challenging. Consequently, the ambient temperature of commercial SLM systems is typically lower than for EBM systems, resulting in a greater potential for thermal distortion. The laser trajectory is determined by numerically controlled optics, allowing rapid and efficient control of the laser beam. Laser technology allows a small spot size, thereby allowing for a manufactured feature resolution that is typically smaller than for EBM systems. SLM systems operate with finer powder distribution and with smaller layer thickness

[1] Note that due to imperfect packing, the density of unfused powder is lower than that of solid material; consequently the thickness of the layer recoat is larger than the screed layer thickness [16].

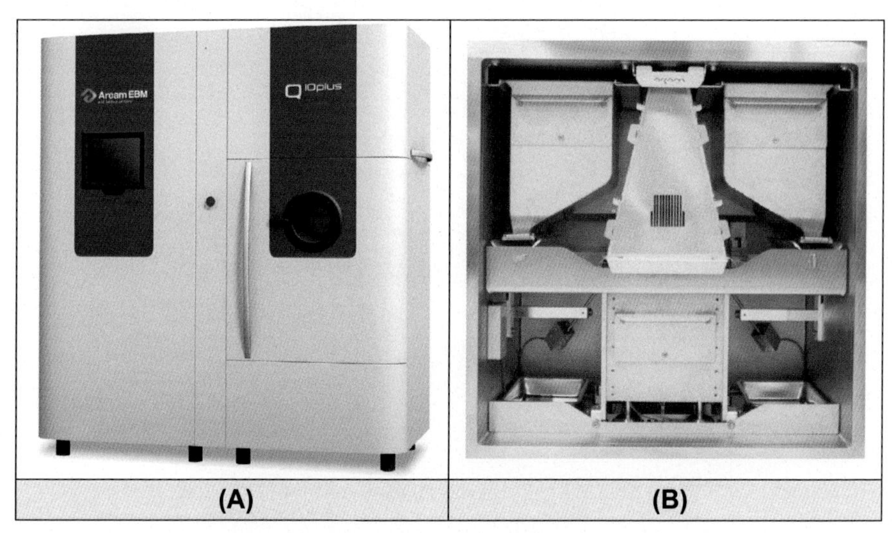

Fig. 11.1 (A) Commercial Electron Beam Melting (EBM) system including (B) enlarged view of vacuum chamber. Courtesy of GE Additive Arcam EBM.

than similar EBM systems, thereby potentially resulting in a finer resolution of manufactured features (Fig. 11.2).

For both EBM and SLM systems, regulated powder fusion is achieved by the precise control of process parameters concurrently with heat source trajectory. Relevant process and toolpath parameters include the system power, P, local toolpath velocity, v, layer thickness, L_t, and hatch spacing, h (Fig. 11.3). Where the *hatch* refers to a series of parallel heat source trajectories typically used to fill the layer *core*; *perimeter* trajectories are

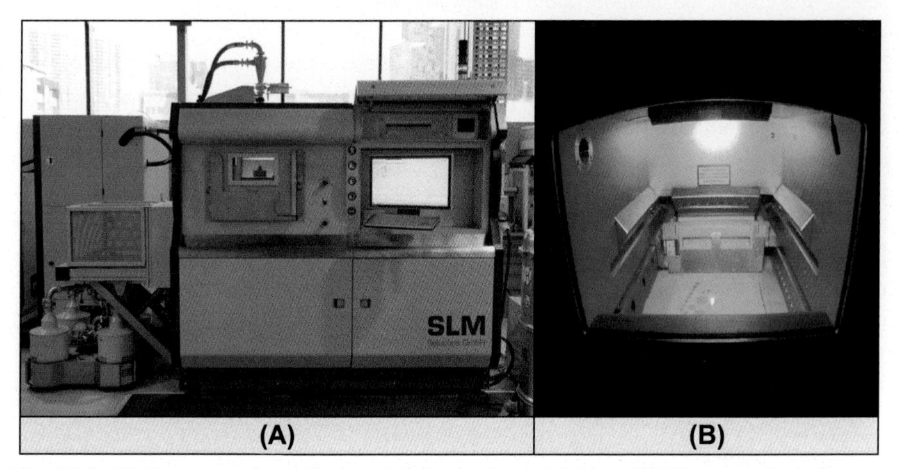

Fig. 11.2 (A) Commercial installation of Selective Laser Melting (SLM) system including (B) enlarged view of processing chamber. Image courtesy of RMIT University Advanced Manufacturing Precinct.

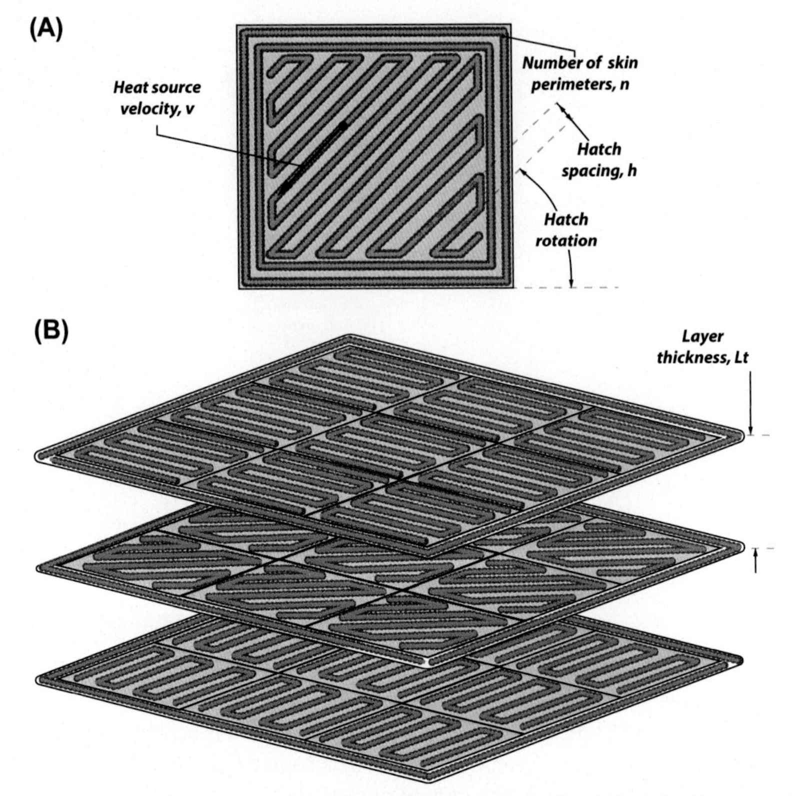

Fig. 11.3 Relevant PBF process and toolpath parameters, including (A) toolpath nomenclature, (B) to alleviate thermal effects an island strategy is commonly applied to sequentially fill manufactured layers; a hatch rotation is specified between consecutive layers to mitigate fusion defects. For clarity, the meltpool width is shown undersized.

then specified to control surface finish. Based on these characteristics, a number of processing parameters may be generated, including the energy density defined in linear, E_l, surface, E_s, and volumetric, E_v, terms (Eqs 11.1–11.3). These processing parameters are commonly reported, and are useful in comparing various input conditions, but are subject to error in calculating the actual absorbed energy. For example, power is typically reported as the net emitted from the energy source, whereas the actual energy received is lower due to reflection and other losses. Of the available energy density measures, volumetric energy density is perhaps the most commonly reported for PBF. However, it assumes that thermal energy is deposited within a precisely defined volume of depth, L_t, which is not typically observed in practice.

$$E_l = \frac{P}{v}, \left[\frac{J}{m}\right] \tag{11.1}$$

$$E_s = \frac{P}{v\ h}, \left[\frac{J}{m^2}\right] \tag{11.2}$$

$$E_v = \frac{P}{v\,L_t\,h}, \left[\frac{J}{m^3}\right] \tag{11.3}$$

The energy density typically associated with commercial PBF systems causes near-instantaneous melting of the input powder. This thermomechanical scenario results in rapid and directional heat transfer, which is fundamental to the PBF metallurgical response. Unlike traditional casting methods where solidification occurs relatively slowly, the cooling rates and thermal gradients for PBF fusion systems are comparatively rapid, resulting in a unique PBF metallographic response. Typical metallographic features include observation of unique melt pool boundaries and anisotropic orientation of crystallographic grains to the build direction (Fig. 11.4).

Commercial PBF applications consist of multiple length scales; for example, the conformal cooling component of Fig. 11.5, has a macroscopic length scale in the build direction, L_z, of 125 mm, with a layer thickness, L_t, of 50 μm, resulting in a total of 2500 unique fused layers,[2] N_z (Eq. 11.4). Furthermore, the nominal component dimension parallel to the platen, L_x, is 45 mm, with a hatch spacing of 100 μm, resulting in 450 unique parallel laser passes, N_x, to completely fuse the layer (Eq. 11.5). Therefore, macroscopic components such as this, consist of many hundreds or thousands of unique layers, each layer consisting of many hundreds or thousands of unique

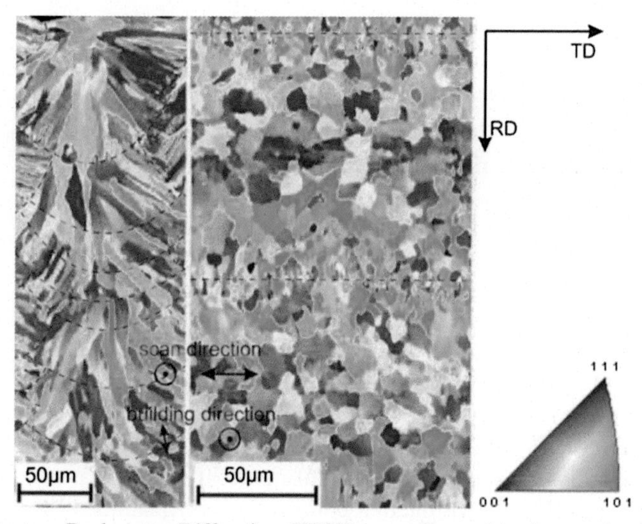

Fig. 11.4 Electron Backscatter Diffraction (EBSD) crystallographic characterization of AlSi10Mg fabricated by SLM indicating anisotropic response to rapid cooling [4]. Observable microstructural attributes include unique melt pool boundaries and grain alignment in build-direction.

[2] To provide an everyday reference, this scenario results in a layer thickness that is less than that of a standard sheet of printer copy paper.

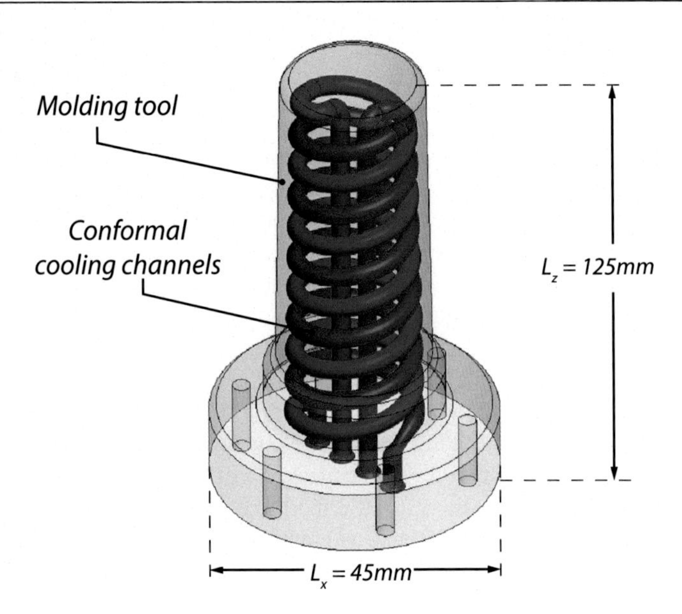

Fig. 11.5 PBF cooling die designed with conformal cooling channels to optimize heat transfer rate.

meltpool passes. Each of these meltpool passes is associated with rapid solidification; consequently, the observed microstructure for PBF components is much more like that of micro-welded structures than of cast specimens. A significant challenge for PBF certification and process control is the management of potential defects that can occur along this extensive network of micro-fused tracks.

$$N_z = \frac{L_z}{L_t} \tag{11.4}$$

$$N_x = \frac{L_x}{h} \tag{11.5}$$

Numerous studies exist on the effect of process parameters on the associated manufactured product, including observations of fused material morphology and measures of *porosity* and *slumping* in order to quantify process setting that result in robustly formable materials (Fig. 11.6). Internal porosity may be quantified by gravimetric or micro-CT methods. Slumping is often referred to qualitatively and refers to the ability of a manufactured test sample to maintain its intended geometry (for example, it is quite feasible to generate specimens with low porosity, but very high distortion due to overheating). Once process parameters are optimized for generic specimens, the manufacturability of specific geometry of interest can be quantified, for example the fabrication of lattice structural elements.

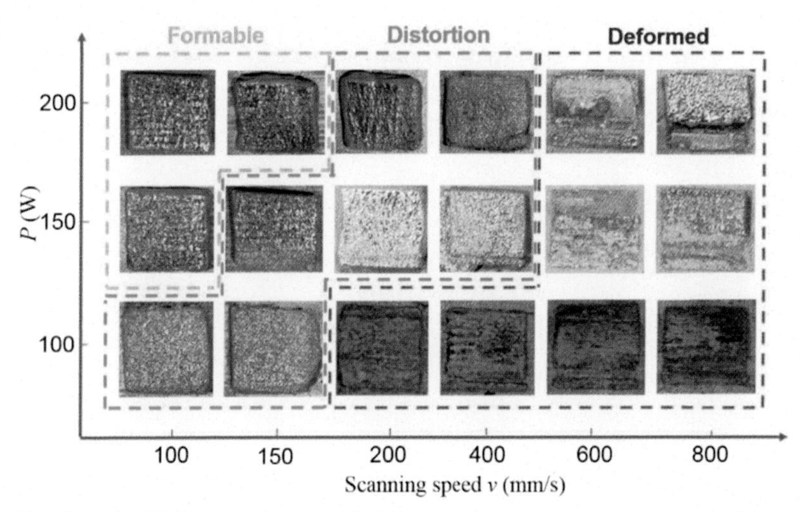

Fig. 11.6 Sample SLM processing map. In this scenario, manufacturing power and scanning speed are applied to quantify zones of robust manufacturability [5].

Defect morphology provides insight into the basis for observed failures and includes potential defects such as balling (a phenomenon where fusion is incomplete); gas-entrapped porosity (associated with gas entrapped during powder manufacture); defects induced by sub-optimal toolpaths, especially associated with core track end-points and skin/core interactions; keyholing porosity (where deep laser penetration induces porosity); and lack of fusion defects where planar defects are induced due to incomplete fusion (Fig. 11.7).

11.2 Materials of interest

As to be expected for an emerging technology, commercial alloys utilized for PBF systems are typically repurposed from established manufacturing technologies, notably from casting and powder-metallurgy applications. These commercially available alloys enable PBF technology acceleration given that alloy composition and processing are technically understood and potential applications are familiar to potential end-users. However, these potential advantages are offset somewhat because these alloys are not tuned to the specific thermo-mechanical processing inherent to the PBF process. The development of PBF-optimised materials remains an open commercial research opportunity.[3]

Modern PBF systems can process a range of candidate metals, including challenging materials such as copper and aluminium, which, for example, can be highly

[3] This research opportunity is challenging, as the certification of novel alloys for high-value applications, such as medical and aerospace, necessitates significant commercial investment.

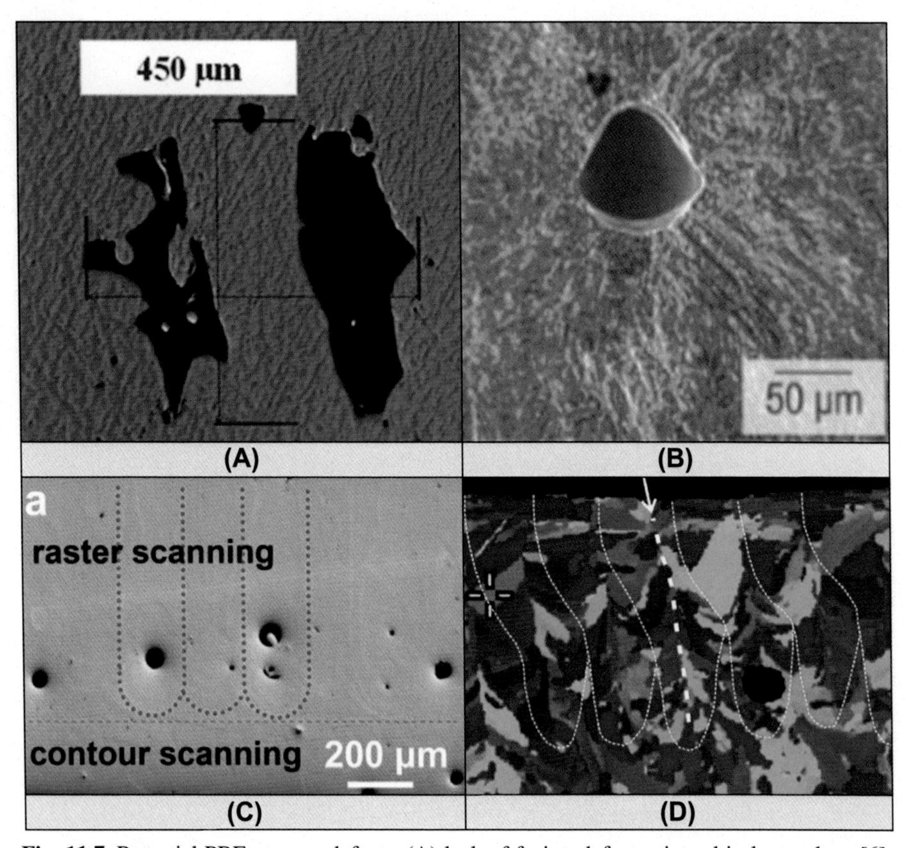

Fig. 11.7 Potential PBF process defects: (A) lack of fusion defects viewed in layer plane [6], (B) entrapped gas pore [7], (C) toolpath induced defects [8] (D) porosity defect associated with keyholing mode [9].

reflective to laser power; as well as materials that are challenging for traditional manufacture, including titanium and Inconel alloys. To be commercially feasible, PBF source materials must not only satisfy the associated technical processing requirements, but also be available in commercial quantities at an acceptable cost. Table 11.1 describes the basic properties of a series of potential PBF alloys, including the following:

- Titanium alloys provide high strength with relatively low density as well as fatigue resistance and high biocompatibility. Furthermore, titanium alloys are challenging to fabricate with traditional methods, but due to their low thermal conductivity are highly suitable to the fabrication of fine resolution PBF geometry. Commercial titanium alloys provide exceptional opportunities for high-value commercial applications; consequently, the Ti64 alloy is the predominantly reported PBF material in the research literature.

Table 11.1 Candidate PBF materials, relevant material properties and classifications.

Material	ρ	T_m	k	C_p	CTE	Comments	Applications
Titanium (Ti64)	4430	1604–1660	6.7	526	8.6	Light-alloy with excellent biocompatibility.	Medical implants, aerospace
Inconel (624)	8440	1290–1350	9.8	410	12.8	Excellent properties at elevated temperature	Gas turbines, rocket engines.
Aluminium (AlSi10Mg)	2670	577	110	915	21	High castability light-alloy with relatively high mechanical properties.	Complex cast geometry.
Copper (CuCr1Zr)	8900	1075–1080	320	380	17	Precipitation hardening copper alloy	Electrodes and electrical contacts.
Tool steel (H13)	7800	1427	24	460	11	High hardenability, wear-resistance and toughness.	Die casting and forging tools.

Generic material properties reported at room temperature for the candidate PBF alloy. Material properties: Density ρ (kg/m^3), Melting temperature, T_m (°C), Thermal conductivity, k (W/Mk), Specific heat capacity, C_p (J/kgK), Thermal expansion coefficient (CTE) (μm/mk).

- Inconel is a technically important engineering alloy for high-value thermal structural applications, including rocket engine nozzles and high temperature gas turbine applications. Even more so than titanium, Inconel is challenging to fabricate with traditional methods and is therefore an exceptional candidate for PBF processing. Various alloys are of technical importance, including Inconel 624 and 718.
- Aluminum alloys enable lightweight design at low cost and are consequently an important engineering alloy. Aluminum is important for AM applications due to the low melting temperature and high thermal conductivity, thereby reducing build time by allowing for relatively large layer thickness, albeit with a relatively large minimum feature size. The AlSiMg alloy system is an example of an industrially accepted casting alloy that is being extensively applied in commercial PBF applications.
- Copper alloys can provide useful thermal, biological and electrical characteristics. Copper is highly reflective and can therefore be challenging to process with in PBF systems, although it is compatible with contemporary PBF systems that are able to generate a high power density at specific laser wavelengths. Few copper alloys are reported in the PBF literature, although CuCr1Zr is an example of one of the emerging PBF alloys.
- Tool steels, which enable the fabrication of complex tool geometry that facilitates rapid thermal cycling, are an important commercial opportunity for PBF systems. H13 is an example of such a material that is reported in the literature for high-value tooling applications.

Powdered materials vary significantly in their fundamental morphology, resulting in significant differences in flowability and packing density within the powder bed. Various relevant powder properties exist, including distributions of particle size (Fig. 11.8), dynamic powder response, and experimental measures of flowability. High packing density is typically achieved by the selection of a broad distribution of powder sizes (in order that pores of various sizes be filled) as well as particles with a high spheroidicity (to avoid interlocking of non-spherical powder).

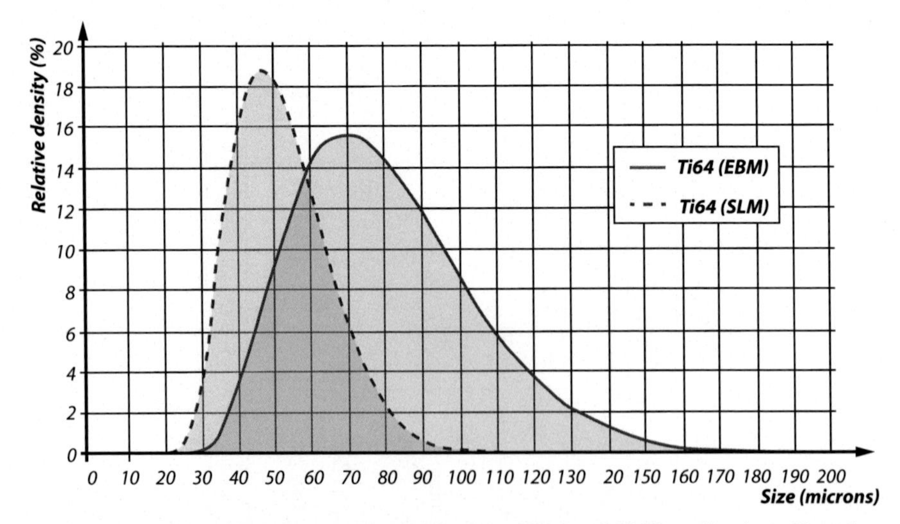

Fig. 11.8 Examples of Ti64 powder distributions for EBM and SLM applications. Note that EBM powder is typically of larger mean size than for SLM.

11.3 Technical challenges to PBF implementation

PBF provides a technologically unique opportunity for the inexpensive fabrication of high-value functional engineered systems with high geometric complexity utilizing technically robust engineering alloys. Despite this profound techno-economic opportunity, there remain significant research challenges that must be addressed for this opportunity to translate into routine commercial engineering practice. These Design for AM (DFAM) challenges include the need for:

- toolpath optimization to eliminate unnecessary porosity defects
- robust tools for sophisticated support structure deployment
- a deep understanding of the underlying structural mechanics and observed mechanical response of high-value PBF structures
- thermal-mechanical models of PBF manufacture, especially idealized models that allow for the rapid approximation of thermo-mechanical response for initial design studies
- experimental quantification of microstructural anisotropy and influence on mechanical response
- quantification of the as-manufactured geometry for common structural elements

11.3.1 Toolpath optimization

At the heart of the PBF technology lies an extensive network of fused meltpool that, over time, integrates to the consolidation of a manufactured product. Extensive DFAM tools exist to enable the automated generation of these toolpaths; however, there exist significant opportunities for enhanced toolpath deployment strategies. Potential challenges to optimized toolpath generation include the potential for a disconnection between the functional requirements of the AM system and the specified toolpath, as well as the obfuscation or encryption of explicit toolpath data. These challenges are significant: the former is relevant to technical process optimization and the latter is a particular challenge for the manufacture of safety-critical and high-value componentry that requires process certification.

For example, the toolpaths of Fig. 11.9 are commercially generated for an SLM process to fabricate inclined cylinders for high-value lattice structures. These toolpaths consist of the typically applied skin-and-core strategy (Fig. 11.3). This strategy is well suited to bulk structures but is potentially challenged by scenarios that do not conform well to the standard layout, including small features, features that transition from skin only to skin-and-core, and features with high aspect ratio, such as inclined cylindrical features. These scenarios can result in sub-optimal toolpaths that introduce unintended porosity defects: a particularly relevant challenge for the fabrication of complex geometry (such as lattice structures) that enable the techno-economic benefits of PBF.

This challenge presents an advanced DFAM opportunity in the development of toolpath generation methods that allow the intended function of the PBF structure to be accommodated, especially for high-value structures such as lattice and Triply

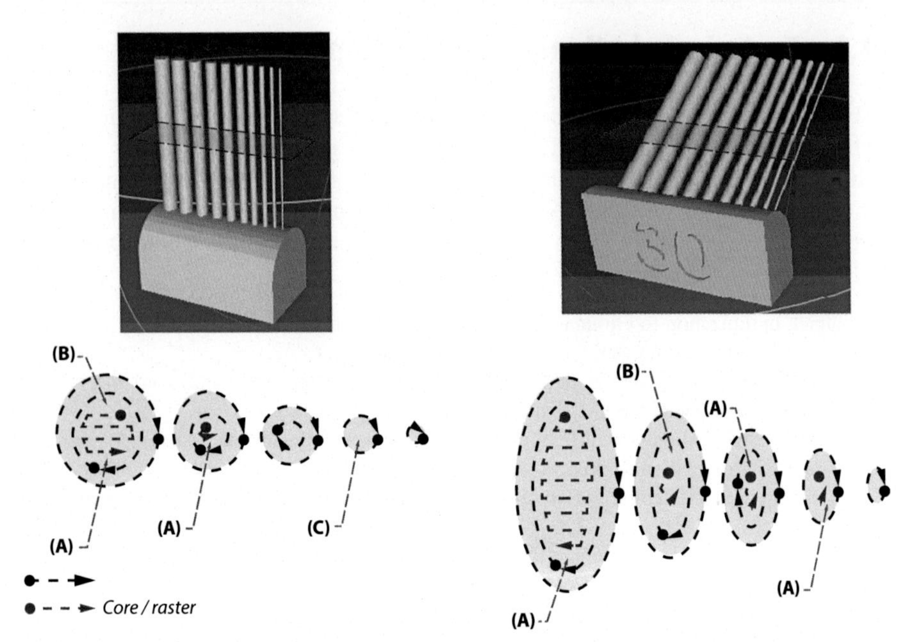

Fig. 11.9 Example SLM toolpath generated for vertical ($\alpha = 0°$, left) and inclined ($\alpha = 30°$, right) cylindrical specimens. Arrow indicates sites of recurring porosity defects due to (A) interaction between toolpath endpoints, (B) insufficient core fill, (C) insufficient skin fill.

Periodic Minimal Surfaces (TPMS) (Chapter 5). These opportunities can be integrated into the design of AM-capable CAD systems (Chapter 3).

11.3.2 Support structure deployment

PBF support structures are subject to complex and often conflicting design requirements. PBF is an inherently thermal process and thermal loads from the solidifying meltpool must be grounded in a controlled manner to ensure robust manufacture. Excessive local temperatures result in manufacturing failure either by thermal overload or excessive thermal distortion (Fig. 11.10). These challenges can be met in PBF systems by the deployment of supporting structures such that thermal flow is increased, and thermal distortion is physically restrained. These support structures are typically deployed manually, based on the experience of the designer; however, this intuitive approach can lead to overdesign or flawed manufacturing outcomes. The latter being a particular challenge for economical low-volume production with PBF systems. In responses to this challenge, prescriptive DFAM methodologies are emerging by which the thermo-mechanical response of specific AM support materials is quantified (Fig. 11.11); however, such data are sparsely reported in the literature, resulting in a significant emerging research opportunity.

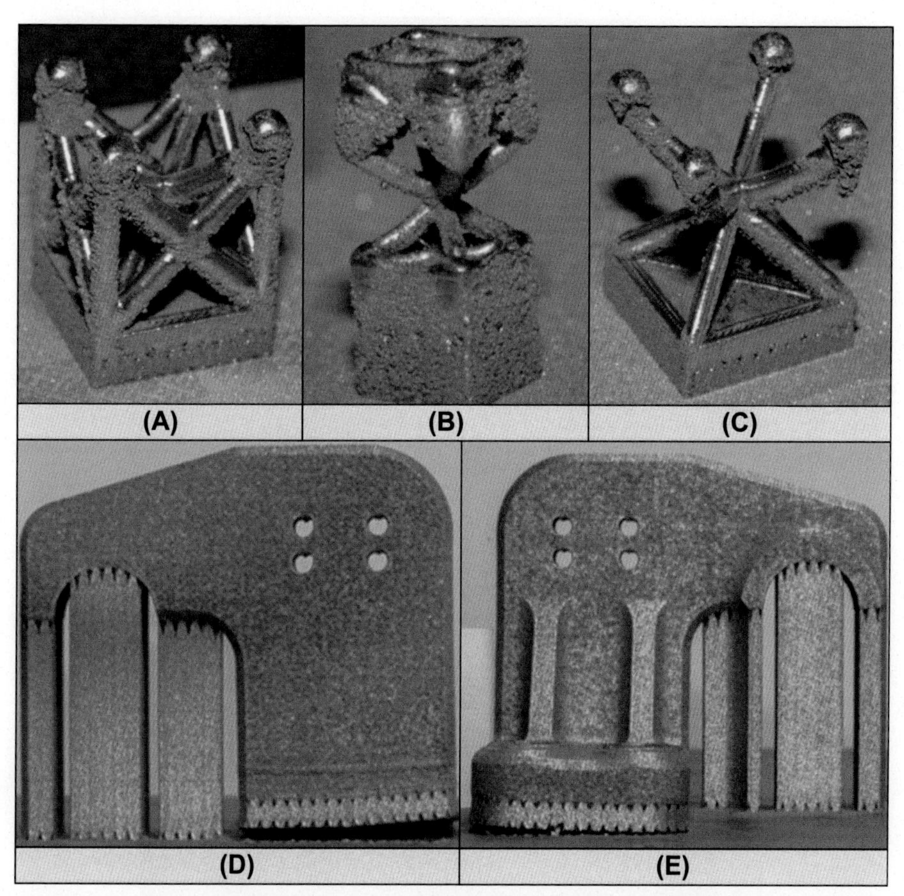

Fig. 11.10 PBF structures are subject to intense thermal loading. This loading can result in a failure to achieve robust manufacturing outcomes due to excessive local temperatures, as for the SLM aluminium lattices structures of (A–C). PBF are typically defined manually for PBF systems due to their inherent complexity in providing both support structures and thermal load paths. This manual DFAM outcome can result in failed manufacturing outcomes (D and E).

11.3.3 Structural mechanics of high-value PBF structures

PBF systems are eminently compatible with the fabrication complexity necessary for the manufacture of complex structures such as Triply Periodic Minimal Surfaces (TPMS) and sophisticated lattice structures. Despite this opportunity, the structural response of such structures is somewhat uncertain, especially for conformal structures with non-uniform periodicity and structures that intent exploit nuances in mechanical response that are not directly compatible with the failure modes prescribed by the Gibson-Ashby model (Fig. 11.12). Furthermore, much of the structural simulation

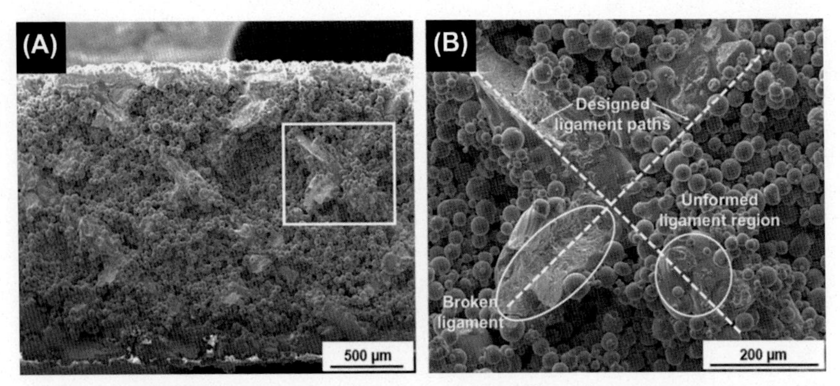

Fig. 11.11 Emerging research, such as this study on the effect of support structure design on mechanical and thermal response completed for SLM Ti64 [10].

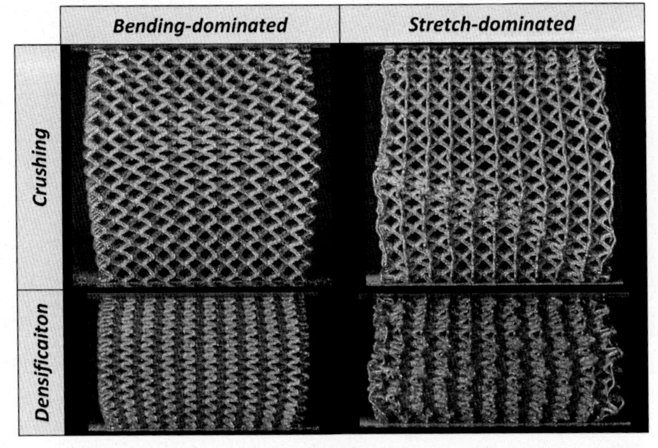

Fig. 11.12 Observed mechanical response of SLM lattice specimens fabricated from Inconel 625 displaying bending-dominated (left) and stretch-dominated response (right). (See Chapter 5).

research is focussed on highly idealized representations of the PBF structure. This idealization is due to the combined challenges of uncertainty in the actual geometry of as-manufactured PBF structures, as well as the computational challenges inherent to the simulation of complex multi-scale structures. Despite these challenges, advanced DFAM contributions are emerging that more correctly accommodate the complexities of PBF manufactured structures (Fig. 11.13). The opportunities and challenges associated with AM design of high-value structures are presented in detail in Chapter 5.

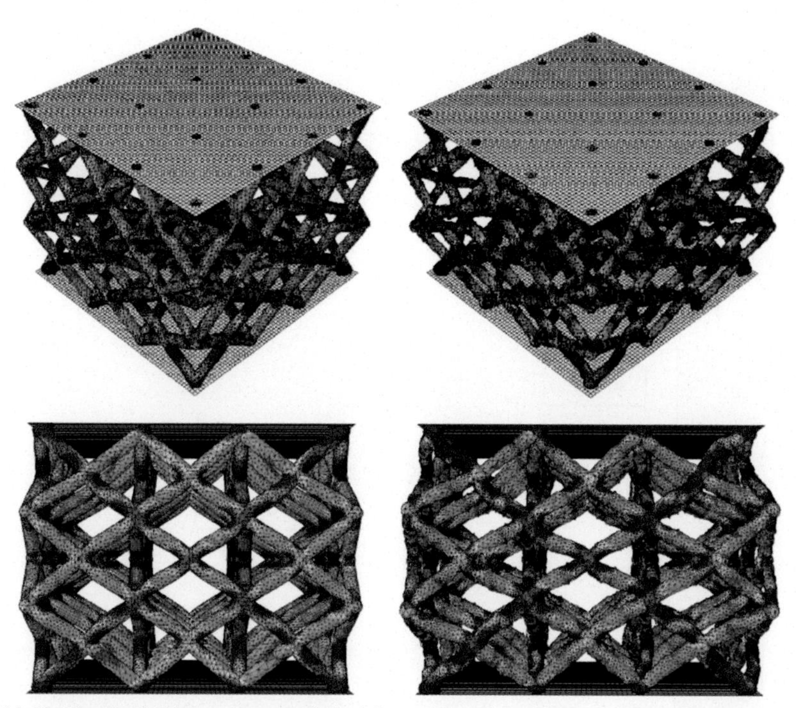

Fig. 11.13 Idealized CAD geometry (left) fails to capture the nuanced mechanical response of the pseudo-random strut elements (right). (See Chapter 5).

11.3.4 Mechanical response and influence of microstructural anisotropy

The rapid solidification inherent to the PBF process results in a microstructure that is highly textured and anisotropic. This microstructure includes unique overlapping melt-pool boundaries with variable grain structures, microsegregation and associated porosity defects. These anisotropic features align with the specimen loading direction according to the alignment of the specimen during manufacture (Fig. 11.14). This fundamental microstructural anisotropy and alignment to the loading direction can result in variability in reported strength and ductility. This response is a function of process parameters, alloy system and local thermal profile. DFAM design data to robustly quantify this response is emerging; however, much valuable research understanding remains to be acquired, especially as commercial PBF systems increase their productivity and therefore vary the associated process parameters.

11.3.5 As-manufactured feature quantification

The manufacturability of real world PBF structures is complex, and geometric artefacts are challenging to predict. In order to confidently understand these as-manufactured structures, coupon level structures must be fabricated (Fig. 11.15). These data create

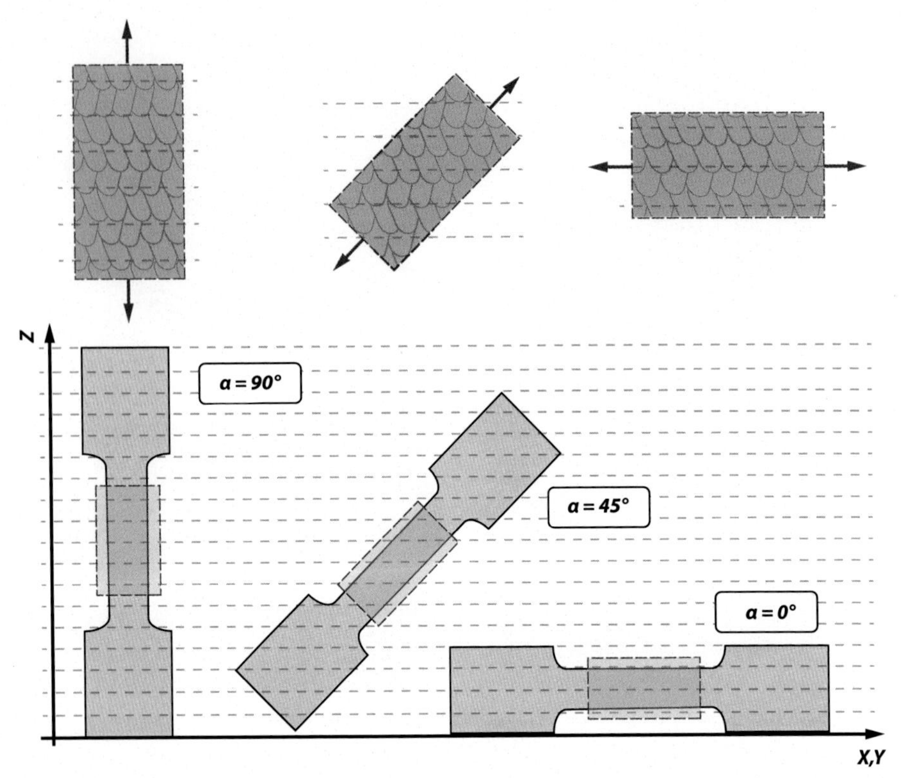

Fig. 11.14 Influence of specimen inclination to platen, α, on microstructural texture. Red vectors indicate tensile loading direction.

a DFAM database of as-manufactured structure response, which then enables increased confidence in manufacturability, the application of generative design methods (Chapter 7), and design input for advanced numerical simulation methods (Chapter 5). Such data are emerging within the DFAM literature, for example [11,12]; however, comprehensive databases of as-manufactured structural response remains relatively limited for PBF systems.

11.4 Application case studies

Powder bed fusion systems provide an exceptional opportunity for the commercial and clinical implementation of high-value systems that are not technically or economically feasible with traditional methods. The successful application of PBF requires the application of economic, digital and design DFAM strategies (Chapters 2, 3 and 4 respectively) such that high-value PBF product is reliably and repeatably produced to the intended functional requirements. These high-value applications benefit from generative methods that reduce the manual design burden and aid in product certification (Chapter 7). The following case studies provide insight into the application of PBF technologies for the commercial development of medical and aerospace systems.

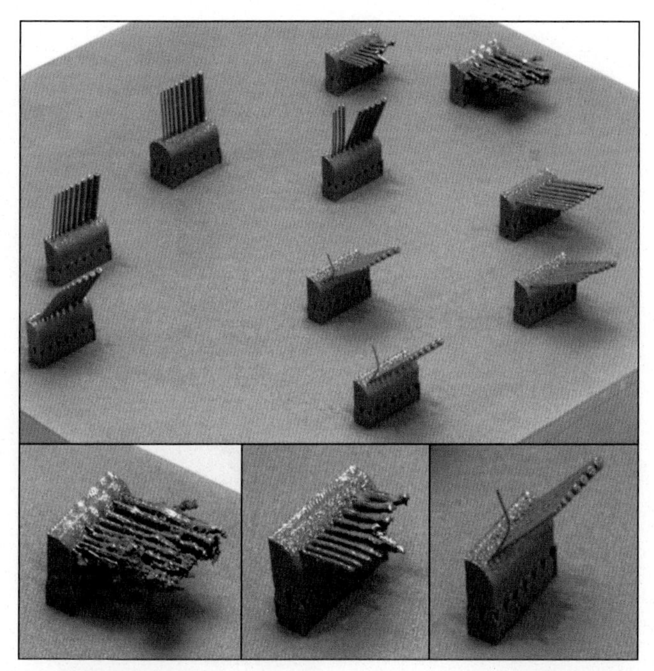

Fig. 11.15 Example of manufactured structures to assess, in this case, the manufacturability of inclined lattices strut elements.

11.4.1 PBF case study: medical applications

The techno-economic attributes of PBF are eminently suited to the fabrication of both patient-specific and series production medical devices. Both scenarios are strictly governed by medical certification requirements; however, the technical requirements of patient-specific and series production are somewhat unique. Series production requires that design and manufacturing methods be demonstrated; however, each manufactured product has the financial and time resources required for extensive physical testing and certification. The design of patient-specific devices is typically in response to an urgent or otherwise inoperable patient requirement and would rarely allow the same level of component testing associated with a series manufactured product. For patient-specific applications, these design challenges are specifically associated with robust algorithmic design, reference to approved series production devices, and non-destructive physical testing of the manufactured device. Case studies for both patient-specific and series production medical applications are demonstrated below.

PBF series production provides a unique opportunity for the fabrication of moderate to high production volumes[4] of medical devices with clinically valuable features, such as prescribed roughness surfaces for cellular attachment, which would be technically

[4] The potential for incorporating a range of unique size-variants as is enabled by the flexibility of AM processes and generative design methods (Chapters 2 and 7).

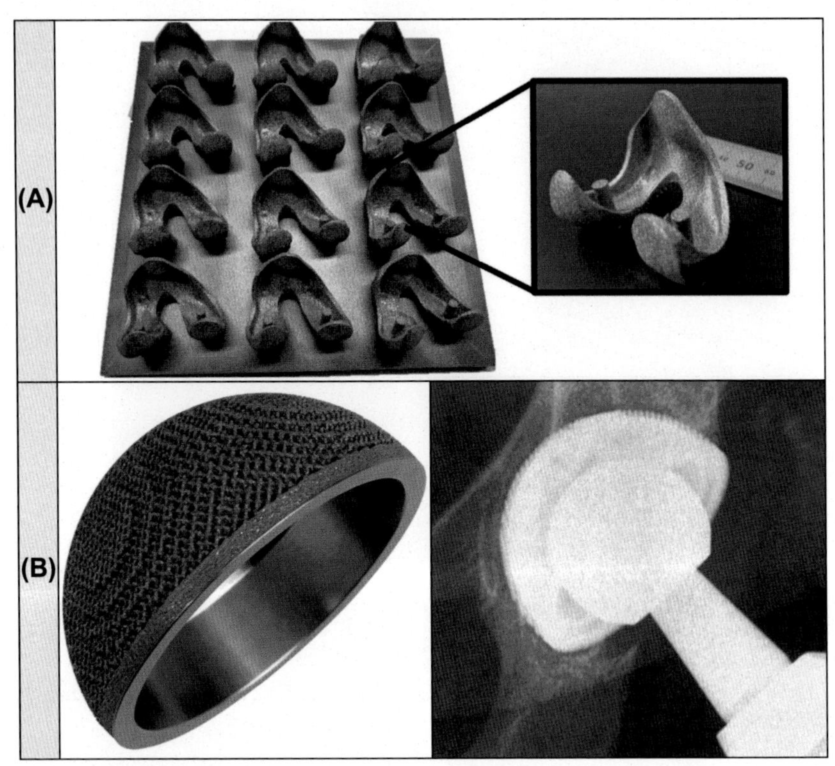

Fig. 11.16 (A) Technology demonstration of series production of the femoral component of a total knee reconstruction implant [13], (B) clinical application of acetabular cup component with engineered surface texture integrated within the EBM process and in-vivo implantation [14]. Internal features are machined post AM fabrication.

challenging for traditional manufacturing methods. For example, Fig. 11.16 illustrates the potential for series production of the femoral component of a total knee replacement system, as well as the potential for the integration of bulk and cellular geometry, for structure and cell ingrowth respectively, in a functionally integrated single-piece acetabular implant. The fabrication of a single piece implant, rather than the addition of cellular material as an additional coating step, provides commercial advantages for regulatory scenarios where each unique manufacturing method must be uniquely certified.

The commercial and clinical application of such AM fabricated implants is increasing exponentially[15]. These clinical devices utilize the flexibility of PBF systems to allow enhanced functionality within a single manufacturing process. These opportunities include surfaces with engineered surface roughness, combined with machined or post-AM modified surfaces as required; integrated threaded holes for fastening; lattice infill for bone ingrowth and to allow insertion of bone growth factors; and uniquely scribed identification markings (Fig. 11.17).

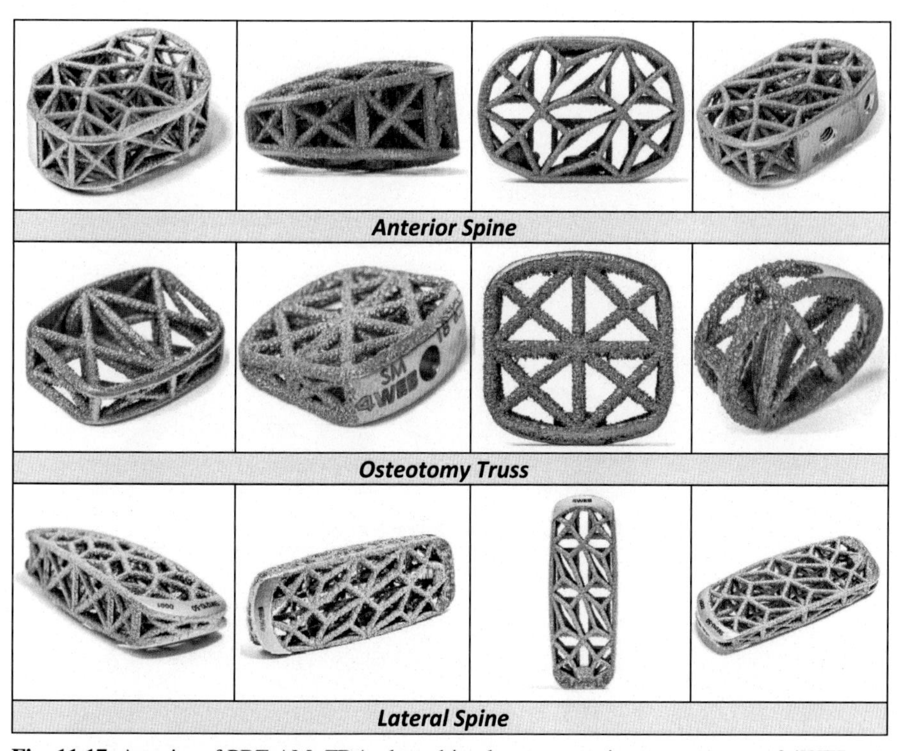

Fig. 11.17 A series of PBF AM, FDA-cleared implant systems,images courtesy of 4WEB Medical, USA.

Patient-specific medical implant design enables material and geometry to be bespoke designed for specific clinical needs; however, this additional effort is typically warranted only in scenarios that are unsuited to existing series production devices. Fig. 11.18 graphically summarises a selection of the potential design phases for the deployment of a bespoke medical device. To assist in efficient structural deployment, generative methods are applied to rapidly generate multiple design concepts; these concepts can then be assessed, and the preferred structure further optimized and specified for manufacture. This generative design methodology is further described in Chapter 7.

11.4.2 PBF case study: aerospace and space applications

Aerospace and space applications are highly suited to the technical and economic advantages of PBF systems. The functional optimization enabled by geometrically complex PBF structures enables mass reduction that allows increased payload and reduced self-loading, as well as the potential elimination of failure modes associated with interacting components. Furthermore, a particular challenge associated with the fabrication of aerospace materials (including titanium and Inconel) is their technical processing

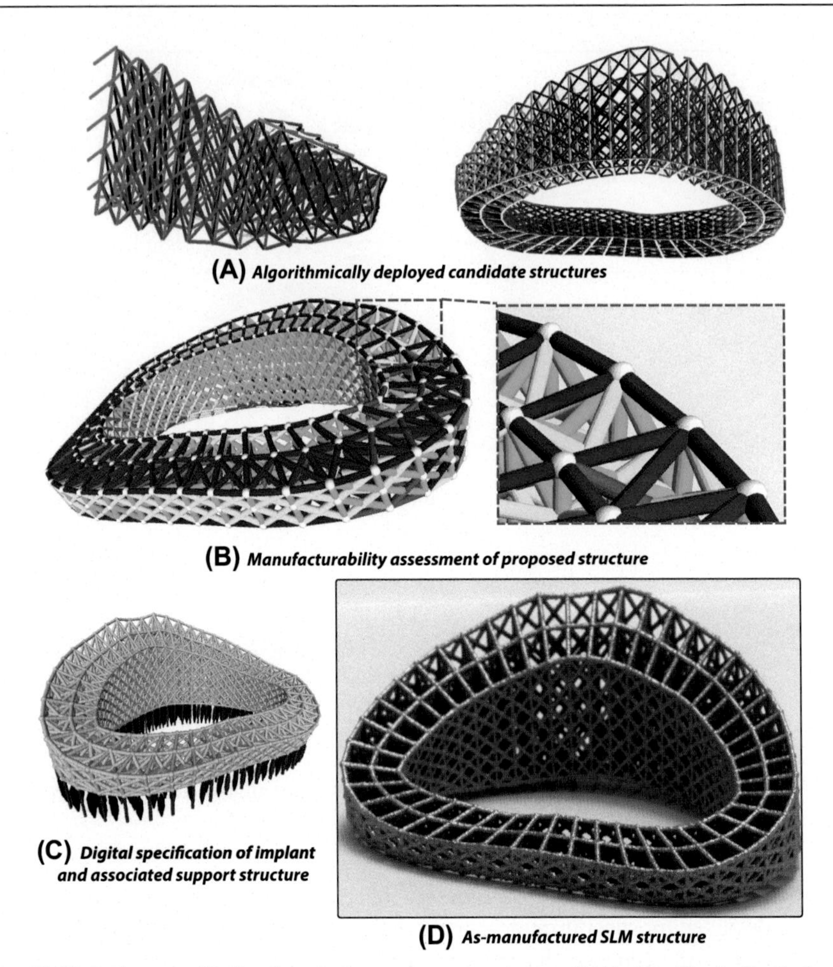

Fig. 11.18 Patient-specific implant design process, consisting of (A) algorithmically deployed candidate structures (B) from which the preferred embodiment can be specified and generated as a (C) digital structure for (D) SLM manufacture.

difficulty with traditional manufacture, notably forging and machining. Conversely, the low conductivity of titanium results in a relatively small PBF meltpool, enabling the fabrication of highly complex titanium structures.

The design flexibility of the PBF classification of AM technologies is demonstrated in the customized rocket engine inlet manifold of Fig. 11.19. This high-value structure is exposed to substantial thermal and mechanical loads while being subject to significant weight restrictions. In response to this challenging design specification, a near net-shape structure was conceived as a single piece structural element with continuously contoured geometry to efficiently accommodate local stresses while minimizing frictional losses through the manifold structure. A conformal iso-grid was specified to increase stiffness within the available mass budget; local grid geometry was filleted to

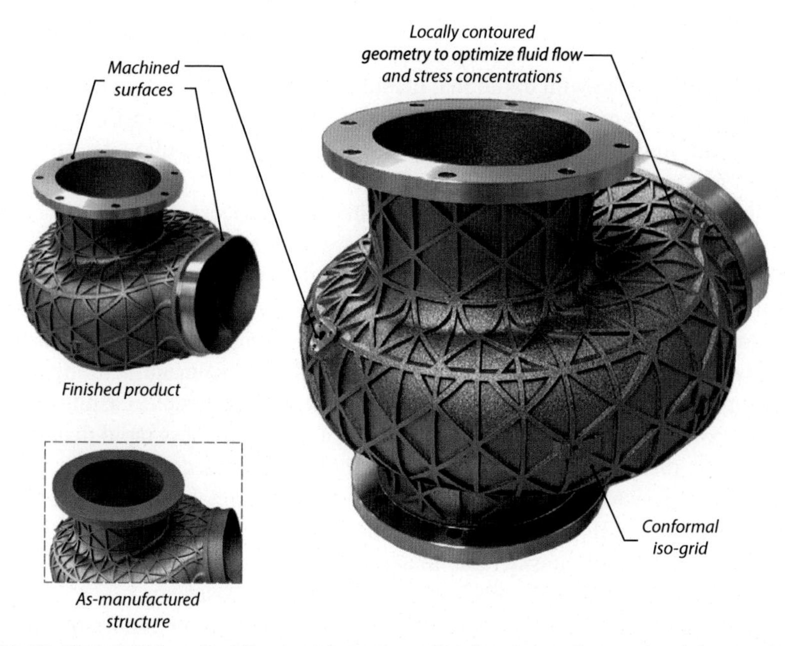

Fig. 11.19 Ti64 EBM applied for the fabrication of high-value rocket engine inlet manifold designed and manufactured by Zenith Tecnica Ltd., New Zealand.

avoid stress concentrations. Bulk geometry was optimized for orientation within the Electron Beam Melting (EBM) system (overall size: 330 × 300 × 250 mm – weight: 5 kg).

11.5 PBF DFAM summary

Powder Bed Fusion (PBF) is an advanced AM technology that has matured over an extended period of research and industrial development. This technology development has enabled wide uptake of commercial applications of PBF technology, particularly for the production of high-value products that are not technically feasible with traditionally manufacturing processes. In particular, commercial opportunities exist for the manufacture of highly complex geometry at small production volume; for example, as required for medical and aerospace applications.

For both medical and aerospace applications, titanium is one of the preferred materials: for medical applications this preference is primarily due to biocompatibility while for aerospace applications this light alloy is structurally desirable. Consequently, the application of titanium (in particular Ti64) is relatively well developed within the research literature.

Alternative alloys also provide commercial benefit, and much research opportunity exists for commercially focussed researchers who can quantify robust process parameters for these materials. Alternative materials of interest include Inconel alloys, which have high importance for high-temperature engineering applications and are challenging to fabricate with traditional manufacturing methods; copper alloys, which are compatible with contemporary PBF systems, but have received limited research attention; and aluminium alloys that provide a low cost light-alloy feedstock with numerous technical applications.

For these alloy systems it is logical that initial PBF process development be made on alloys that are commercially available and are known to the engineering community. This is especially important for alloy systems that require functional certification; for example, the biocompatibility of medical grade alloys and for aerospace applications. However, the application of extant alloys does impose a technical challenge, especially as these alloys have been optimized for alternative manufacturing methods that display fundamentally distinct process parameters. For example, casting alloys are developed for solidification rates that are orders of magnitude less rapid than for PBF. Consequently, these alloys are unlikely to provide an optimized mechanical response, and significant research opportunities exist for the commercial development of alloys that are optimal for PBF processing. Decisions regarding the specific alloys of interest should also consider the considerable financial barriers to the application of novel alloys to high-value applications (including medical and aerospace).

Optimization of the digital aspects of PBF design provides an important, and often overlooked, opportunity for functional optimization. In particular, the development of advanced toolpath scheduling algorithms that mitigate fusion defects and optimize local microstructures are necessary for the commercial application of PBF technologies to high-value systems.

The prediction of the thermo-mechanical response of PBF systems is commercially necessary; however, contemporary technologies and experimental data present a barrier to application. The development of rapid numerical modelling tools and the validation of experimental data remains an open research opportunity. In particular, DFAM tools that can predict structural anisotropy as a function of intended production geometry and processing parameters are highly in demand.

References

International standard reference for principles and terminology of AM

[1] ISO/ASTM 52900:2015(en). Additive manufacturing — general principles — terminology. Geneva, Switzerland: International Organization for Standardization (ISO); 2015.

Internationally regarded annual review of AM technology

[2] Wohlers T. Wohlers report. Wohlers Associates Inc.; 2019.http://www.wohlersassociates.com.

Initial PBF patent developed at the Fraunhofer institute

[3] Meiners W, Wissenbach K, Gasser A, Fraunhofer Gesellschaft zur Forderung der Angewandten Forschung. Selective laser sintering at melting temperature. 2001. U.S. Patent 6,215,093.

Characterization of the anisotropic microstructure of SLM fabricated AlSi10Mg

[4] Thijs L, Kempen K, Kruth JP, Van Humbeeck J. Fine-structured aluminium products with controllable texture by selective laser melting of pre-alloyed AlSi10Mg powder. Acta Materialia 2013;61(5):1809–19.

Characterization of the microstructure of Ti6Al4V SLM fabricated in vacuum environment

[5] Zhou B, Zhou J, Li H, Lin F. A study of the microstructures and mechanical properties of fabricated by SLM under vacuum. Materials Science and Engineering: A 2018;724:1–10.

Experimental observation of meltpool defects

[6] Coeck S, Bisht M, Plas J, Verbist F. Prediction of lack of fusion porosity in selective laser melting based on melt pool monitoring data. Additive Manufacturing 2019;25:347–56.

Fatigue life of EBM Ti64 including experimental observations of crack initiating defects

[7] Günther J, Krewerth D, Lippmann T, Leuders S, Tröster T, Weidner A, Biermann H, Niendorf T. Fatigue life of additively manufactured Ti–6Al–4V in the very high cycle fatigue regime. International Journal of Fatigue 2017;94:236–45.

Effect of processing parameters and toolpath on surface roughness and porosity

[8] Yu W, Sing SL, Chua CK, Tian X. Influence of re-melting on surface roughness and porosity of AlSi10Mg parts fabricated by selective laser melting. Journal of Alloys and Compounds 2019;792:574–81.

EBSD characterization of SLM fabricated IN738LC including keyhole porosity

[9] Guraya T, Singamneni S, Chen ZW. Microstructure formed during selective laser melting of IN738LC in keyhole mode. Journal of Alloys and Compounds 2019;792:151–60.

Effect of support structure design on mechanical and thermal response

[10] Bobbio LD, Qin S, Dunbar A, Michaleris P, Beese AM. Characterization of the strength of support structures used in powder bed fusion additive manufacturing of Ti-6Al-4V. Additive Manufacturing 2017;14:60–8.

Characterization of the manufacturability of Ti64 EBM lattice elements

[11] Zhang XZ, Tang HP, Leary M, Song T, Jia L, Qian M. Toward manufacturing quality Ti-6Al-4V lattice struts by selective electron beam melting (SEBM) for lattice design. JOM 2018;70(9):1870–6.

Characterization of the manufacturability of SLM fabricated Ti6Al4V and AlSi12Mg lattice elements

[12] Mazur M, Leary M, McMillan M, Sun S, Shidid D, Brandt M. Mechanical properties of Ti6Al4V and AlSi12Mg lattice structures manufactured by Selective Laser Melting (SLM). In: Laser additive manufacturing. Woodhead Publishing; 2017. p. 119–61.

Technology demonstration of potential for PBF series production of medical devices

[13] Narra SP, Mittwede PN, Wolf SD, Urish KL. Additive manufacturing in total joint arthroplasty. Orthopedic Clinics 2019;50(1):13–20.

EBM fabricated acetabular cup and in-vivo implantation

[14] Marin E, Fusi S, Pressacco M, Paussa L, Fedrizzi L. Characterization of cellular solids in Ti6Al4V for orthopaedic implant applications: trabecular titanium. Journal of the Mechanical Behavior of Biomedical Materials 2010;3(5):373–81.

Review of EBM fabricated lattice structures

[15] Zhang XZ, Leary M, Tang HP, Song T, Qian M. Selective electron beam manufactured Ti-6Al-4V lattice structures for orthopaedic implant applications: current status and outstanding challenges. Current Opinion in Solid State and Materials Science 2018;22(3): 75–99.

Novel insight into the actual powder bed height

[16] Wischeropp TM, Emmelmann C, Brandt M, Pateras A. Measurement of actual powder layer height and packing density in a single layer in selective laser melting. Additive Manufacturing; 2019.

Directed Energy deposition

12

Directed Energy deposition (DED) is an ISO/ASTM classification for AM technologies 'in which focused thermal energy is used to fuse materials by melting as they are being deposited' [1]. DED enables the production of physically large structures with high deposition rates using standard and novel materials in either powder or wire form. Protection from oxidizing environments is provided by either local gas shielding or by physical enclosure in a vacuum environment.

Commercial DED systems are technically robust and are typically based on the integration of existing commercial machine elements, such that sophisticated DED systems are inexpensively implemented. These machine elements include industrial robotics; standalone laser and welding systems; and shielding gas delivery hardware. Commercial systems range in size from small semi-portable systems that can be transported to allow in-place refurbishment, for example in rail and energy applications, to physically large installations for the series-production of high-value product. Commercial integration challenges include the development of robust toolpath planning methodologies that accommodate the unique design challenges and opportunities of DED, as well as process and material optimisation to manage induced thermal stresses. Technical responses to these challenges include implementations that increase manufacturability by allowing controlled workpiece orientation, as well as effective feedback control systems that allow material deposition to be optimized for the production of high-density solid structures with controlled thermal stresses and low surface roughness. In addition, inspection and certification are particular challenges, especially for DED applications that attempt to compete with traditional production methods such as forging and machining; existing and emerging DFAM strategies to accommodate these challenges are presented.

The specific production attributes of DED technologies combined with the high availability of commercial implementations enables a unique set of techno-economic outcomes for DED technologies. For example, DED systems allow the fabrication of very large-scale structural elements utilizing known structural metal alloys, as well as allowing novel alloy combinations. These alloys may be blended in-situ to allow functionally graded materials and geometry. Furthermore, DED is commercially well established in *cladding* scenarios, whereby a relatively thin skin of material is metallurgically bound to a substrate material of interest.

Despite the extensive commercial implementation of DED systems, there remain extensive DFAM research opportunities. In fact, the opportunities for generalisable research outcomes are potentially higher for DED systems because much of the fundamental technical knowledge exists as in-house knowledge that is not generally accessible. Consequently, research opportunities exist within many disciplines, including the robust characterization of standard engineering alloys; the design of novel alloys

Design for Additive Manufacturing. https://doi.org/10.1016/B978-0-12-816721-2.00012-9

that are optimized for the solidification conditions inherent to DED; the development of inspection and certification methodologies; and protocols for functionally graded deposition, for example, to allow the structural integration of low-cost bulk materials with high value functional alloys as required, such that highly functional product can be inexpensively manufactured. Emerging research includes the extension of DED to the processing of challenging metallic alloys as well as ceramics. These emerging applications provide a significant opportunity for commercial reward; however, significant research effort and associated DFAM tools are required for these opportunities to be translated to commercial practice.

12.1 Technology history and current commercial methods

The Directed Energy Deposition (DED) classification refers to AM systems that utilize some thermal energy to fuse powdered materials as they are deposited (Fig. 12.1). The functional DED system therefore requires that the materials and thermal energy source coexist at the intended location, and that this energy and material delivery be managed by some control system (either open or closed-loop). DED system architectures vary somewhat, although in general these systems utilize the existence of mature machine tools and systems to achieve powder delivery, process control and meltpool shielding. The core enabling technology for this commercial implementation is the associated thermal energy source.

Contemporary DED systems are predominantly laser based and utilize inexpensive commercial laser systems. The term *laser* is an acronym for 'Light Amplification by Stimulated Emission of Radiation' and refers to a device that emits coherent light. Coherence allows the emitted light to be of a narrow spectrum that can be collimated to a narrowly focussed beam with very high energy density. The first practical laser system was reported by Theodore Maiman of Hughes Research Laboratories in 1960 [2]. Extensive commercial and fundamental development has evolved the laser from these early laboratory-based implementations to a commercially robust machine tool. Notable milestones in this technology development include the development of industrially robust optics and laser control technologies; energy efficient optical fibre delivery systems; and the development of inexpensive high-energy laser implementations such as the disk laser. Laser technologies are so well developed that multiple kilowatt systems are routinely implemented for commercial DED applications.

DED requires that material be delivered to the thermally induced meltpool. This input material may be of either wire or powder form. Wire provides advantages, including relatively low cost (for commercially available grades) and compatibility with existing welding hardware and associated control systems. Powdered materials are available as a manufacturing feedstock for powdered metallurgy and other technology applications, and custom AM powder feedstocks are becoming increasingly available. Furthermore, powder feedstock provides valuable DFAM opportunities, including in-situ alloying and functional gradients in deposited materials.

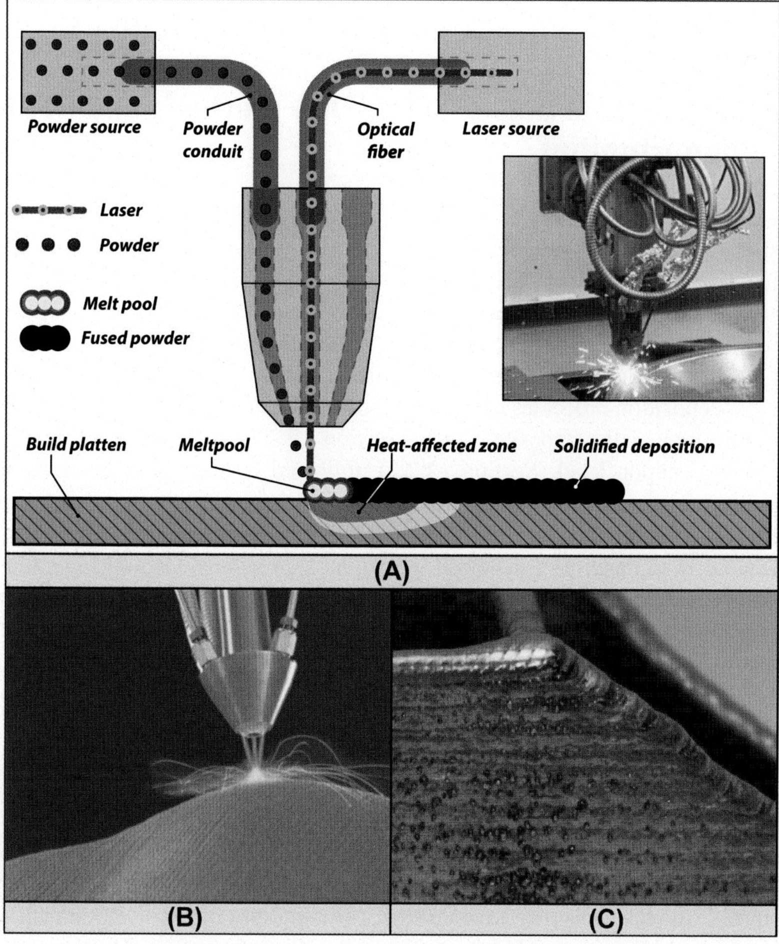

Fig. 12.1 (A) Schematic representation of Directed Energy Deposition (DED) implementation, inset shows commercial implementation at the RMIT Advanced Manufacturing Precinct, (B) high-speed photography of DED system with central laser power source and (three) radially aligned concentric powder nozzles [3], (C) typical surface finish detail for laser DED product.

The system integration for powder-based DED systems requires the fabrication of custom *powder deposition nozzles*. Various technical implementations are feasible, including the off-axis and coaxial delivery of powder, each of which provides unique opportunities and challenges associated with implementation, control and dexterity in material deposition.

Commercial practice for deposition toolpath planning is typically based on the re-purposing of existing CNC-derived tool path planning software. Custom DED toolpath planning software is emerging; however, the development such DFAM tools remains a commercially relevant research opportunity. In particular, the development of advanced toolpath planning software that can accommodate thermal data acquisition systems and advanced dexterity DED applications discussed in the following sections.

12.2 Commercial best practice DED methods and DFAM opportunities

Direct energy deposition is a well-developed industrial process that is routinely applied in industrial settings. Despite DED being a relatively mature technology, especially in cladding applications, much technology development is required to allow its robust commercial application. This requirement is due to the highly dimensional physics associated with this thermal AM process, as well as the commercial in-confidence nature of much of the industrial research in this sector. In order to obtain robust industrial outcomes, deposition processes are required to be statistically characterized, including the optimization of the deposited track (Section 12.2.1) and associated contiguous layers at the coupon level (Section 12.2.2), followed by the optimization of three-dimensional component level structures (Section 12.2.3).

Once these fundamental process requirements are acceptably under control, advanced DED applications can be considered, including the prediction and mitigation of thermal stresses; the application of thermal data acquisition systems for certification and closed-loop process control; and the application of DED systems that allow advanced workpiece orientation.

12.2.1 Deposition track optimization

The commercial application of new or otherwise unquantified alloys requires experimental process optimization, such that a robust metallurgical bond is achieved between the substrate, and that the deposition be acceptably crack free, and that excessive track dilution is avoided. This characterization is typically achieved by the full factorial experimental assessment of the relevant process parameters, including system power, P, toolpath velocity, v, and powder flow rate, f. The deposited track is then characterized by the geometric properties, width, w, height, h, and depth, d. These parameters may be represented in terms of the *dilution*, or fractional penetration of the track into the substrate material, A_p, to the total track area including the area of track above the substrate, A_b (Eq. 12.1).

$$\text{dilution} = \frac{A_p}{A_p + A_b} \approx \frac{d}{d + h} \tag{12.1}$$

An example of initial process parameter optimization for the development of a specific alloy for laser-based DED is provided in Fig. 12.2.

12.2.2 Coupon level optimization

Once track deposition parameters are optimized, the optimization of a manufactured layer at the coupon level is required. Typical challenges include the robust fabrication

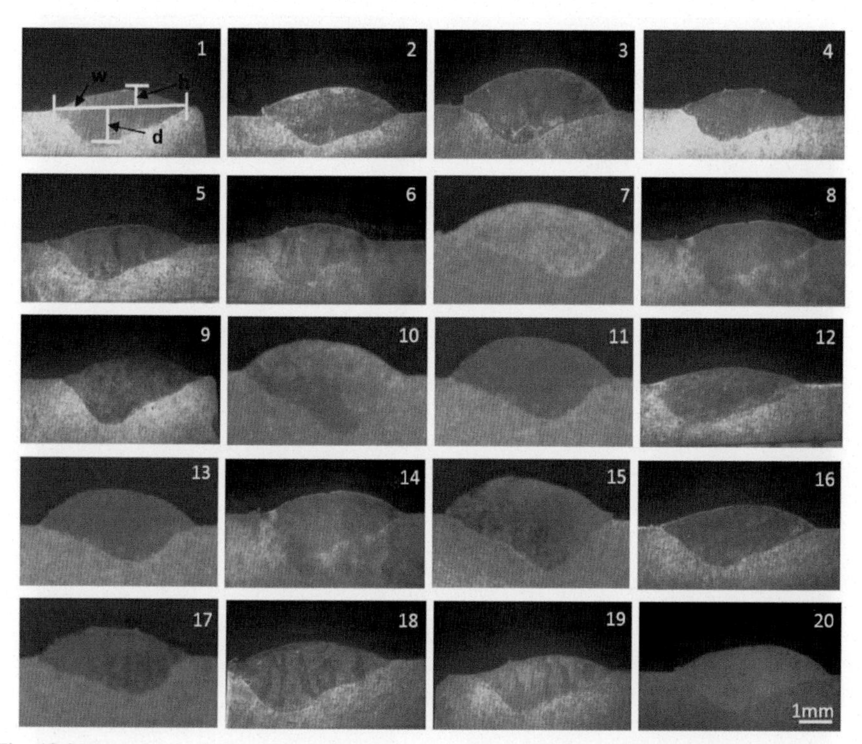

Fig. 12.2 Experimental characterization of the track cross-section (width, height and depth), and associated dilution for various process parameters permutations (power, velocity and powder feed rate) [4].

of track endpoints and track overlap. Therefore, layer manufacturability is typically characterized by the fabrication of planar structures with interacting individual tracks (Fig. 12.3). The fabrication of clad surfaces is typically associated with a single layer only, with post-fabrication machining used to achieve the final manufactured geometry. For cladding applications, it is therefore only necessary that a microstructurally robust and defect free layer be achieved: there is no necessity that excessive local height be avoided.

However, for 3D AM fabrication, excessive deposition height can result in structures that incrementally exceed the intended manufactured geometry with an increasing number of layers. In the extreme case, excessive distortion can occur and track quality can be compromised, or collision can occur between the deposition head and part. To avoid such defects, the interacting effects of both process parameters and toolpath design on the manufactured geometry much be characterized at the component level (Fig. 12.3). Toolpath optimization may include the manipulation of laser start and end positions, as well as the local variation of powder flow rates and velocity at these challenging locations.

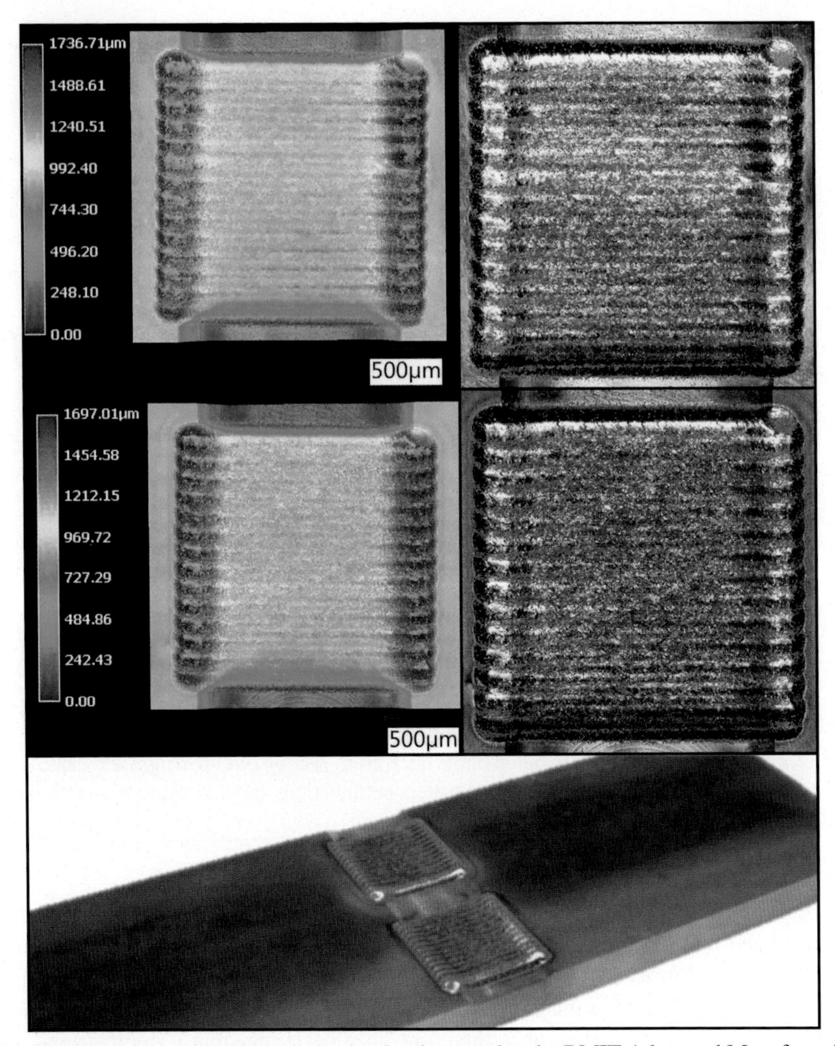

Fig. 12.3 Coupon level DED optimization implemented at the RMIT Advanced Manufacturing Precinct to assess the interaction between track deposition parameters and the effect of track interaction. Three-dimensional colourized imaging confirms that the clad layer is effective in filling the required groove depth.

12.2.3 Component level structures

Once individual layers are robustly implemented, the optimisation of DED parameters can then be extended to the manufacture of component level structures. DED allows deposition rates that are significantly higher than for competing metallic AM systems such as Powder Bed Fusion (PBF) systems but with a proportionately higher surface roughness. This surface roughness potentially introduces geometric stress concentrations that significantly compromise the fatigue resistance of structural components. Furthermore, as-manufactured DED structures typically contain slumping defects, whereby the intended geometry is incompatible with the material deposition profile and may contain oxidization defects due to excessive local temperatures and inadequate local shielding (Fig. 12.4). Consequently, for high-value structures this dimensionally compromised, high roughness surface is typically removed by a post-manufacture machining operation.

This production approach, whereby materials are deposited near-net with DED and then machined to generate the final production geometry provides a highly efficient path to the inexpensive fabrication of high-value product. This advantage is especially pronounced for materials that are challenging to machine, such as Inconel and titanium, as these materials typically show a poor *buy-to-fly* ratio for aerospace applications - implying that for typical manufacture, oversize billets are purchased and then extensively machined to fabricate the final product. Figs 12.5 and 12.6 indicate a potential application of a component level structure design for laser-based DED applications with low buy-to-fly ratio.

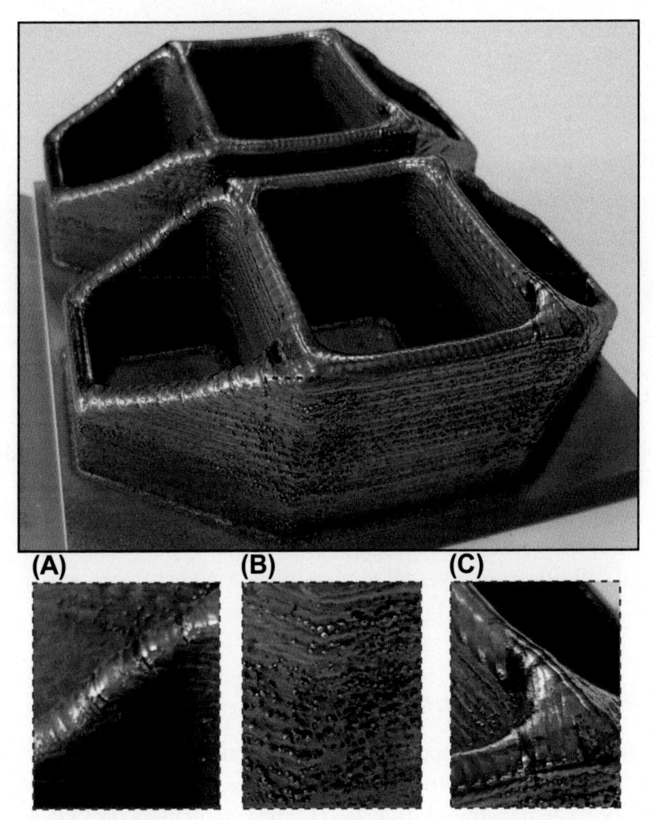

Fig. 12.4 Typical as-manufactured laser-based DED product, including (A) slumping defect, (B) layerwise geometry and adhered particles, and (C) potential overheating and oxidization.

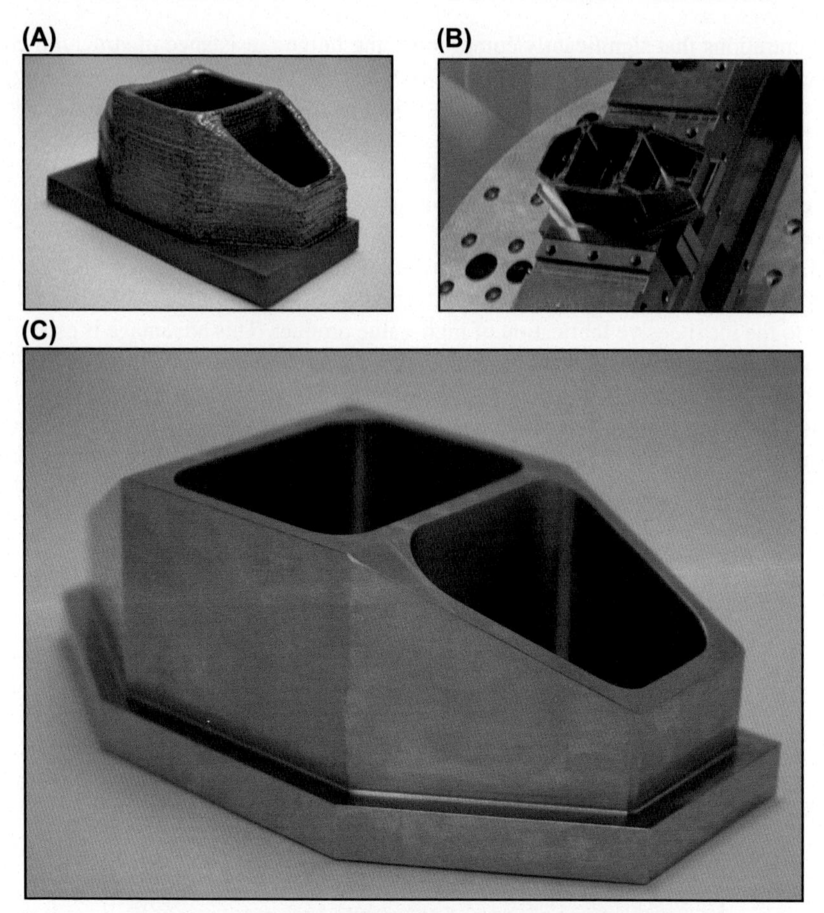

Fig. 12.5 Design and fabrication of laser-based DED product for titanium aerospace application completed at the RMIT Advanced Manufacturing Precinct, including (A) as-manufactured DED product, (B) machining operation, and (C) final product.

This production approach requires that additional DED material be deposited to accommodate a machining allowance to remove the surface defects inherent to DED systems. Furthermore, when compared with homogeneous billet, the anisotropy of DED materials must also be considered. This anisotropy is associated with the rapid and directional solidification inherent to metal fusion AM systems, and is discussed in the context of PBF in Chapter 11.

12.2.4 DED support structures

The development of DED support structures remains a commercially relevant but underutilized DFAM research opportunity; however, the technical feasibility of such structures has been demonstrated (Fig. 12.7). These structures provide an opportunity for cost-reduction in commercial DED production, as well as providing innovative

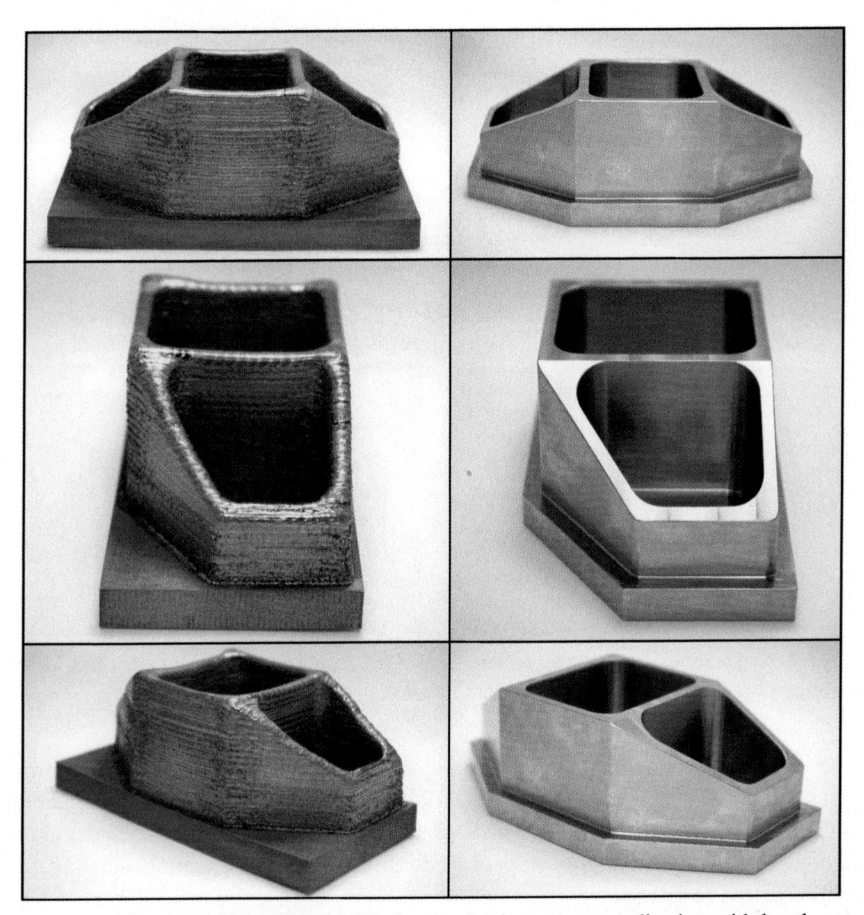

Fig. 12.6 Application of laser-based DED for titanium aerospace application with low buy-to-fly ratio. Material is deposited near-net with DED (left) and then machined to generate the final production geometry (right). Build platen is integrated into the final design to reduce certification challenges and product cost.

opportunities in design and certification. For example, frangible DED support structures provide an opportunity for the limitation of internal residual forces within manufactured structures, thereby providing a mechanism for certification and process determination, as well as limiting the (potentially damaging) stresses applied to the platen structure. Furthermore, unlike powder-bed systems, DED systems do not provide inherrent support of overhanging structures, and therefore the lack of robust supporting structures restricts the achievable manufacturability of the DED process.

12.2.5 Ceramic DED applications

The processing of ceramic materials is highly challenging due to their high hardness and high melting temperature. Emerging research is demonstrating the feasibility of

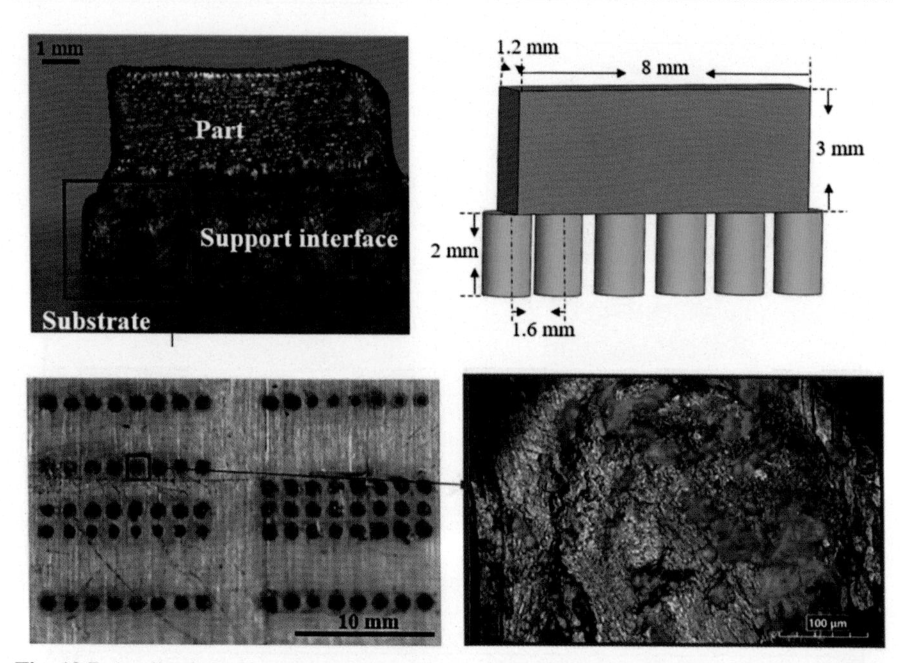

Fig. 12.7 Application of supporting structures for DED application [5]. This emerging DFAM capability provides a commercially relevant research opportunity.

DED processing for ceramic materials. Such materials include, for example (Fig. 12.8), lead zirconate titanate (PTZ), an important piezoelectric material; Al_2O_3, a refractory metal that is also used as an industrial catalyst; and zirconia alumina (ZrO_2-Al_2O_3), a biocompatible, high-temperature structural ceramic with high corrosion resistance.

These emerging research outcomes provide an enabling capability for the industrial application of high-value thermal structures such as gas-turbines with ultra-high inlet temperatures. These DED ceramic structures provide an alternative technology to the current nickel-based superalloys by providing a significant increase in allowable operating temperature.

12.2.6 Multi-axis applications

Historically, the industrial application of DED systems has been predominantly within the cladding role, an application that does not require sophisticated toolpath planning. As DED emerges as a 3D AM technology in its own right, a commercial need for bespoke DED toolpath planning strategies and associated DFAM tools has emerged. These tools must accommodate the specific challenges required of commercial DED application, including the planning of complex trajectories in 3D space; allowing workpiece orientation such that deposition attributes are optimized; avoiding workpiece collisions, especially for highly dexterous 5-axis robotics; accommodating the

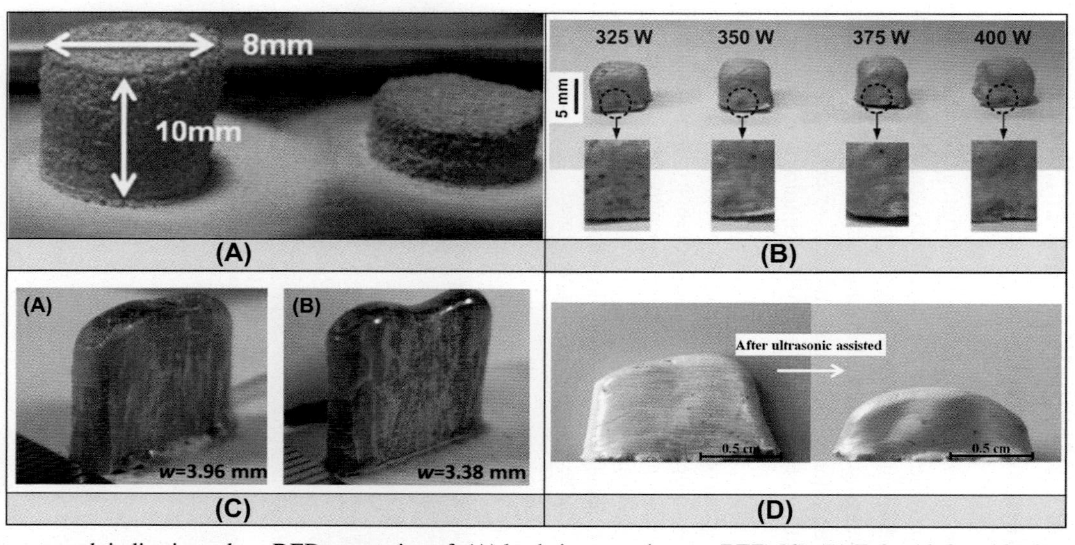

Fig. 12.8 Emerging research indicating robust DED processing of: (A) lead zirconate titanate (PTZ) [6], (B) ZrO_2-Al_2O_3 with ultrasonic vibration [7], (C) Al_2O_3 [8] (D) ZrO_2-Al_2O_3 with ultrasonic vibration [9].

deposition nozzle inclination to optimize meltpool geometry; and accommodating advanced workpiece orientation, such that complex structures can be fabricated without necessitating support structures or reductant materials (Fig. 12.9).

12.3 Technical opportunities and challenges to DED implementation

DED is a robust commercial AM technology that integrates existing commercial machine elements to provide inexpensive and highly capable manufacturing systems. These systems can fabricate robust structures at high build-rates with commercial alloys including challenging grades and even ceramic materials. Despite these existing commercial outcomes, there remain significant research DFAM opportunities to enhance the application of this important AM technology, including:

- toolpath optimization tools to accommodate the specific challenges and opportunities of DED fabrication including complex 3D trajectories, deposition nozzle inclination, and the accommodation of workpiece orientation to allow the fabrication of complex support-free structures
- integration of DED systems with in-situ process monitoring to enable closed-loop process control and product validation
- integration of DED manufacturing methodologies with generative DFAM tools such that high-value products can be algorithmically implemented

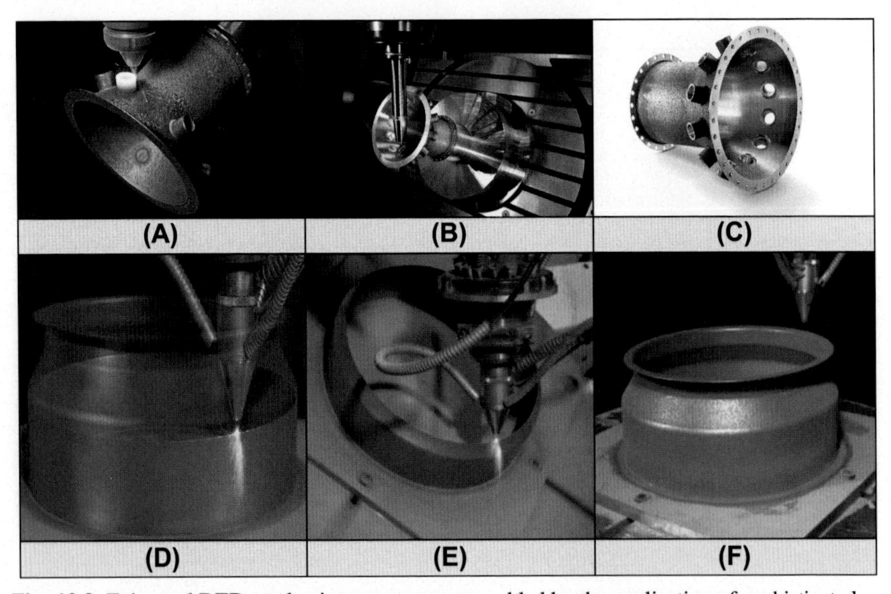

Fig. 12.9 Enhanced DED production outcomes are enabled by the application of sophisticated toolpath optimization in conjunction with highly dextrous robotics and workpiece articulation [10].

- enhancement of the limited understanding of the technical methods for the deployment of frangible DED support structures, including DFAM tools that correlate process and toolpath inputs to a specific support strength
- the extension of existing DED systems to ceramic materials, especially the extension of existing coupon level testing to the fabrication of functional structures.

References

International standard reference for principles and terminology of AM

[1] ISO/ASTM 52900:2015(en). Additive manufacturing — general principles — terminology. Geneva, Switzerland: International Organization for Standardization (ISO); 2015.

The first documented experimental observation of laser light

[2] Maiman TH. Stimulated optical radiation in ruby Nature 1960;187:493–4.

Review of contemporary laser methods

[3] Löffler K. Developments in disk laser welding. In: Handbook of laser welding technologies. Woodhead Publishing; 2013. p. 73–102.

Experimental DED reference including dilution reference

[4] Bhardwaj T, Shukla M, Paul CP, Bindra KS. Direct energy deposition-laser additive manufacturing of titanium-molybdenum alloy: parametric studies, microstructure and mechanical properties. Journal of Alloys and Compounds 2019;787:1238–48.

Initial research in the field of DED support structure design

[5] Jiang J, Weng F, Gao S, Stringer J, Xu X, Guo P. A support interface method for easy part removal in directed energy deposition. Manufacturing Letters 2019;20:30–3.

Initial research in ceramic DED processing – lead zirconate titanate (PTZ)

[6] Bernard SA, Balla VK, Bose S, Bandyopadhyay A. Direct laser processing of bulk lead zirconate titanate ceramics. Materials Science and Engineering: B 2010;172(1):85–8.

Initial research in ceramic DED processing – ZrO_2-Al_2O_3 with ultrasonic vibration

[7] Hu Y, Ning F, Cong W, Li Y, Wang X, Wang H. Ultrasonic vibration-assisted laser engineering net shaping of ZrO_2-Al_2O_3 bulk parts: effects on crack suppression, microstructure, and mechanical properties. Ceramics International 2018;44(3):2752–60.

Initial research in ceramic DED processing – Al_2O_3

[8] Niu F, Wu D, Zhou S, Ma G. Power prediction for laser engineered net shaping of Al_2O_3 ceramic parts. Journal of the European Ceramic Society 2014;34(15):3811–7.

Initial research in ceramic DED processing – ZrO_2-Al_2O_3 with ultrasonic vibration

[9] Yan S, Wu D, Niu F, Ma G, Kang R. Al_2O_3-ZrO_2 eutectic ceramic via ultrasonic-assisted laser engineered net shaping. Ceramics International 2017;43(17):15905–10.

Commercial best practice DED application for high-value product

[10] Liu R, Wang Z, Sparks T, Liou F, Newkirk J. Aerospace applications of laser additive manufacturing. In: Laser additive manufacturing. Woodhead Publishing; 2017. p. 351–71.

Binder jetting

13

Binder jetting (BJT) is an ISO/ASTM classification for AM technologies 'in which a liquid bonding agent is selectively deposited to join powder materials' [1]. Commercial BJT implementations apply layerwise binding of material in a powder bed. This implementation has many technical and economic advantages, allowing relatively low-cost fabrication of complex geometry with no thermal distortion and high production yield. BJT components in the as-bound state are relatively fragile, but are useful for aesthetic prototyping and form-and-fit applications. Post-processing can be applied to enhance mechanical properties, and although these methods were challenged by the emergence of high-powered PBF systems, the commercial opportunities for post-processed BJT components is emerging as a commercially significant opportunity for the inexpensive fabrication of high-value structures. Much commercial research is required to translate this emerging manufacturing capability into robust commercial practice.

13.1 Technology history and current commercial methods

Binder jetting technologies require the controlled deposition of a binding agent to locally adhere powdered materials. Commercial jetting technologies are highly complex, but are technically well understood due to intense technology development for the paper printing market, as described in their application to Material Jetting technologies (MJT, Chapter 9). These jetting technologies are applied in commercial BJT systems to precisely deposit binding materials within a powder bed. A rake or wiper precedes the traversal of the binder jetting system such that the jetted binding agent integrates with the freshly deposited powder. This mix of powder and binding agent may then be cured by ambient air exposure or by the aid of thermal or UV-curing (Fig. 13.1).

This fabrication method has a number of commercial and technical advantages. The powder bed provides active support for the deposited binder, and there is no direct physical contact between the jetting assembly and the deposited powder. Therefore, commercial BJT technologies allow acute build inclinations and even isolated island features with no direct connection to other components or to the build platen. The binder jetting process is essentially non-thermal, allowing unlimited three-dimensional nesting of fabricated components within the available build volume. Commercial jetting systems are technically sophisticated and allow high precision and throughput, thereby enabling

Design for Additive Manufacturing. https://doi.org/10.1016/B978-0-12-816721-2.00013-0

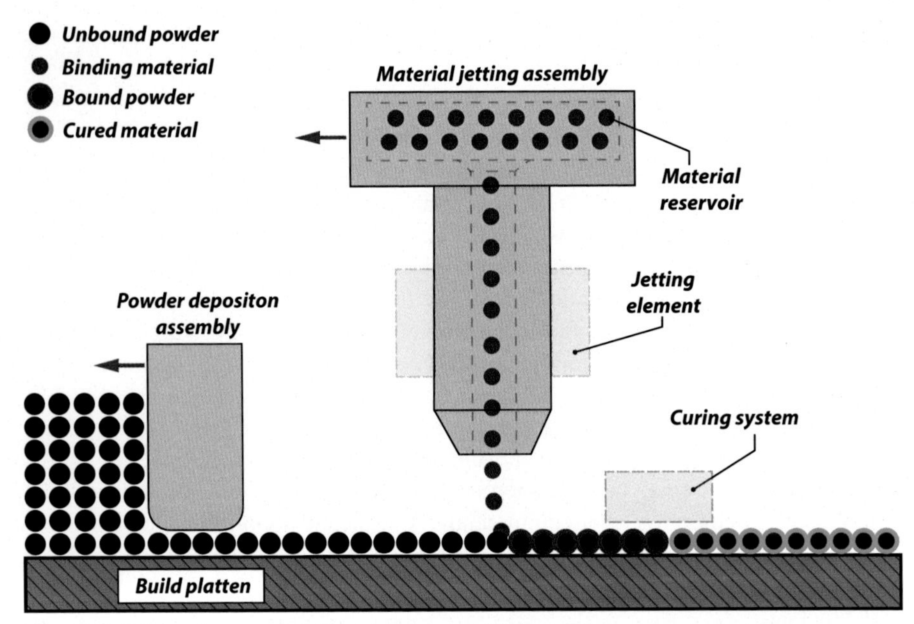

Fig. 13.1 Schematic representation of commercial BJT technology implementation.

the rapid generation of high-resolution features. Furthermore, these jetting systems are inexpensive due to their extensive technology development for paper printing applications. These combined capabilities provide a unique techno-economic profile for BJT technologies in the inexpensive manufacture of high-precision structures.

The as-bound component is referred to as *green*, a metal casting term that refers to the relatively fragile material used for sand casting molds. Similarly, the BJT component in the as-bound state relies on the physical strength of the binding agent alone and is therefore relatively fragile. In response, BJT is often implemented as a two-step process whereby the green components is then post-processed to provide additional strengthening, typically by a sintering process whereby material is heated such that neighbouring particles locally fuse, thereby increasing mechanical properties.

Historically, sintered BJT components provided the only powder bed AM technology for the robust fabrication of metallic structures. However, intense technology development of Powder Bed Fusion (PBF, Chapter 11) enabled the fusion of structurally robust components directly within the powder bed, a capability that eclipsed that of sintered BJT components at the time. More recently, BJT technologies have experienced a commercial revival. This revival is based on the fundamental advantages of second-generation BJT systems to inexpensively deploy complex geometry with robust mechanical properties. The commercial application of BJT technologies requires the development of DFAM tools that accommodate a formal understanding of the mechanical and geometric performance of these structures.

13.2 Technical challenges and research contributions

Although fundamentally based on commercially robust jetting hardware, binder jetting technologies are complex commercial systems, unlike, for example, the relative simplicity of Material Extrusion technologies (MEX, Chapter 8). Consequently, much of the fundamental BJT technology development occurs within commercial research and development facilities. Furthermore, commercial BJT systems are typically not directly compatible with the development of custom DFAM processing algorithms. Despite these restrictions, valuable research contributions are being made into the novel application of BJT, as well as the

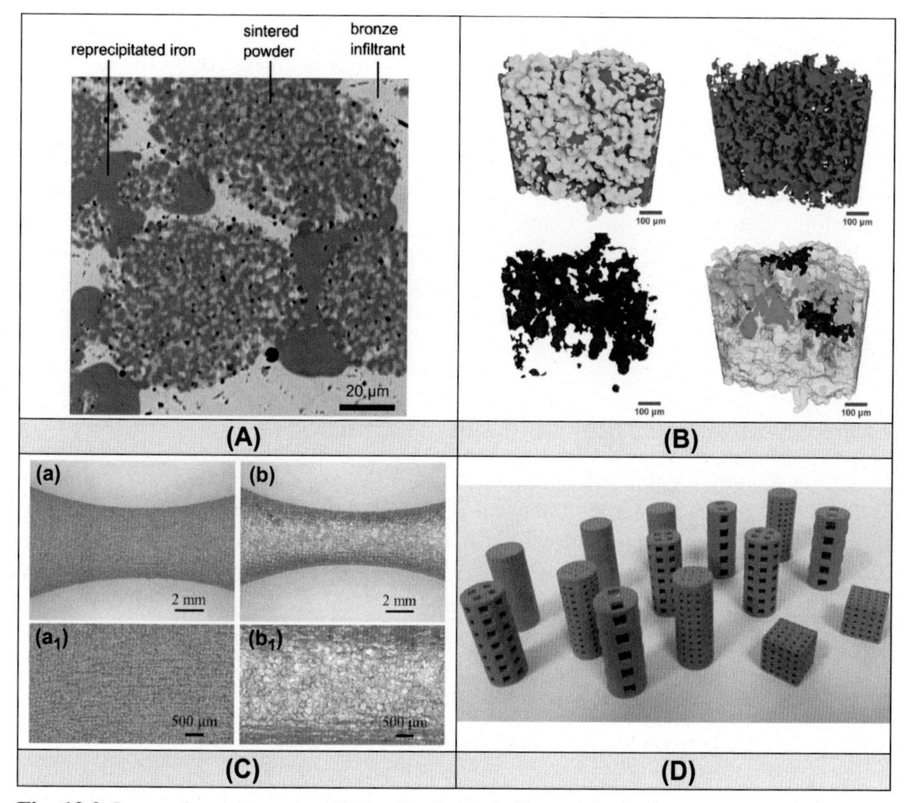

Fig. 13.2 Innovative response to BJT technology challenges, including: (A) molten bronze infiltration [2], (B) the deployment of silicone materials in a custom BJT system [3], (C) characterization of mechanical response of BJT Inconel 625 [4], (D) characterization of geometric capabilities for custom-medical implant applications [5].

characterization of the geometric and mechanical response of the as manufactured product (Fig. 13.2).

13.2.1 Melt infiltration to improve mechanical properties

Binder jetting allows the manufacture of complex geometry but is potentially challenged by the associated mechanical properties of the sintered preform. The infiltration of these structures with alternate materials remains an active research field. For example, the infiltration of BJT fabricated components with molten bronze has been shown to increase part-density and improve strength by mitigating the stress concentrations at particle connections [2]. Advanced application of material infiltration provides a unique commercial research opportunity for the fabrication of high-performance product with microstructure and micro-topology that is optimized for the BJT technology.

13.2.2 Novel processing techniques and materials

Despite the technical challenges associated with the development of custom BJT systems in a research environment, valuable research is emerging on the development of custom BJT systems and associated materials. For example, the utilizaiton of silicone materials enabled by custom developed BJT systems [3]. These emerging technical opportunities potentially enable the application of modified BJT systems for the manufacture of high-value structures for pharmaceutical and biological applications. Commercially robust BJT systems are becoming available with the potential to fabricate ceramics and other materials are that are potentially challenging for alternate AM technologies – the development of DFAM tools that accommodate these novel material processing capabilities are required.

13.2.3 Characterization of mechanical response

Emerging technical capabilities are allowing the application of BJT technologies to the design of functional technical products. For design engineers to confidently apply BJT for these scenarios requires that the associated mechanical response be formally characterized. In response, design data is being generated, especially associated with high-value materials such as Inconel [4]. The lack of robust data in this space provides a sound research focus for commercially engaged DFAM researchers.

13.2.4 Characterization of geometric response

The geometric capability of BJT systems is impressive, but not unlimited, especially when combined with the effect of thermal post-processing. The robust design of functional products requires that the geometric response of the manufacturing process be statistically characterized such that Geometric Dimension and

Tolerancing (GD&T) methods can be systematically applied. In response to the technical capabilities of BJT technologies, engineering data is emerging to aid in the characterization of the associated process capabilities [5]. This data is commercially valuable and provides a robust DFAM research opportunity.

13.3 Technical opportunities and challenges to BJT implementation

Binder jetting provides a unique AM technology platform for the inexpensive fabrication of high-value product with structurally robust materials; including the potential to fabricate ceramics and other materials that are potentially challenging for alternate AM technologies. The commercial applicability of BJT to these material systems has matured significantly in recent years. However, there remains a need for commercially focussed DFAM research teams to generate the material, geometry and processing data required for the robust implementation of this technology. A nascent opportunity exists in the application of generative design methods to the BJT technology classification. This opportunity will become commercially viable as the requisite DFAM tools for BJT design become increasingly available.

References

International standard reference for principles and terminology of AM

[1] ISO/ASTM 52900:2015(en), Additive manufacturing — general principles — Terminology. Geneva, Switzerland: International Organization for Standardization (ISO); 2015.

Strengthening of binder jetting structures

[2] Cordero ZC, et al. Strengthening of ferrous binder jet 3D printed components through bronze infiltration. Additive Manufacturing 2017;15:87–92.

Binder jetting of silicone for biomedical applications

[3] Liravi F, Vlasea M. Powder bed binder jetting additive manufacturing of silicone structures. Additive Manufacturing 2018;21:112–24.

Binder jetting surface finish and fatigue properties

[4] Mostafaei A, et al. Characterizing surface finish and fatigue behavior in binder-jet 3D-printed nickel-based superalloy 625. Additive Manufacturing 2018;24:200–9.

Lattice structures manufactured with binder jetting

[5] Vangapally S, et al. Effect of lattice design and process parameters on dimensional and mechanical properties of binder jet additively manufactured stainless steel 316 for bone scaffolds. Procedia Manufacturing 2017;10:750–9.

Index

'*Note*: Page numbers followed by "f" indicate figures and "t" indicates tables.'

Printed in the United States
By Bookmasters